SYMMETRIES IN SUBATOMIC PHYSICS

Related Titles from AIP Conference Proceedings and the Subseries on High Energy Physics

533 Next Generation Nucleon Decay and Neutrino Detector: NNN99
Edited by Milind V. Diwan and Chang Kee Jung, August 2000, 1-56396-956-4

531 Particles and Fields: Seventh Mexican Workshop
Edited by Alejandro Ayala, Guillermo Contreras, and Gerardo Herrera, July 2000, 1-56396-954-8

530 Colliders and Collider Physics at the Highest Energies: Muon Colliders at 10 TeV to 100 TeV, HEMC'99 Workshop
Edited by Bruce J. King, August 2000, 1-56396-953-X

512 Nuclear Physics at Storage Rings: Fourth International Conference: STORI99
Edited by Hans-Otto Meyer and Peter Schwandt, June 2000, 1-56396-928-9

508 Hadron Physics: Effective Theories of Low Energy QCD
Edited by A. H. Blin, B. Hiller, M. C. Ruivo, C. A. Sousa, and E. van Beveren, March 2000, 1-56396-927-0

494 New Directions in Quantum Chromodynamics
Edited by Chueng-Ryong Ji and Dong-Pil Min, November 1999, 1-56396-908-4

490 Particles and Fields: Eighth Mexican School
Edited by Juan Carlos D'Olivo, Gabriel López Castro, and Myriam Mondragón, November 1999, 1-56396-895-9

488 High Energy Physics at the Millennium: MRST'99
Edited by Pat Kalyniak, Stephen Godfrey, and B. Kamal, October 1999, 1-56396-902-5

453 Particles, Fields, and Gravitation
Edited by Jakub Rembieliński, December 1998, 1-56396-837-1

452 Toward the Theory of Everything: MRST'98
Edited by James M. Cline, Marcia E. Knutt, Gregory D. Mahlon, and Guy D. Moore, November 1998, 1-56396-845-2

424 Twenty Beautiful Years of Bottom Physics: Proceedings of the b20 Symposium
Edited by R. A. Burnstein, D. M. Kaplan, and H. A. Rubin, March 1998, 1-56396-745-6

423 Fundamental Particles and Interactions: Frontiers in Contemporary Physics
Edited by Robert S. Panvini and Thomas J. Weiler, February 1998, 1-56396-725-1

To learn more about these titles, or the AIP Conference Proceedings Series, please visit the webpage **http://www.aip.org/catalog/aboutconf.html**

SYMMETRIES IN SUBATOMIC PHYSICS

3rd International Symposium

Adelaide, Australia   13–17 March 2000

EDITORS
X.-H. Guo
A. W. Thomas
A. G. Williams
University of Adelaide, Australia

Melville, New York, 2000
AIP CONFERENCE PROCEEDINGS ■ VOLUME 539

Editors:

Xin-Heng Guo
Anthony W. Thomas
Anthony G. Williams

CSSM
University of Adelaide
Adelaide, SA 5005
AUSTRALIA

E-mail: xhguo@physics.adelaide.edu.au
athomas@physics.adelaide.edu.au
awilliam@physics.adelaide.edu.au

Authorization to photocopy items for internal or personal use, beyond the free copying permitted under the 1978 U.S. Copyright Law (see statement below), is granted by the American Institute of Physics for users registered with the Copyright Clearance Center (CCC) Transactional Reporting Service, provided that the base fee of $17.00 per copy is paid directly to CCC, 222 Rosewood Drive, Danvers, MA 01923. For those organizations that have been granted a photocopy license by CCC, a separate system of payment has been arranged. The fee code for users of the Transactional Reporting Service is: 1-56396-964-5/00/$17.00.

© 2000 American Institute of Physics

Individual readers of this volume and nonprofit libraries, acting for them, are permitted to make fair use of the material in it, such as copying an article for use in teaching or research. Permission is granted to quote from this volume in scientific work with the customary acknowledgment of the source. To reprint a figure, table, or other excerpt requires the consent of one of the original authors and notification to AIP. Republication or systematic or multiple reproduction of any material in this volume is permitted only under license from AIP. Address inquiries to Office of Rights and Permissions, Suite 1NO1, 2 Huntington Quadrangle, Melville, N.Y. 11747-4502; phone: 516-576-2268; fax: 516-576-2450; e-mail: rights@aip.org.

L.C. Catalog Card No. 00-108019
ISBN 1-56396-964-5
ISSN 0094-243X
Printed in the United States of America

Contents

Preface ix
Organizing Committee xi

Latest Experimental Information on ϵ'/ϵ 1
    Y. B. Hsiung
CP Violation 2000: Status and Perspectives 13
    A. J. Buras
Latest Results from Super-Kamiokande 31
    T. Kajita for the Super-Kamiokande Collaboration
Additional Isospin-Breaking Effects in ϵ'/ϵ 41
    S. Gardner
Isospin Breaking in the Extraction of Isovector and Isoscalar Spectral Functions from $e^+e^- \to$ hadrons 48
    K. Maltman and C. E. Wolfe
Neutrino Mixing and CP Violation in Matter 54
    Z. Xing
Tritium Decay, Neutrino Mixing and Neutrino Interactions 60
    B. H. McKellar, G. J. Stephenson, Jr., and T. Goldman
Constraints on a Parity-even/Time-reversal-odd Interaction 66
    W. T. H. van Oers
What can we Learn from QED at Large Couplings? 73
    A. W. Schreiber, R. Rosenfelder, and C. Alexandrou
CP Violation in $\Lambda \to p\pi^-$: SM vs New Physics 80
    G. Valencia
Quark Mixing Angles and CP-violating Phase from Flavour Permutational Symmetry Breaking 86
    A. Mondragón and E. Rodríguez-Jáuregui
Atomic Theory and Tests of the Standard Model in Atomic Experiments 92
    V. V. Flambaum
Precision Tests of the Standard Model at Electron Colliders 99
    D. Muller
Lorentz and CPT Tests in Atomic Systems 109
    R. Bluhm
New Clock Comparison Searches for Lorentz and CPT Violation 119
    R. L. Walsworth, D. Bear, M. Humphrey, E. M. Mattison,
    D. F. Phillips, R. E. Stoner, and R. F. C. Vessot
Charges, Parity Doublets and Parity-non-conservation 130
    B. Desplanques
Enhancement of Parity and Time Invariance Violation in Heavy Atoms 136
    V. A. Dzuba, V. V. Flambaum, and J. S. M. Ginges
Symmetry Motivated Estimation of Nucleon Delta-Excitation 142
    S. I. Sukhoruchkin

*Italicized name indicates author who presented the paper.

Pion-Baryon Couplings and SU(3) 148
    A. J. Buchmann and *E. M. Henley*

Direct CP Violation in Charmed Hadron Decays via ρ-ω Mixing 153
    X.-H. Guo and A. W. Thomas

Large N_c QCD Sum Rules 160
    P. W.-Y. Hwang

Is Time Reversal Invariance Violated in Muon Decay?
A Measurement of the Transverse Positron Polarization 167
    W. Fetscher, K. Bodek, A. Budzanowski, N. Danneberg,
    C. Hilbes, L. Jarczyk, K. Kirch, S. Kistryn, J. Klement,
    K. Köhler, A. Kozela, J. Lang, G. Llosá Llácer,
    M. Markiewicz, X. Morelle, T. Schweizer, J. Smyrski,
    J. Sromicki, E. Stephan, A. Strzałkowski, and J. Zejma

Test of Time Reversal Invariance in $K_{\mu 3}$ Decay 177
    Y. Kuno

T Violation and CPT Tests at CPLEAR 187
    L. A. Schaller

Latest Results from BaBar 197
    B. Abbott for the BaBar Collaboration

Symmetries in Parton Distributions 207
    J. T. Londergan

Studies of Nucleon Spin Structure at HERMES 217
    B. Tipton

The Nucleon's Strange Form Factors 228
    M. L. Pitt

Charge Symmetry Violation in np$\rightarrow$dπ^0 238
    A. K. Opper

Isospin Symmetry Breaking in Nuclei—ONS Anomaly 245
    K. Saito

Charge Symmetry Violation in Parton Distributions 255
    C. Boros, F. M. Steffens, J. T. Londergan, and A. W. Thomas

A Precision Measurement of the Michel Parameter ξ''
in Polarized Muon Decay 261
    P. van Hove, N. Danneberg, J. Deutsch, J. Egger, W. Fetscher,
    F. Foroughi, J. Govaerts, M. Hadri, C. Hilbes, K. Kirch, P. Knowles,
    J. Lang, M. Markiewicz, R. Medve, X. Morelle, O. Naviliat,
    A. Ninane, R. Prieels, T. Schweizer, and J. Sromicki

Nonleptonic Decays of Supermultiplets 265
    R. Delbourgo and D. Liu

Distinguished Features of Muon Colliders Physics Potential 273
    F. F. Tikhonin

New Facility for Fundamental Physics with Polarized Cold Neutrons 280
    J. Sromicki, K. Bodek, P. Böni, N. Danneberg, W. Fetscher,
    C. Hilbes, S. Kistryn, J. Lang, M. Lüthy, M. Markiewicz,
    A. Pusenkov, A. Schebetov, and A. Serebrov

---

*Italicized name indicates author who presented the paper.

A Proposed Measurement of the β Asymmetry in Neutron Decay with the Los Alamos Ultra-Cold Neutron Source 286

B. *Tipton*, A. Alduschenkov, K. Asahi, T. Bowles, B. Filippone,
M. Fowler, P. Geltenbort, F. Hartmann, R. Hill, A. Hime,
M. Hino, S. Hoedl, G. Hogan, T. Ito, C. Jones, T. Kawai,
A. Kharitonov, K. Kirch, T. Kitagaki, S. Lamoreaux, M. Lassakov,
C.-Y. Liu, M. Makela, J. Martin, R. McKeown, C. Morris,
A. Pichlmaier, M. Pitt, Y. Rudnev, A. Saunders, S. Seestrom,
A. Serebrov, D. Smith, K. Soyama, M. Utsuro, A. Vasilev,
B. Vogelaar, P. Walstrom, J. Wilhelmy, A. R. Young, and J. Yuan

Current Status of the CHORUS Experiment at CERN 292

K. *Kodama*

Parity Violation in p-p Scattering: the TRIUMF Experiment 298

S. A. *Page*, A. R. Berdoz, J. Birchall, J. B. Bland, J. D. Bowman,
J. R. Campbell, C. A. Davis, A. A. Green, P. W. Green, A. A. Hamian,
D. C. Healey, R. Helmer, Y. Kuznetsov, L. Lee, C. D. P. Levy,
R. E. Mischke, W. D. Ramsay, S. D. Reitzner, G. Roy, P. W. Schmor,
A. M. Sekulovich, J. Soukup, G. M. Stinson, T. Stocki, V. Sum,
N. A. Titov, W. T. H. van Oers, R. J. Woo, and A. N. Zelenski

Asymmetric Quarks in the Proton 305

W. *Melnitchouk*

Status Report and Preliminary Results of the KLOE Detector at the DAΦNE ϕ-factory 311

G. *Venanzoni* for the KLOE Collaboration

Neutrino Oscillation Results from CERN 321

K. *Varvell*

Rare Kaon Decays 333

T. *Numao*

Experiments Searching for Lepton Number Violation 341

K. P. *Jungmann*

List of Participants 355
Author Index 359

*Italicized name indicates author who presented the paper.

Preface

The 3rd International Symposium on Symmetries in Subatomic Physics (Symm2000) was held at the Adelaide Hilton International from March 13-17, 2000. The Special Research Centre for the Subatomic Structure of Matter (CSSM) at the University of Adelaide took responsibility for the organisation.

There were 85 attendees at Symm2000, from 14 countries. The study of the symmetry principles, which govern the Universe in which we live, is absolutely fundamental to modern subatomic physics. Our quantum field theories are built around these symmetries while their occasional violation not only surprises or delights us but can also offer deep insight into the dynamics of complicated systems. In this meeting the practitioners of the improbable and unlikely came from around the world to present a wonderful status report of recent results and future plans. From neutrino oscillations to B-factories, from beta decay to colliders to masers, we were privileged to hear the latest theoretical and experimental developments in this field. These proceedings present a valuable snapshot of the state of the art.

IUPAP as well as the CSSM and the National Institute for Theoretical Physics sponsored Symm2000. On behalf of all the participants we would like to express our thanks to these institutions. The International Advisory Committee provided excellent advice on possible speakers and topics while the Program Committee did a great job of turning these ideas into reality.

Much of the organisational work fell onto the capable shoulders of Sharon Johnson and Sara Boffa. They did a marvellous job, at times under pressure of an enormous workload and everyone who attended expressed admiration for their work. Ramona Adorjan did a thorough and professional job of ensuring that delegates had effective Internet access. Alicia Thomas also contributed effectively to the organisation on a casual basis. The Local Organising Committee and the students and staff of the CSSM all contributed beyond the call of duty to ensure that the meeting functioned smoothly. Everyone who attended will long remember the excursion with dinner at Hardy's Winery (McLaren Vale). Hopefully it will be remembered together with the charm of the city of Adelaide and many of the overseas delegates will return in future.

Finally, we wish Klaus Jungmann and his team all the best for a successful meeting in Heidelberg in three years time.

A.W. Thomas
Conference Chairman

International Advisory Committee:

Eric Adelberger (Seattle)
Guido Altarelli (Geneva)
Gerald Brown (Stony Brook)
J. David Bowman (Los Alamos)
Andrzej J. Buras (Munich)
Jules Deutsch (Louvain la Neuve)
Hiro Ejiri (Osaka)
Ernest Henley (Seattle)
Pauchy Hwang (Taipei)
Dan Kaplan (Illinois)
Iosif Khriplovich (Novosibirsk)
Alan Kostelecky (Indiana)
Heiri Leutwyler (Bern)
V.M. Lobashev (Moscow)
Bruce McKellar (Melbourne)
Jerry Miller (Washington)

M. Moshinsky (Mexico)
Lev Okun (Moscow)
Roberto Peccei (California)
Norman Ramsey (Harvard)
Patrick Sandars (Oxford)
Lukas Schaller (Fribourg)
Stephen R. Sharpe (Seattle)
Wim van Oers (Winnipeg)
Erich Vogt (TRIUMF)
Hans Weidenmueller (Heidelberg)
Bruce Winstein (Chicago)
Yoshio Yamaguchi (KEK-Tanashi)
Chen Ning Yang (Stony Brook)

Program Committee:

J. Deutsch (Louvin la Neuve)
E. Henley (Seattle)
W.T. van Oers (Winnipeg)
H. Ejiri (Osaka)
B.H.J. McKellar (Melbourne)
L. Schaller (Fribourg)
A.W. Thomas (CSSM)

Latest Experimental Information on ϵ'/ϵ

Yee B. Hsiung*

*Fermi National Accelerator Laboratory
P.O.Box 500, Batavia Illinois 60510

Abstract. We review the latest experimental results in searching for "direct" CP-violation by measuring the CP-violating parameters $Re(\epsilon'/\epsilon)$ in neutral kaon decays. Recent result from Fermilab-KTeV $Re(\epsilon'/\epsilon) = (28.0 \pm 4.1) \times 10^{-4}$, and new preliminary result from CERN-NA48 $Re(\epsilon'/\epsilon) = (14.0 \pm 4.3) \times 10^{-4}$, are presented. Both experiments, though using very different techniques, have now performed very well by collecting millions of events for all four relevant decay modes of $K_{L,S}$ to $\pi^+\pi^-$ and $\pi^0\pi^0$ simultaneously. The current world average on this important measurement is $Re(\epsilon'/\epsilon) = (19.3 \pm 2.4) \times 10^{-4}$ with a $\chi^2/ndf = 11.1/5$, establishing the existence of "direct" CP-violation. The experimental status of such crucial measurements and the future prospects are also discussed here.

INTRODUCTION

Studying symmetry or the lack of symmetry in nature is a powerful tool in modern physics to understand many of its underlying fundamental laws of nature. The big bang is thought to have created equal amounts of matter and antimatter, but to the best of our knowledge, all the antimatter has disappeared along with most of the matter. The answer is not known yet, but the clue lies in understanding the symmetry, or the lack of it, between the basic interaction of matter and antimatter. Three of the most important symmetry operations in physics are: charge conjugation, C, in which particles are replaced by their anti-particles; parity inversion, P, in which all three spatial coordinates are reversed; and time reveral, T. We believe that the "violation" of two of these symmetry operations – charge conjugation and parity inversion, combined as CP – is intimately involved in the dominance of matter over anti-matter in the universe.

CP violation was first discovered in 1964 by Cronin and Fitch [1]. They observed that if you waited long enough for only the long-lived kaon were present, you will see occasionally two-pion decays at a rate of 1000 times smaller than the short-lived kaon decay rate. This could be explained by the mixing phenomena that a small amount (0.23%) wrong CP states was mixed in both long-lived kaon, K_L and short-lived kaon, K_S. For more than 35 years, the CP violation has only been observed in weak decays and so far only in the neutral kaon system, e.g. charge asymmetry

TABLE 1. Experimental measurements on Re(ϵ'/ϵ) since 1986.

Experiments	Year Published	Re(ϵ'/ϵ) ($\times 10^{-4}$)
E731A [7]	1988	(32 ± 30)
NA31 ('86) [8]	1988	(33 ± 11)
E731B (20%) [9]	1990	(-4 ± 15)
E731B (final) [10]	1993	(7.4 ± 5.9)
NA31 (final) [11]	1993	(23.0 ± 6.5)
KTeV (23% '96-'97) [12]	1999	(28.0 ± 4.1)
NA48 ('97) [13]	1999	(18.5 ± 7.3)
NA48 ('98 prelim.) [14]	2000	(12.2 ± 4.9)
Average		(19.3 ± 2.4)
		($\chi^2/ndf = 11.1/5$)

δ_l in K_{e3} and $K_{\mu3}$; η_{+-}, η_{00} and $\eta_{+-\gamma}$ in $K_L \to 2\pi$ and $K_L \to \pi^+\pi^-\gamma$ decays [2]; as well as the recent CP-odd and T-odd angular asymmetry in $K_L \to \pi^+\pi^- e^+ e^-$ [3].

While CP violation can be accommodated within the Standard Model with three generation of quark families [4], we still do not fully understand the *origin* of this violation and do not know whether the Standard Model provides the *sole* source of CP violation or not [5]. The search for a more complete understanding of CP violation has been the driving force behind a variety of recent kaon experiments, such as KTeV, NA48 and KLOE, as well as B-factory experiments. Besides a small amount of unequal mixture can give such tiny effect, called indirect CP-violation parametrized by ϵ; there is also other decay processes [6] in the Standard Model can give a new kind of CP-violation directly, smaller than ϵ, which has only been established very recently [11] [12]. Such effect referred to as "direct" CP-violation parametrized by ϵ' which contributes differently to the decay rates of $K_L \to \pi^+\pi^-$ versus $K_L \to \pi^0\pi^0$ (relative to the corresponding K_S decays), and would be observed as a nonzero value in the ratio of Re(ϵ'/ϵ).

Experimentally we measure the double ratio R of the four 2π decay rates,

$$R = \frac{\Gamma(K_L \to \pi^0\pi^0)/\Gamma(K_S \to \pi^0\pi^0)}{\Gamma(K_L \to \pi^+\pi^-)/\Gamma(K_S \to \pi^+\pi^-)} \approx 1 - 6\text{Re}(\epsilon'/\epsilon). \quad (1)$$

Table 1 lists the Re(ϵ'/ϵ) measurements since 1986 including the most recent results.

The standard Cabbibo-Kobayashi-Maskawa (CKM) model accomodates CP violation with a complex phase in the quark mixing matrix. However, the theoretical calculations of Re(ϵ'/ϵ) are still uncertain depending on several input parameters and on the method used to estimate the hadronic matrix elements [15], though the recent estimates had favored non-zero values somewhat below 10^{-3}. Alternatively, a "superweak" interaction [16] could also produce the observed CP-violating mixing effect (ϵ) but would give Re(ϵ'/ϵ) = 0. Therefore, a non-zero measurement of Re(ϵ'/ϵ) would rule out the possibility that a superweak interaction is the sole source of CP violation, and would establish the "direct" CP-violation phenomenon from the decay process itself.

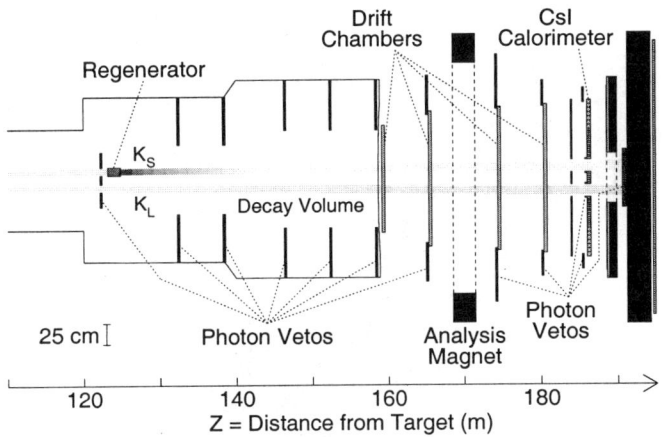

FIGURE 1. Plan view of the KTeV apparatus with double kaon beam as configured to measure Re(ϵ'/ϵ). The evacuated decay volume ends with a thin vacuum window at $Z = 159\ m$ followed by charged spectrometer and CsI calorimeter.

If the direct CP violation exist, not only we would observe a non-zero value for Re(ϵ'/ϵ), but also we would observe very rare direct CP-violating kaon decay modes, such as $K_L \to \pi^0 e^+ e^-$, $K_L \to \pi^0 \mu^+ \mu^-$ and $K_L \to \pi^0 \nu \bar{\nu}$. The probability to observe such rare decays is quite small, less than 10^{-10} or 10^{-11}, current experiments are barely reaching such sensitivity [17] and the search is still on-going.

Experimental Challenges and Methods

To measure the double ratio R we need both K_L and K_S decays with high statistics. High precision electromagnetic (EM) calorimeter is required to measure $\pi^0 \pi^0$ mode which decays to 4γ's, to match the resolution of charged spectrometer for $\pi^+ \pi^-$ mode. Since the γ's and charged pions can not be measured with the same detector element, experiments are looking for techniques would cancel the systematics, such as double kaon beam technique used in E731, KTeV and NA48, moving K_S target station used in NA31, as well as K_S lifetime wighting method used in NA48. Various systematic sources have to be controlled and understood well in such high-rate experiments, such as backgrounds, detector or reconstruction inefficiencies, accidental losses and acceptance corrections, as well as the energy scales and non-linearity of the detector response.

The KTeV experiment (shown in Fig. 1) was designed to improve on the previous experiments and ultimately to have the sensitivity to establish "direct" CP-violation if Re(ϵ'/ϵ) is on the order of 10^{-3} with a sensitivity of 10^{-4}. The experimental technique was essentially the same as in E731 [18] with many improvements in beam and detector performance. Double kaon beams from a BeO target (pro-

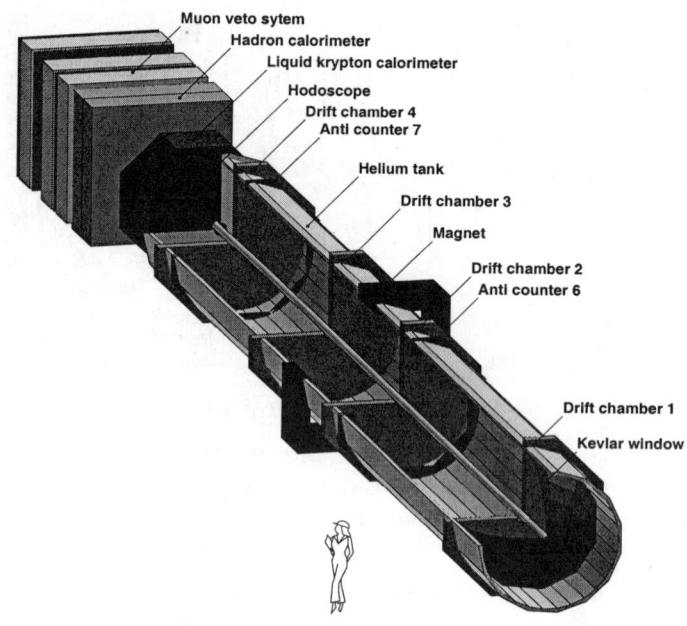

FIGURE 2. NA48 detector layout after the vacuum decay region.

duced from an 800 GeV/c proton beam striking at 4.8 $mrad$ angle with an intensity of 4.5×10^{12} per pulse) was used to enable the simultaneous collection of K_L and K_S decays to minimize the systematics due to time variation of beam flux and detector inefficiencies. A precision magnetic spectrometer (with 412 MeV/c p_T^2 kick in KTeV but 200 MeV/c for E731) was used to minimize backgrounds in the $\pi^+\pi^-$ samples and to allow *in situ* calibration of the calorimeter with electrons. A high precision EM calorimeter, 3100-crystal Cesium Iodide (CsI) array, was used in KTeV instead of lead-glass calorimeter in E731 for $\pi^0\pi^0$ reconstruction and better background suppression. Superb mass resolutions (1.5 MeV/c^2 for $\pi^0\pi^0$ and 1.6 MeV/c^2 for $\pi^+\pi^-$) and photon energy resolution (better than 0.7% above 20 GeV/c^2) were achieved. Nearly hermetic photon vetoes were employed for further background reduction for $\pi^0\pi^0$ mode. A new beamline was constructed for KTeV with cleaner beam collimation and improved muon sweeping. While the method of producing a K_S beam (by passing a K_L beam through a "regenerator") was also the same as E731, the KTeV regenerator was made of scintillator and was fully active to reduce the scattered background to the coherently regenerated K_S. Unlike E731, both $\pi^+\pi^-$ and $\pi^0\pi^0$ data were taken simultaneously in KTeV.

On the other hand, NA31/NA48 started with quite orthogonal techniques. First, NA31 collected data alternately between K_L and K_S mini-run periods with separate

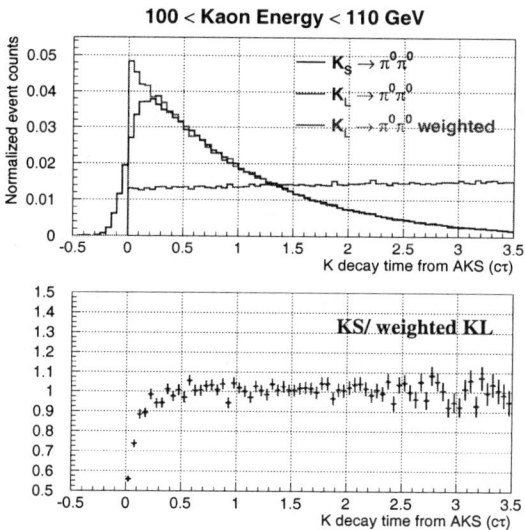

FIGURE 3. NA48 weighting method as shown for $2\pi^0$ decay time distribtion and the ratio.

targets for a single K_L or K_S beam and with different proton beam energy and intensity. In the K_S runs, there were a K_S target train moving along the 50 m decay region to provide 41 K_S target stations mimicking the K_L decay distribution to minimize the acceptance differences and corrections. Non-magnetic spectrometer was used to collect $\pi^+\pi^-$s and a liquid argon calorimeter was used for detecting $2\pi^0$s. The results of NA31 (shown in Table 1) gave a first indication of a possible 3σ non-zero $\text{Re}(\epsilon'/\epsilon)$. The result of E731, consistent with zero, did not confirm such finding.

The NA48 experiment (shown in Fig. 2) was designed to measure $\text{Re}(\epsilon'/\epsilon)$ with an accuracy of 2×10^{-4}. The principle of the NA48 is to use nearly collinear K_L and K_S beams pointing at the center of the detector with similar momentum spectra and to detect all 4 decay modes at the same time. By applying a K_S lifetime weighting procedure to the events of K_L decays in both modes, the difference of acceptance between K_L and K_S can be largely reduced (as shown in Fig. 3).

The K_L beam was produced from a 450 GeV/c proton beam (with intensity 1.5×10^{12} per pulse) striking a Be target at 2.4 $mrad$. A small amount of protons behind the target were channeled and deflected back along the K_L beam after being identified by a tagging hodoscope. The proton beam of intensity 3×10^7 per pulse striking a second target located 120 m downstream and 72 mm above the K_L beam to produce the K_S beam. The beginning of the decay volume was defined for K_S by an anti-counter as a hardware veto to remove K_S decays upstream. The decay region extended over 90 m inside a vacuum tank terminated by a thin kevlar window. Downstream of the vacuum decay region, the magnetic spectrometer (with 265 MeV/c p_t kick) was used for $\pi^+\pi^-$ with a mass resolution of 2.5 MeV/c^2. A quasi-homogeneous liquid krypton (LKr) calorimeter with 13212 projective towers was used for the precision measurement of neutral mode. The reconstructed π^0 mass resolution for $2\pi^0$ decays was about 1 MeV/c^2. Only events within 3.5 τ_S lifetime would be accepted for the analysis as shown in Fig. 3. High speed, high bandwidth data acquisition system and pipeline readout systems were built for collecting large quantity of good data (e.g. 50 Tbytes per year).

KTEV RESULTS AND STATUS

KTeV has been taking high statistics 2π data in two periods, 1996-1997 and 1999 fixed target runs at Fermilab. The statistics for 1996-1997 data sample gives about 4 million $K_L \to \pi^0\pi^0$ (the limiting statistics mode) and for 1999 run about 4.5 million $K_L \to \pi^0\pi^0$ decays.

First result of Re(ϵ'/ϵ) based on a sub-sample from 23% of 1996-1997 data sample of KTeV were published in 1999 [12]. The Re(ϵ'/ϵ) was extracted from the background-subtracted data using a fitting program which analytically calculates regeneration and decay distributions accounting for $K_S - K_L$ interference. The net yields after background subtraction are 2.607M $\pi^+\pi^-$ events in the vacuum beam, 4.516M $\pi^+\pi^-$ in the regenerator beam, 862K $\pi^0\pi^0$ in the vacuum beam and 1.434M $\pi^0\pi^0$ in the regenerator beam. After the monte carlo acceptance correction, the resulting prediction for each decay mode is integrated over Z and compared to data in 10 GeV bins of kaon energy. CPT symmetry is assumed, and the values of $K_S - K_L$ mass difference (Δm) and K_S lifetime (τ_S) are fixed to PDG values [2]. The regeneration amplitude is allowed to float in the fit, but constrained to have a power law dependence on kaon energy, with the phase determined by analyticity [18] [19]. The kaon energy distribution are also allowed to float for $\pi^+\pi^-$ and $\pi^0\pi^0$ modes in each energy bin (24 fit parameters in all).

Fitting was done "blind", by hiding the value of Re(ϵ'/ϵ) with an unknown offset between η_{+-} and η_{00}, until after the analysis and systematic error evaluation were finalized. The final fit result is Re(ϵ'/ϵ) = $(28.0 \pm 3.0) \times 10^{-4}$, where the error is statistical only with a χ^2 equals 30 for 21 degrees of freedom. Table 2 summarize the studies of various systematics where the details can be found in reference [12]. The total systematic error is 2.8×10^{-4}. Effects due to accidental activities were taking into account in the monte-carlo acceptance simulation by overlaying the random accidental triggers taken during the run on top of monte-carlo events. Figure 4 shows the data vs monte-carlo comparisons for the systematic studies of acceptance. Several cross-checks on this Re(ϵ'/ϵ) result have been performed in this analysis. Consistent values were obtained at all kaon energies (see Fig. 5), and there was no significant variation as a function of time or beam intensity.

KTeV has measured Re(ϵ'/ϵ) = $(28.0 \pm 3.0\ (stat) \pm 2.8\ (syst)) \times 10^{-4}$; combining errors in quadrature, Re(ϵ'/ϵ) = $(28.0 \pm 4.1) \times 10^{-4}$. This result, nearly 7σ above

TABLE 2. Systematic uncertainties for KTeV $\mathrm{Re}(\epsilon'/\epsilon)$.

Source of Uncertainty	Uncertainty ($\times 10^{-4}$)	
	$\pi^+\pi^-$	$\pi^0\pi^0$
Trigger and Level 3 filter	0.5	0.3
Energy scale, nonlinearity	0.1	0.9
Detector calibration, alignment	0.3	0.4
Analysis cut variations	0.6	0.8
Background subtraction	0.2	0.8
Detector aperture, resolution	0.5	0.5
Drift chamber simulation	0.6	-
Z dependence of acceptance	1.6	0.7
Monte Carlo statistics	0.5	0.9
Kaon flux and physics parameters	0.35	
TOTAL	2.8	

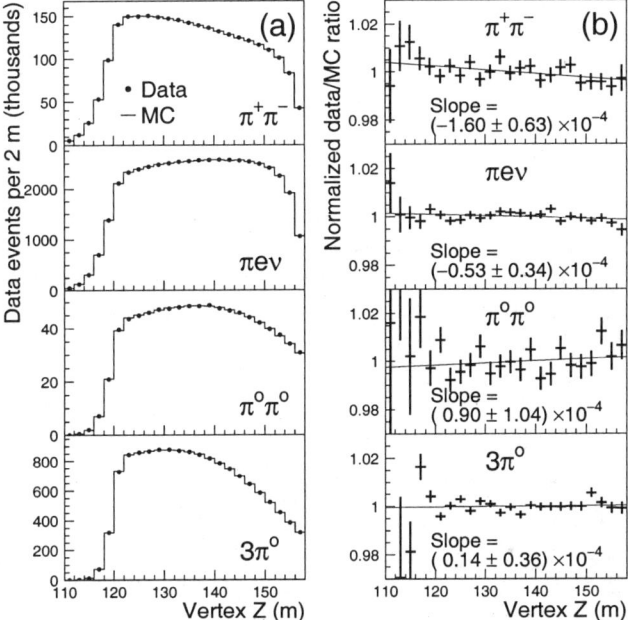

FIGURE 4. (a) Data versus Monte Carlo comparisons of vacuum-beam decay vertex Z distributions for $\pi^+\pi^-$, $\pi e\nu$, $\pi^0\pi^0$, and $3\pi^0$ decays in KTeV. (b) Linear fits to the data/MC ratio of Z distributions for each of the four decay modes.

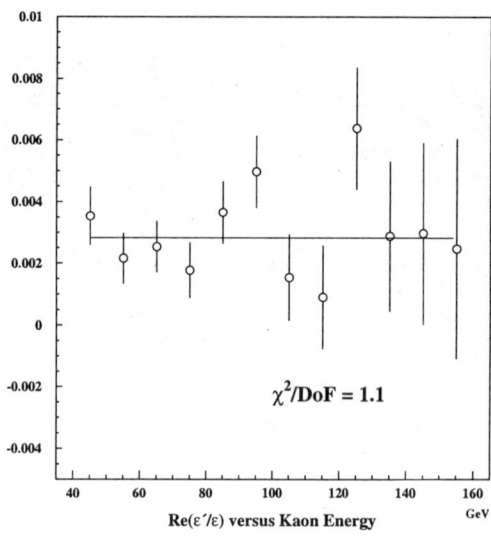

FIGURE 5. A systematic check of Re(ϵ'/ϵ) vs kaon energy for KTeV.

zero, firmly establishes the existence of CP-violation in a "decay process", agreeing better with the earlier measurement from NA31 than with E731 and shows that a superweak interaction cannot be the sole source of CP-violation in the K meson system.

The rest of data from 1997 KTeV run are currently being analyzed to reduce both statistical and systematic uncertainties. More data have been taken in 1999 run with an aim of doubling the statistics with much improved detector performance and additional systematic checks, such as drift chamber efficiency in the beam region, reliable CsI readout electronics and better calibration, intensity studies and regeneration studies. We expect KTeV will reduce the Re(ϵ'/ϵ) statistical uncertainty to $\sim 1 \times 10^{-4}$ and lower the systematics to a similar level.

NA48 RESULTS AND STATUS

NA48 at CERN has also started data taking in 1997, since then they have acquired good statistics in three running periods. The data sample for 1997 run gave 0.49 million $K_L \to \pi^0\pi^0$ decays and for 1998 run 1.14 million, for 1999 run about 2 million $K_L \to \pi^0\pi^0$. First result of Re(ϵ'/ϵ) based on the smaller 1997 data sample from NA48 has been published in 1999 [13]. New preliminary result with the larger 1998 data sample has been anounced recently [14]. The detector performance has been improved over the years, such as fixing the dead stripes and the blocking capacitors in LKr calorimeter, level-2 charged trigger efficiencies as well as stable

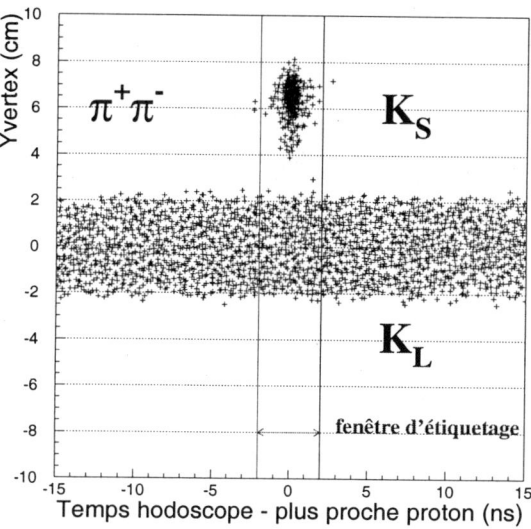

FIGURE 6. Distribution of minimum tagging time-of-flight versus decay y-vertex for the $\pi^+\pi^-$ data sample from NA48 experiment.

detector running and data collection.

The K_L and K_S decays were separated by the tagging system by requiring events fall into $\pm 2ns$ time-of-flight window between the event time in the detector and the time in the tagger as K_S. Events outside this window were treated as K_LS. This can be measured quite well with the charged mode data sample as shown in Fig. 6. The mistagging probability for a K_L to be counted as K_S due to accidentals in the tagging counter was measured to be $\alpha_{LS}^{+-} = (11.05 \pm 0.01)\%$ and the tagging inefficiency for K_S was $\alpha_{SL}^{+-} = (1.97 \pm 0.05) \times 10^{-4}$. Tagging differences between $\pi^0\pi^0$ and $\pi^+\pi^-$ mode would bias the Re(ϵ'/ϵ) measurement. For neutral mode, events with π^0 Dalitz decays or photon conversion as well as a special K_S only run limited the difference between α_{SL}^{00} and α_{SL}^{+-} to be less than 0.5×10^{-4}. The mistagging probability for $\pi^0\pi^0$ mode can only be checked by comparing the side bands in untagged K_L time window between $\pi^0\pi^0$ and $\pi^+\pi^-$ and comparing the $K_L \to 3\pi^0$ in the tagging window with the untagged $K_L \to \pi^+\pi^-$ in the side bands. The corrections and uncertainties are shown in Table 3.

The remaining acceptance correction (about -5×10^{-4} after K_S lifetime weighting were estimated by monte-carlo simulation. The correction for charged trigger was due to the level-2 trigger efficiency $(91.68 \pm 0.09)\%$ in 1997 data and $(97.75 \pm 0.05)\%$ in 1998.

The statistics for 1998 data sample after background subtraction and mistagging corrections were 4.87M $K_L \to \pi^+\pi^-$ events, 7.46M $K_S \to \pi^+\pi^-$, 1.14M $K_L \to \pi^0\pi^0$

TABLE 3. Corrections and systematic uncertainties for Re(ϵ'/ϵ) for 1997 data analysis and 1998 preliminary result from NA48. Units on Re(ϵ'/ϵ) are (10^{-4}).

Source of Uncertainty	1997 data correction	error	1998 preliminary correction	error
Charged trigger	-1.5	±3.8	+0.2	±1.8
Mistagging probability	-3.0	±1.5	-0.2	±1.3
Tagging efficiency	-	±1.0	-	±0.5
Neutral scale	-	±2.0	-	±1.7
Charge vertex	-	±0.8	-0.3	±0.3
Acceptance	-4.8	±2.0	-5.2	±1.5
Neutral background	+1.3	±0.3	+1.2	±0.3
Charged background	-3.8	±0.7	-3.2	±0.5
Beam scattering	+2.0	±0.5	+1.7	±0.5
Accidental activity	+0.3	±2.3	-0.3	±2.0
TOTAL	-9.5	±5.8	-6.2	±4.0

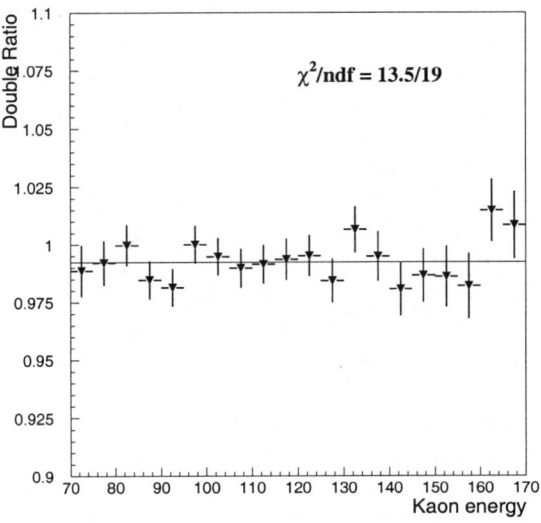

FIGURE 7. A systematic check of double ratio R vs kaon energy for NA48.

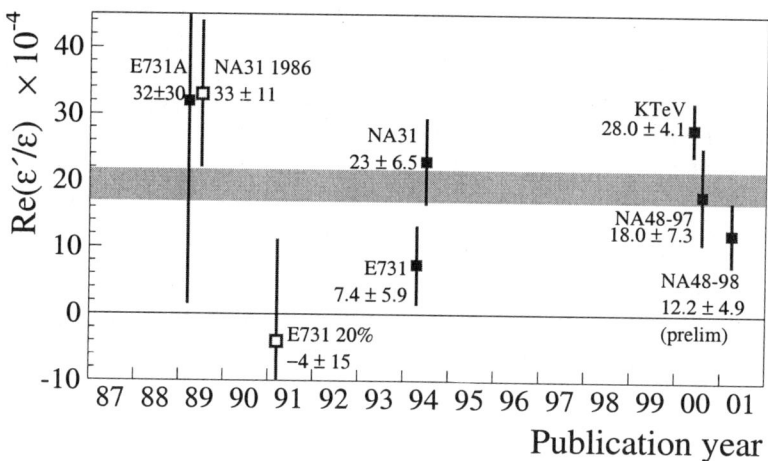

FIGURE 8. Comparison of recent Re(ϵ'/ϵ) measurements.

and 1.80M $K_S \to \pi^0\pi^0$. The final result was obtained after applying a series of small corrections as shown in Table 3 with systematic errors. A systematic check of double ratio R versus kaon energy is shown in Fig. 7. Consistent values were obtained at all kaon energies and there was no significant variation as a function of time and cut variations. Several cross checks has been performed in this analysis.

The result of 1997 data was $Re(\epsilon'/\epsilon) = (18.5 \pm 4.5(stat) \pm 5.8(syst)) \times 10^{-4}$ and of the preliminary 1998 measurement was $Re(\epsilon'/\epsilon) = (12.2 \pm 2.9(stat) \pm 4.0(syst)) \times 10^{-4}$. Combining both together, the new NA48 result was $Re(\epsilon'/\epsilon) = (14.0 \pm 4.3) \times 10^{-4}$, though a 3σ above zero it agrees better with E731 result than KTeV (2.4σ difference). Clearly more precise measurements in the near future are needed from both NA48 and KTeV to clarify such discrepancy.

NA48 will take more data (another 25-30%) in year 2001 at CERN after rebuilding all the damaged drift chambers due to an accident of carbon fiber vacuum pipe implosion at the end of 1999. More systematic studies will be done during this run.

SUMMARY

The average of all the measurements from KTeV, NA48, NA31 and E731 is $Re(\epsilon'/\epsilon) = (19.3 \pm 2.4) \times 10^{-4}$ with a not so great $\chi^2/ndf = 11.1/5$. While this result is at the high end of standard-model predictions which supports the notion of a nonzero phase in the CKM matrix, further theoretical and experimental advances are needed before one can say whether or not there are other sources of CP violation beyond the standard model. Figure 8 shows the trend of recent experimental results on $Re(\epsilon'/\epsilon)$ since 1986. In next few years we expect $Re(\epsilon'/\epsilon)$ to be precisely measured by experiments (such as KTeV, NA48 and KLOE) to 5-10% of itself, which would challenge the theorists to further refine their calculations to understand the origin of direct CP violation. Such result may well be the most precise measurement in search for "direct" CP-violation in the next 5 to 10 years before upcoming B-physics experiments and the next generation $K_L \to \pi^0 \nu \bar{\nu}$ experiments. The $K_L \to \pi^0 \nu \bar{\nu}$ decay though very challenging experimentally, is essentially pure direct CP violation and can be calculated theoretically very cleanly and precisely [20]. Its branching ratio depends directly on the CP-violating phase of the Standard CKM Model with little theoretical uncertainty. Therefore, an observation of $K_L \to \pi^0 \nu \bar{\nu}$ events in the predicted range would measure directly the magnitude of CP-violating phase in CKM matrix elements. An observation of $K_L \to \pi^0 \nu \bar{\nu}$ outside the range predicted by standard model would indicate interesting new physics [21]. Therefore, by over-constraint the parameters of CKM unitarity triangle through both B and K decays will be the ultimate test to the Standard Model.

REFERENCES

1. J.H. Christenson, J.W. Cronin, V.L. Fitch and R. Turlay, *Phys. Rev. Lett.* **13**, 138 (1964).
2. Particle Data Group, C. Caso et al. *Eur. Phys. J.* C **3**, 1 (1998).
3. A. Alavi-Harati et al. (KTeV collaboration), *Phys. Rev. Lett.* **84**, 408 (2000).
4. M. Kobayashi and T. Maskawa, *Prog. Theo. Phys.* **49**, 652 (1973).
5. B. Winstein and L. Wolfenstein, *Rev. Mod. Phys.* **65**, 1113 (1993).
6. A.J. Buras in *Probing the Standard Model of Particle Interactions*, ed. R. Gupta et al., Amsterdam: Elsevier Science B.V., 1999, pp. 281; hep-ph/9806471.
7. M. Woods et al., *Phys. Rev. Lett.* **60**, 1695 (1988).
8. H. Burkhardt et al., *Phys. Lett.* B **206**, 169 (1988).
9. J.R. Patterson et al., *Phys. Rev. Lett.* **64**, 1491 (1990).
10. L. K. Gibbons et al., *Phys. Rev. Lett.* **70**, 1203 (1993).
11. G. D. Barr et al., *Phys. Lett.* B **317**, 233 (1993).
12. A. Alavi-Harati et al. (KTeV collaboration), *Phys. Rev. Lett.* **83**, 22 (1999).
13. V. Fanti et al. (NA48 collaboration), *Phys. Lett.* B **465**, 335 (1999).
14. M. Lenti, presented at XXXVth Rencontres de Moriond - QCD and High Energy Hadronic Interaction, March 18-25, 2000; see also presentation of CERN Particle Physics Seminar by A. Ceccucci, Feb. 29, 2000.
15. A.J. Buras in this proceedings.
16. L. Wolfenstein, *Phys. Rev. Lett.* **13**, 569 (1964).
17. A. Alavi-Harati et al. (KTeV collaboration), *Phys. Rev. Lett.* **84**, 5279 (2000).
18. L. K. Gibbons et al., *Phys. Rev.* D **55**, 6625 (1997).
19. R.A. Briere and B. Winstein, *Phys. Rev. Lett.* **75**, 402 (1995).
20. G. Buchalla and A. J. Buras, *Phys. Rev.* D **54**, 6782 (1996).
21. Y. Grossman and Y. Nir, *Phys. Lett.* B **398**, 163 (1997).

CP Violation 2000: Status and Perspectives

Andrzej J. Buras

Technische Universität München, Physik Department
D-85748 Garching, Germany

Abstract. We summarize the present status of CP violation discussing in particular the standard analysis of the unitarity triangle and the ratio ε'/ε including most recent developments. The perspectives in this field include in particular B decays, rare K decays such as $K_L \to \pi^0 \nu \bar{\nu}$ and $K_L \to \pi^0 e^+ e^-$ and the improved standard analysis of the unitarity triangle through the measurement of $B_s^0 - \bar{B}_s^0$ mixing.

INTRODUCTION

One of the central issues of elementary particle physics is the question whether the Standard Model (SM) of fundamental interactions is capable of describing the violation of CP symmetry observed in nature. Actually this question has already been answered through the studies of a dynamical generation of the baryon asymmetry in the universe, which is necessary for our existence. It turns out that the size of CP violation in the SM is too small to generate a large enough matter-antimatter asymmetry to produce the baryon number to entropy ratio observed in the universe today.

On the other hand it is conceivable that the physics responsible for the baryon asymmetry involves only very large scales as the GUT scale or the Planck scale and the related CP violation is unobservable in the experiments performed by humans. Yet even such unfortunate situation is a real possibility, it is unlikely that the SM provides an adequate description of CP violation at scales accessible to experiments peformed on our planet in this millennium. On the one hand the Kobayashi-Maskawa (KM) picture of CP violation [1] is so economical that it is hard to believe that it will pass future experimental tests. On the other hand almost any extention of the SM contains additional sources of CP–violating effects. As some kind of new physics is required in order to understand the patterns of quark and lepton masses and generally to understand the flavour dynamics responsible for their mixing, it is very likely that this physics will bring new sources of CP violation modifying KM picture considerably.

This talk gives a brief personal view on the status of CP violation as of June 2000 and outlines some future perspectives. Due to the lack of space the list of references is by far incomplete and I apologize for this from the beginning. They can be found in a number of reviews published during the last years, in particular by Buchalla, Fleischer, Gronau, Nir, Peccei, Rosner and myself.

CP VIOLATION IN THE STANDARD MODEL

According to Kobayashi and Maskawa [1], CP violation in the Standard Model is supposed to arise from a single phase in the unitary 3×3 matrix $\hat{V}_{\mathrm{CKM}}$ which parametrizes the charged current interactions of quarks

$$J_\mu^{cc} = (\bar{u}, \bar{c}, \bar{t})_L \gamma_\mu \begin{pmatrix} V_{ud} & V_{us} & V_{ub} \\ V_{cd} & V_{cs} & V_{cb} \\ V_{td} & V_{ts} & V_{tb} \end{pmatrix} \begin{pmatrix} d \\ s \\ b \end{pmatrix}_L. \tag{1}$$

The matrix $\hat{V}_{\mathrm{CKM}}$, the Cabibbo-KM matrix [1,2], connects the mass eigenstates of the down quarks (d, s, b) to the corresponding flavour eigenstates. It can be parametrized in the standard manner [3] in terms of three sines of mixing angles and one complex phase δ:

$$s_{12} = |V_{us}|, \quad s_{13} = |V_{ub}|, \quad s_{23} = |V_{cb}|, \quad \delta \tag{2}$$

where the first three entries are excellent approximations.

A more transparent but approximate parametrization due to Wolfenstein [4] uses as parameters the set $(\lambda, A, \varrho, \eta)$ with $\lambda = |V_{us}| = 0.22$ playing the role of an expansion parameter and η representing the CP violating phase:

$$\hat{V}_{CKM} = \begin{pmatrix} 1 - \frac{\lambda^2}{2} & \lambda & A\lambda^3(\varrho - i\eta) \\ -\lambda & 1 - \frac{\lambda^2}{2} & A\lambda^2 \\ A\lambda^3(1 - \varrho - i\eta) & -A\lambda^2 & 1 \end{pmatrix} + O(\lambda^4) \tag{3}$$

In order to study CP violation it is necessary to include $\mathcal{O}(\lambda^4)$ and $\mathcal{O}(\lambda^5)$ terms. As in any perturbative expansion these higher order terms are not unique. A particular prescription for finding these terms, adopted by many authors in the literature, has been proposed in [5]. It gives, in particular, up to $\mathcal{O}(\lambda^6)$ corrections

$$V_{us} = \lambda, \quad V_{cb} = A\lambda^2, \tag{4}$$

$$V_{ub} = A\lambda^3(\varrho - i\eta), \quad V_{td} = A\lambda^3(1 - \bar{\varrho} - i\bar{\eta}), \tag{5}$$

$$\mathrm{Im} V_{ts} = -\eta A\lambda^4, \quad \mathrm{Im} V_{cd} = -\eta A^2 \lambda^5 \tag{6}$$

where

$$\bar{\varrho} = \varrho(1 - \frac{\lambda^2}{2}), \qquad \bar{\eta} = \eta(1 - \frac{\lambda^2}{2}). \qquad (7)$$

With these formulae at hand it is an easy matter to construct the unitarity triangle that is obtained by using the unitarity relation

$$V_{ud}V_{ub}^* + V_{cd}V_{cb}^* + V_{td}V_{tb}^* = 0, \qquad (8)$$

rescaling it by $\mid V_{cd}V_{cb}^* \mid = A\lambda^3$ and depicting the result in the complex $(\bar{\rho}, \bar{\eta})$ plane as shown in fig. 1. The lenghts CB, CA and BA are equal respectively to 1,

$$R_b \equiv \sqrt{\bar{\varrho}^2 + \bar{\eta}^2} = (1 - \frac{\lambda^2}{2})\frac{1}{\lambda}\left|\frac{V_{ub}}{V_{cb}}\right| \quad \text{and} \quad R_t \equiv \sqrt{(1-\bar{\varrho})^2 + \bar{\eta}^2} = \frac{1}{\lambda}\left|\frac{V_{td}}{V_{cb}}\right|. \qquad (9)$$

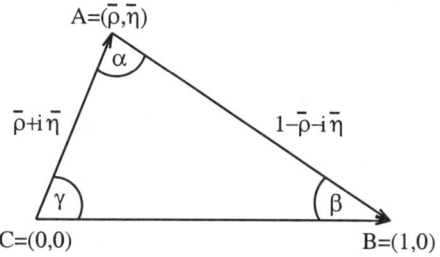

FIGURE 1. Unitarity Triangle.

The triangle in fig. 1, $\mid V_{us} \mid$ and $\mid V_{cb} \mid$ give the full description of the CKM matrix. We observe that the measurements of four CP *conserving* decays sensitive to $|V_{us}|$, $|V_{cb}|$, $|V_{ub}|$ and $|V_{td}|$ can tell us whether CP violation ($\bar{\eta} \neq 0$) is predicted in the SM. This fact is often used to determine the angles of the unitarity triangle without the study of CP violating quantities.

Indeed, measuring the ratio $|V_{ub}/V_{cb}|$ in tree-level decays and $|V_{td}|$ through $B_d^0 - \bar{B}_d^0$ mixing allows to determine R_b and R_t respectively. If so determined R_b and R_t satisfy

$$1 - R_b < R_t < 1 + R_b \qquad (10)$$

then $\bar{\eta}$ is predicted to be non-zero on the basis of CP conserving transitions in the B-system alone without any reference to CP violation discovered in $K_L \to \pi^+\pi^-$ in 1964 [6]. Moreover one finds

$$\bar{\eta} = \pm\sqrt{R_b^2 - \bar{\varrho}^2}, \qquad \bar{\varrho} = \frac{1 + R_b^2 - R_t^2}{2}. \qquad (11)$$

The natural question then arises whether the values of $\bar{\eta}$ and $\bar{\varrho}$ extracted in this manner are compatible with future measurements of CP asymmetries in B decays, with CP violation observed in K_L decays and with future measurements of $K_L \to \pi^0 \nu \bar{\nu}$ and $K_L \to \pi^0 e^+ e^-$. More generally using the language of the unitarity triangle the question is whether the various curves in the $(\bar{\varrho}, \bar{\eta})$ plane extracted from different decays and transitions will cross each other at a single point as shown in fig. 2 and whether the angles (α, β, γ) in the resulting triangle will agree with those extracted from CP-asymmetries in B decays and CP-conserving B decays. It is truly exciting that during the present decade we should be able to answer all these questions and in the case of the inconsistencies in the $(\bar{\varrho}, \bar{\eta})$ plane get some hints about the physics beyond the SM. One obvious inconsistency would be the violation of the constraint (10).

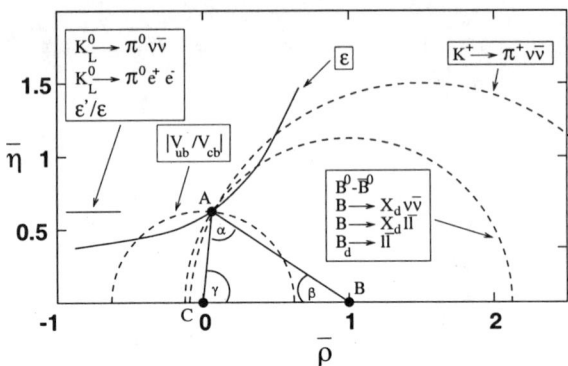

FIGURE 2. The ideal Unitarity Triangle.

In order to extract the curves in fig. 2 from measured branching ratios and CP asymmetries one needs not only very accurate experiments but also sufficiently accurate theoretical calculations. Now the amplitude for an *exclusive* decay of a meson M into a final state F can be written with the help of the Operator Product Expansion (OPE) and the renormalization group techniques as follows

$$A(M \to F) = \frac{G_F}{\sqrt{2}} V_{\text{CKM}} \sum_i C_i(\mu) \langle F \mid Q_i(\mu) \mid M \rangle \qquad (12)$$

where V_{CKM} denotes the relevant CKM factor. Here Q_i are local operators generated by QCD and electroweak interactions. They are built out of quark and lepton fields and govern "effectively" the decays in question. $C_i(\mu)$, the Wilson coefficients of Q_i, give the strength with which a given operator enters the expansion in (12). They can be considered as scale dependent couplings related to the vertices Q_i and can be calculated using perturbative techniques as long as μ is not too small. They are known in the Standard Model for all the relevant operators including leading and next-to-leading QCD and QED corrections [7]. Typically $\mu = \mathcal{O}(m_b)$ and $\mu = \mathcal{O}(1-2\,\text{GeV})$ for B- and K-decays respectively.

Unfortunately the evaluation of the hadronic matrix elements $\langle Q_i(\mu) \rangle$ is a non-perturbative question. As presently available non-perturbative methods are subject to sizable uncertainties, the main theoretical uncertainties in the determination of $\hat{V}_{\text{CKM}}$ come from the evaluation of $\langle Q_i(\mu)\rangle$. This is in particular the case of non-leptonic decays, which involve four-quark operators. In this case it is a common practice to express $\langle Q_i(\mu)\rangle$ in terms of non-perturbative factors B_i. For instance:

$$\langle \bar{K}^0 \mid (\bar{s}d)_{V-A}(\bar{s}d)_{V-A} \mid K^0 \rangle = \frac{8}{3} B_K(\mu) F_K^2 m_K^2, \qquad \hat{B}_K = B_K(\mu)\left[\alpha_s(\mu)\right]^{-2/9} \quad (13)$$

where the dimensionless factor $\hat{B}_K$ is μ-independent. In the so-called vacuum insertion method $\hat{B}_K = 1$ but in QCD $\hat{B}_K \neq 1$. We will encounter other B_i factors below. The accuracy of the extraction of V_{CKM} from the data depends sensitively on the accuracy with which the factors B_i can be calculated.

Fortunately in certain situations the factors B_i can be experimentally determined or eliminated by taking the ratios of branching ratios and using flavour-symmetry arguments. This can most easily be done in leptonic and semi-leptonic decays for which only the matrix elements of weak currents are needed, but occassionally also in non-leptonic decays.

PRESENT STATUS OF CP VIOLATION

Experiment

CP violation has been observed undisputably only in the decays of the K_L meson. One can distinguish between two types of CP violation: the *indirect* CP violation and the *direct* CP violation. The indirect CP violation is the manifestation of the fact that K_L is not exactly a CP=−1 eigenstate but a linear combination of K_2 (CP=−1) and K_1 (CP=+1): $K_L = K_2 + \bar{\varepsilon}K_1$ with $\bar{\varepsilon} = \mathcal{O}(10^{-3})$. This implies that K_L can decay through the K_1 component into CP=+1 state $\pi\pi$. The complex parameter $\varepsilon \approx \bar{\varepsilon}$ is a measure of this type of CP violation. For the discussion of the difference between ε and $\bar{\varepsilon}$ we refer to [8]. The direct CP violation, on the other hand, is realized via a direct transition between states of different CP parity. In the case at hand through the transition $K_2 \to \pi\pi$. The complex parameter ε' is a measure of this type of CP violation.

Experimentally ε and ε' can be found by measuring the ratios

$$\eta_{00} = \frac{A(K_L \to \pi^0\pi^0)}{A(K_S \to \pi^0\pi^0)}, \qquad \eta_{+-} = \frac{A(K_L \to \pi^+\pi^-)}{A(K_S \to \pi^+\pi^-)}. \quad (14)$$

Indeed, assuming ε and ε' to be small numbers one finds

$$\eta_{00} = \varepsilon - \frac{2\varepsilon'}{1 - \sqrt{2}\omega}, \qquad \eta_{+-} = \varepsilon + \frac{\varepsilon'}{1 + \omega/\sqrt{2}} \quad (15)$$

where $\omega = \text{Re}A_2/\text{Re}A_0 = 0.045$ with $A_{0,2}$ being isospin amplitudes.

In the absence of direct CP violation $\eta_{00} = \eta_{+-}$. The ratio ε'/ε can then be measured through

$$\mathrm{Re}(\varepsilon'/\varepsilon) = \frac{1}{6(1+\omega/\sqrt{2})}\left(1 - \left|\frac{\eta_{00}}{\eta_{+-}}\right|^2\right) \tag{16}$$

It should be remarked [9] that the experimental groups in giving their results for $\mathrm{Re}(\varepsilon'/\varepsilon)$ omitt the term $\omega/\sqrt{2}$ in (16). Yet in order to be consistent with the definitions used by theorists:

$$\varepsilon = \frac{A(K_L \to (\pi\pi)_{I=0})}{A(K_S \to (\pi\pi)_{I=0})}, \tag{17}$$

$$\varepsilon' = \frac{1}{\sqrt{2}}\mathrm{Im}\left(\frac{A_2}{A_0}\right)\exp(i\Phi_{\varepsilon'}), \qquad \Phi_{\varepsilon'} = \frac{\pi}{2} + \delta_2 - \delta_0 \approx \frac{\pi}{4}, \tag{18}$$

this term should be kept. Here $\delta_{0,2}$ are the strong phases corresponding to $A_{0,2}$. Consequently all existing experimental results for $\mathrm{Re}(\varepsilon'/\varepsilon)$ quoted below should be rescaled down by 3.2%. Clearly at present this rescaling is academic as the experimental error in $\mathrm{Re}(\varepsilon'/\varepsilon)$ is roughly ±20% and the theoretical one at least ±50%. We will therfore omit this rescaling in what follows.

The indirect CP violation has been discovered in 1964 in $K_L \to \pi^+\pi^-$ [6] and subsequently measured in $K_L \to \pi^0\pi^0$, $\pi l\nu$, $\pi^+\pi^-\gamma$ and recently in $K_L \to \pi^+\pi^- e^+ e^-$. All these CP violating effects can be described by

$$\varepsilon_{exp} = (2.280 \pm 0.013) \cdot 10^{-3}\ \exp(i\Phi_\varepsilon), \qquad \Phi_\varepsilon = \frac{\pi}{4}. \tag{19}$$

The most recent experimental results for the ratio ε'/ε read

$$\mathrm{Re}(\varepsilon'/\varepsilon) = \begin{cases} (28.0 \pm 4.1) \cdot 10^{-4} & (\text{KTeV})\ [10] \\ (14.0 \pm 4.3) \cdot 10^{-4} & (\text{NA48})\ [11], \end{cases} \tag{20}$$

Together with the older NA31 measurement $((23 \pm 7) \cdot 10^{-4})$ [12], these data confidently establish direct CP violation in nature and taking also the E731 result $((7.4 \pm 5.9) \cdot 10^{-4})$ [13] into account one finds the grand average [11]

$$\mathrm{Re}(\varepsilon'/\varepsilon) = (19.2 \pm 4.6) \cdot 10^{-4} \tag{21}$$

close to the NA31 result but with a smaller error. Here a scale factor [3] of 1.86 has been included in the error to account for the large spread in quoted results. In my opinion even this enlarged error is questionable as the KTeV result appears to be inconsistent with the E731 result. In addition the substantial difference between KTeV and NA48 is disturbing. Let us hope that these issues will be clarified in the coming years. In this context an independent measurement of ε'/ε by KLOE at Frascati will be very important.

Theory

Standard Analysis of the Unitarity Triangle

In the Standard Model ε receives the dominant contributions from the box diagrams with internal $W^\pm$ and top quark exchanges and smaller contributions involving charm quark exchanges. Equating the theoretical result for ε with (19) determines the following hyperbola in the $(\bar{\varrho}, \bar{\eta})$ plane:

$$\bar{\eta} \left[(1 - \bar{\varrho}) A^2 \eta_2 S_0(x_t) + P_c(\varepsilon) \right] A^2 \hat{B}_K = 0.226, \qquad (22)$$

with $\hat{B}_K = 0.80 \pm 0.15$ from various non-perturbative calculations. $P_c(\varepsilon) = 0.31 \pm 0.05$ summarizes the contributions of box diagrams with two charm quark exchanges and the mixed charm-top exchanges. $S_0(x_t) = 0.784 \cdot x_t^{0.76}$ with $x_t = m_t^2/M_W^2$ describes the m_t dependence of the top quark contribution. Finally $\eta_2 = 0.57$ is a short distance QCD correction. The hyperbola (22) is depicted in fig. 2. The main uncertainties in the constraint (22) reside in $\hat{B}_K$ and to some extent in A^4 which multiplies the leading term.

With $A = 0.82 \pm 0.04$ corresponding to $|V_{cb}| = 0.040 \pm 0.002$ and $m_t(m_t) = (165 \pm 5)$ GeV the constraint (22) is consistent with

$$R_b = \sqrt{\bar{\varrho}^2 + \bar{\eta}^2} = 4.44 \cdot \left| \frac{V_{ub}}{V_{cb}} \right| = 0.40 \pm 0.07, \qquad (23)$$

$$R_t = \frac{1}{\lambda} \frac{|V_{td}|}{|V_{cb}|} = 1.0 \cdot \left[\frac{|V_{td}|}{8.8 \cdot 10^{-3}} \right] \left[\frac{0.040}{|V_{cb}|} \right] \qquad (24)$$

where

$$|V_{td}| = 8.8 \cdot 10^{-3} \left[\frac{200 \text{ MeV}}{\sqrt{\hat{B}_{B_d}} F_{B_d}} \right] \left[\frac{170 \text{ GeV}}{\overline{m}_t(m_t)} \right]^{0.76} \left[\frac{\Delta M_d}{0.50/\text{ps}} \right]^{0.5} \sqrt{\frac{0.55}{\eta_B}} \qquad (25)$$

and also with

$$\frac{|V_{td}|}{|V_{ts}|} = \xi \sqrt{\frac{m_{B_s}}{m_{B_d}}} \sqrt{\frac{\Delta M_d}{\Delta M_s}}, \qquad \xi = \frac{F_{B_s} \sqrt{\hat{B}_{B_s}}}{F_{B_d} \sqrt{\hat{B}_{B_d}}}. \qquad (26)$$

Here $\Delta M_{d,s}$ are the mass differences related to $B^0_{d,s} - \bar{B}^0_{d,s}$ mixings and measured to be

$$\Delta M_d = (0.471 \pm 0.016)/\text{ps}, \qquad \Delta M_s > 14.3/\text{ps}. \qquad (27)$$

Finally $\sqrt{\hat{B}_{B_d}} F_{B_d} = (200 \pm 40)$ MeV and $\xi = 1.14 \pm 0.08$ from lattice and QCD sum rules calculations.

Using $(\Delta M_d)_{exp}$ and (26) one derives a useful approximate formula

$$(R_t)_{\max} = 1.0 \cdot \xi \sqrt{\frac{10.2/ps}{(\Delta M_s)_{\min}}}. \tag{28}$$

Using simultaneously (22)–(28) one finds the allowed range for the apex of the unitarity triangle, that is the allowed area in the $(\bar{\varrho}, \bar{\eta})$ plane. There are many analyses of this type done in the literature, in particular those by Ali and London, Schune et al and Stocchi et al. Here for a change I show in fig. 3 the result of an analysis by Stefan Schael. Only the dark region is allowed. From this figure one extracts

$$\alpha = 90° \pm 9°, \qquad \beta = 23° \pm 4° \qquad \gamma = 67° \pm 8° \tag{29}$$

$$\sin 2\beta = 0.73 \pm 0.10, \qquad |V_{td}| = (8.3 \pm 0.5) \cdot 10^{-3} \tag{30}$$

In this analysis Gaussian errors for all input parameters have been used. My own, more conservative analysis that uses scanning for all input parameters gives

$$\alpha = 90° \pm 23°, \qquad \beta = 23° \pm 6° \qquad \gamma = 71° \pm 26° \tag{31}$$

$$\sin 2\beta = 0.72 \pm 0.13, \qquad |V_{td}| = (8.4 \pm 1.4) \cdot 10^{-3} \tag{32}$$

The "true" errors are probably between these two estimates.

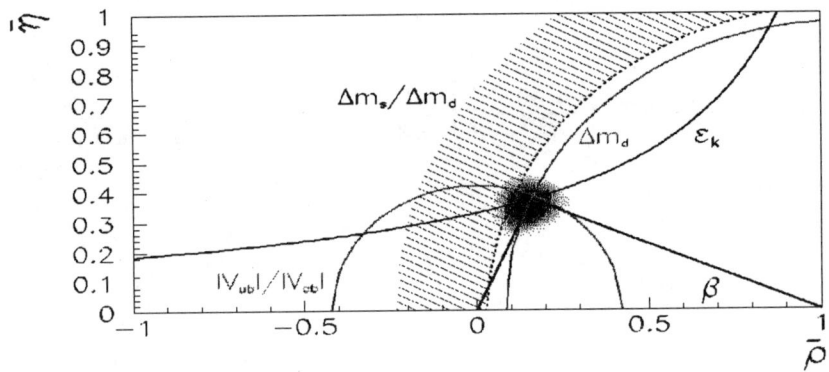

FIGURE 3. The Unitarity Triangle 2000.

We conclude that the Standard Model is capable of describing the observed indirect CP violation in K_L decays, taking into account the data on $B^0_{d,s} - \bar{B}^0_{d,s}$ mixings, $|V_{cb}|$ and $|V_{ub}/V_{cb}|$. We also observe that R_b and R_t from $|V_{ub}/V_{cb}|$ and

$B^0_{d,s} - \bar{B}^0_{d,s}$ mixings alone satisfy the condition (10). Taking $(R_b)_{min} = 0.33$, this condition reads $0.67 < R_t < 1.33$. From (27) and (28) one has $R_t < 1.03$. On the other hand $|V_{td}|_{min}$ is governed by $B^0_d - \bar{B}^0_d$ mixing. From (24) and (32) one has then $R_t > 0.75$. Consequently CP violation in B-decays is predicted on the basis of $|V_{ub}/V_{cb}|$ and $B^0_{d,s} - \bar{B}^0_{d,s}$ mixings alone. Indeed the first results for the CP asymmetry in $B_d \to \psi K_S$ with $\sin 2\beta = 0.73 \pm 0.43$ from CDF establish CP violation in B-decays at 93% C.L. and moreover are consistent with (30) and (32).

$$\varepsilon'/\varepsilon$$

With the values of all CKM parameters at hand we are in the position to calculate $\mathrm{Re}(\varepsilon'/\varepsilon)$ in the SM. Here ε'/ε is governed by QCD-penguin and electroweak penguin operators which contribute to ε'/ε with positive and negative signs respectively. While the Wilson coefficients of these operators are known including NLO corrections, their hadronic matrix elements are subject to large theoretical uncertainties.

In order to appreciate these uncertainties let me recall an approximate formula for ε'/ε presented in [14]

$$\mathrm{Re}(\varepsilon'/\varepsilon) = \mathrm{Im}\lambda_t \cdot F_{\varepsilon'} \tag{33}$$

where $\mathrm{Im}\lambda_t = V_{td}V^*_{ts}$ and

$$F_{\varepsilon'} \approx 13 \left[\frac{110\,\mathrm{MeV}}{m_s(2\,\mathrm{GeV})}\right]^2 \left(\frac{\Lambda^{(4)}_{\overline{MS}}}{340\,\mathrm{MeV}}\right) \left[B^{(1/2)}_6(1 - \Omega_{\eta+\eta'}) - 0.4 B^{(3/2)}_8 \left(\frac{m_t}{165\,\mathrm{GeV}}\right)^{2.5}\right] \tag{34}$$

Here $B^{(1/2)}_6$ and $B^{(3/2)}_8$ are hadronic parameters corresponding to the dominant QCD penguin and electroweak penguin operator respectively. The matrix elements of these operators evaluated in the large N limit exhibit $1/m_s^2$ dependence seen in (34). We note also a linear dependence on $\Lambda^{(4)}_{\overline{MS}}$ which results from NLO calculation of the Wilson coefficients of the penguin operators. This linear dependence is a good approximation for $\Lambda^{(4)}_{\overline{MS}} = (340 \pm 50)$ MeV. $\Omega_{\eta+\eta'} = 0.16 \pm 0.03$ represents isospin breaking corrections [15]. Finally $\mathrm{Im}\lambda_t = (1.33 \pm 0.30) \cdot 10^{-4}$ [14].

The formulae (33) and (34) exhibit very clearly the dominant uncertainties in ε'/ε which reside in the values of $\mathrm{Im}\lambda_t$, m_s, $B^{(1/2)}_6$, $B^{(3/2)}_8$, $\Lambda^{(4)}_{\overline{MS}}$ and $\Omega_{\eta+\eta'}$. Moreover, the partial cancellation between QCD penguin ($B^{(1/2)}_6$) and electroweak penguin ($B^{(3/2)}_8$) contributions requires accurate values of $B^{(1/2)}_6$ and $B^{(3/2)}_8$ for an acceptable estimate of ε'/ε. Because of the accurate value $m_t(m_t) = 165 \pm 5$ GeV, the uncertainty in ε'/ε due to the top quark mass amounts only to a few percent.

There are different opinions whether the grand average in (21) can be accommodated within the SM. We list the results from various groups in table 1. The

TABLE 1. Results for ε'/ε in units of 10^{-4}.

Reference	ε'/ε $[10^{-4}]$
Munich [14]	$7.7^{+6.0}_{-3.5}$ (MC)
Munich [14]	$1.1 \to 28.8$ (S)
Rome [16,17]	$8.1^{+10.3}_{-9.5}$ (MC)
Rome [16,17]	$-13.0 \to 37.0$ (S)
Trieste [18]	23 ± 6 (MC)
Trieste [18]	$13 \to 39$ (S)
Dortmund [19]	$6.8 \to 63.9$ (S)
Montpellier [20]	24.2 ± 8.0
Granada-Lund [21]	34 ± 18
Dubna-DESY [23]	$-3.2 \to 3.3$ (S)
Taipei [22]	$7 \to 16$

labels (MC) and (S) in the second column stand for two error estimates: "Monte Carlo" and "Scanning" respectively. I do not list the values of $B_6^{(1/2)}$ and $B_8^{(3/2)}$ used in these papers as the comparision could be misleading. The reason is that in certain papers these factors are m_s–independent and in other they depend on m_s. Typically $B_6^{(1/2)} = 0.8 - 1.6$ and $B_8^{(3/2)} = 0.6 - 1.2$. Exceptions are the analyses in [20] and [21] where $B_6^{(1/2)}$ in the ballpark of 3 can be found.

In [14,16,17] ε'/ε has been found to be typically by a factor of 2-3 below the data and the KTeV result in (20) could only be accommodated if all relevant parameters were chosen simultaneously close to their extreme values. On the other hand the NA48 result is essentially compatible with [14,16,17] within experimental and theoretical errors. Higher values of ε'/ε than in [14,16,17], in the ballpark of (21), have been found in [18-22]. The result in [23] corresponds to $B_6^{(1/2)} = B_8^{(3/2)} = 1$ as the Dubna-DESY group has no estimate of these non-perturbative parameters. Recent reviews can be found in [18,24,17]. Furthermore it has also been suggested that the final state interactions (FSI) could enhance ε'/ε by a factor of two [18,25,26]. A critical analysis of this suggestion has been presented in [27]. In my opinion the issue of the size of FSI in the evaluation of ε'/ε is still an open question. An interesting recent proposal in [28], if realized, could put the inclusion of FSI in the lattice calculations of hadronic matrix elements in principle under control.

While all theoretical analyses quoted above use the NLO short distance Wilson coefficients calculated by Munich and Rome groups [7], they differ in the evaluation of the hadronic matrix elements of the relevant operators. Being of non-perturbative origin, the latter calculations suffer from large theoretical uncertainties, which at present preclude any firm prediction of ε'/ε within the SM and its extentions. There is a hope, however, that in the coming years the situation may improve through advanced lattice simulations and new analytical studies.

In any case in view of very large theoretical uncertainties and sizable experimental errors there is still a lot of room for non-standard contributions to ε'/ε. Indeed results from NA31, KTeV and NA48 prompted several analyses of ε'/ε within various extensions of the SM, in particular general supersymmetric models [29–32] and models with modified Z^0-penguins [33–35]. Unfortunately these models have many free parameters and are not very conclusive at present. The situation may change in the future when new data from high energy colliders will restrict the possible ranges of parameters involved. For the Minimal Supersymmetric Standard Model (MSSM) this is already the case. A very recent analysis of ε'/ε in MSSM [36] that includes in particular constraints on supersymmetric parameters coming from $B \to X_s \gamma$, $\Delta\varrho$ and the lower bound on the neutral Higgs mass, shows that $(\varepsilon'/\varepsilon)_{\text{MSSM}}$ is hardly distinguishable from $(\varepsilon'/\varepsilon)_{\text{SM}}$. In 1995 deviations of $(\varepsilon'/\varepsilon)_{\text{MSSM}}$ by a factor of two from $(\varepsilon'/\varepsilon)_{\text{SM}}$ were still possible [37].

Finally let us investigate how accurately the SM prediction for ε'/ε could be obtained during this decade. To this end let us assume $B_6^{(1/2)} = 1.2 \pm 0.1$, $B_8^{(3/2)} = 0.80 \pm 0.05$, $m_s(2 \text{ GeV}) = (110 \pm 4)$ MeV and $\hat{B}_K = 0.80 \pm 0.05$. Then Martin Gorbahn finds $\varepsilon'/\varepsilon = (9.9 \pm 1.4) \cdot 10^{-4}$. This exercise is probably optimistic but it shows that on one hand the experimental accuraccy for ε'/ε of $\pm 1.5 \cdot 10^{-4}$ should be aimed for but a better accuracy appears unnecessary.

In summary the SM expectations for ε'/ε are in the ballpark of the experimental data but there is still a lot of room for new physics contributions. In order to reach definite conclusions the calculations of hadronic matrix elements have to be improved considerably. Simultaneously the large spread in experimental results should be clarified. This would help to reduce the experimental error that appears to be dominated by systematics.

PERSPECTIVES

Improved Determinations of the Unitarity Triangle

Here the discovery of $B_s^0 - \bar{B}_s^0$ mixing accompanied by the improved calculations of the non-perturbative ratio ξ in (26) will play the most important role. The measurement of ΔM_s should be achievable in Run II at Fermilab, by HERA-B at DESY and later at LHC. Improved values for ξ should come from unquenched lattice calculations. The latter calculations should also allow more accurate values of $\hat{B}_K$ and $\sqrt{\hat{B}_{d,s}} F_{B_{d,s}}$ required for the ε-hypebola and separate determination of $|V_{td}|$ and $|V_{ts}|$ respectively. It is also expected that the errors on $|V_{cb}|$ and $|V_{ub}/V_{cb}|$ will be decreased through the improved data from B-factories and CLEO III as well as new theoretical, numerical and analytic, efforts. Certainly there is a lot of room for improvements here. While the value of the angle β is already rather constrained, the values of the angles α and γ are still rather uncertain if one wants to be conservative. See (31). The comparision of the more precise shape of the resulting triangle with the results of strategies discussed below will be a very important test of the KM picture of CP violation and will constrain considerably possible patterns of CP violation beyond the SM.

CP Violation in B Decays

During the coming years very important progress in our understanding of CP violation will come through the measurements of CP asymmetries in B-decays as well as various strategies for extracting the angles α, β and γ from two-body B-decays. They are extensively discussed in the working group reports in [39,40] and in [38,41,42].

The CP-asymmetry in the decay $B_d \to \psi K_S$ allows in the SM a direct measurement of the angle β in the unitarity triangle without any theoretical uncertainties [43]. Of considerable interest is also the pure penguin decay $B_d \to \phi K_S$, which is expected to be sensitive to physics beyond the SM. Comparision of β extracted from $B_d \to \phi K_S$ with the one from $B_d \to \psi K_S$ should be important in this respect. An analogue of $B_d \to \psi K_S$ in B_s decays is $B_s \to \psi \phi$. The CP asymmetry measures here η in the Wolfenstein parametrization. It is very small, however, and this fact makes it a good place to look for the physics beyond the SM. In particular the CP violation in $B_s^0 - \bar{B}_s^0$ mixing from new sources beyond the SM should be probed in this decay. Another useful channel for β is $B_d \to D^+ D^-$.

The classic determination of α by means of the time dependent CP asymmetry in the decay $B_d^0 \to \pi^+ \pi^-$ is affected by the "QCD penguin pollution" which has to be taken care of in order to extract α. The recent CLEO results for penguin dominated decays indicate that this pollution could be substantial. The most popular strategy to deal with this "penguin problem" is the isospin analysis of Gronau and London [44]. It requires however the measurement of $Br(B^0 \to \pi^0 \pi^0)$ which is expected to be below 10^{-6}: a very difficult experimental task. For this reason several, rather involved, strategies have been proposed which avoid the use of $B_d \to \pi^0 \pi^0$ in conjunction with $a_{CP}(\pi^+ \pi^-, t)$. They are reviewed in [39,40,42]. For a recent analysis see [45]. It is to be seen which of these methods will eventually allow us to measure α with a respectable precision. It is however clear that the determination of this angle is a real challenge for both theorists and experimentalists.

The two theoretically cleanest methods for the determination of γ are: i) the full time dependent analysis of $B_s \to D_s^+ K^-$ and $\bar{B}_s \to D_s^- K^+$ [46] and ii) the well known triangle construction due to Gronau and Wyler [47] which uses six decay rates $B^\pm \to D_{CP}^0 K^\pm$, $B^+ \to D^0 K^+$, $\bar{D}^0 K^+$ and $B^- \to D^0 K^-$, $\bar{D}^0 K^-$. Both methods are unaffected by penguin contributions. The first method is experimentally very challenging because of the expected large $B_s^0 - \bar{B}_s^0$ mixing. The second method is problematic because of the small branching ratios of the colour suppressed channel $B^+ \to D^0 K^+$ and its charge conjugate, giving a rather squashed triangle and thereby making the extraction of γ very difficult. Variants of the latter method which could be more promising have been proposed in [48,49]. It appears that these methods will give useful results at later stages of CP-B investigations. In particular the first method will be feasible only at LHC-B.

The most recent developments are related to the extraction of the angle γ from the decays $B \to PP$ (P=pseudoscalar). Several of these modes have been observed by the CLEO collaboration [50]. In the future they should allow us to obtain direct information on γ at B-factories (BaBar, BELLE, CLEO III). At present, there are only experimental results available for the combined branching ratios of these modes, i.e. averaged over decay and its charge conjugate, suffering from large uncertainties.

There has been large activity in this field during the last three years. The main issues here are the final state interactions (FSI), SU(3) symmetry breaking effects and the importance of electroweak penguin contributions. Several interesting ideas have been put forward to extract the angle γ in spite of large hadronic uncertainties in $B \to \pi K$ decays [51–55,57]. Recent reviews can be found in [58,54,59].

Three strategies for bounding and determining γ have been proposed. The "mixed" strategy [51] uses $B_d^0 \to \pi^\mp K^\pm$ and $B^\pm \to \pi^\pm K$. The "charged" strategy [57] involves $B^\pm \to \pi^0 K^\pm$, $\pi^\pm K$ and the "neutral" strategy [54] the modes $B_d^0 \to \pi^\mp K^\pm$, $\pi^0 K^0$. General parametrizations for the study of the final state interactions, SU(3) symmetry breaking effects and the importance of electroweak penguin contributions in these channels have been presented in [53–55]. In this context the QCD factorization approach presented in [56] appears to be very promising.

Remarkably, already CP-averaged $B \to \pi K$ branching ratios may imply interesting bounds on γ [51,57,54] which may remove a large portion of the allowed range from the analysis of the unitarity triangle. In particular combining the neutral and charged strategies [54] one finds that the most recent CLEO data favour γ in the second quadrant, which is in conflict with the standard analysis of the unitarity triangle as we have seen in section 3. Other arguments for $\cos\gamma < 0$ using $B \to PP$, PV and VV decays were recently given in [60]. Simultaneously to γ, the CLEO data provide some information on strong phases. The present pattern of these phases indicates either new-physics contributions to the electroweak penguin sector, or a manifestation of large non-factorizable $SU(3)$-breaking effects [54]. There is no doubt that these strategies will be useful in the future.

New strategies for γ using the U-spin symmetry have been proposed in [61,62]. The first strategy involves the decays $B_{d,s} \to \psi K_S$ and $B_{d,s} \to D_{d,s}^+ D_{d,s}^-$ [61]. The second strategy involves $B_s \to K^+ K^-$ and $B_d \to \pi^+\pi^-$ [62]. These strategies are unaffected by FSI and are only limited by U-spin breaking effects. They are promising for Run II at FNAL and in particular for LHC-B.

The Decays $K^+ \to \pi^+ \nu\bar{\nu}$ and $K_L \to \pi^0 \nu\bar{\nu}$

Within the SM these decays are loop-induced semileptonic FCNC processes determined only by Z^0-penguin and box diagrams. While $K^+ \to \pi^+ \nu\bar{\nu}$ is CP conserving, $K_L \to \pi^0 \nu\bar{\nu}$ proceeds in the SM almost entirely through direct CP violation [63].

An important virtue of $K \to \pi \nu \bar{\nu}$ is their clean theoretical character. This is related to the fact that the low energy hadronic matrix elements required are just the matrix elements of quark currents between hadron states, which can be extracted from the leading (non-rare) semileptonic decays. Other long-distance contributions are negligibly small. As a consequence of these features, the scale ambiguities, inherent to perturbative QCD, and the value of m_c in $K^+ \to \pi^+ \nu \bar{\nu}$ essentially constitute the only theoretical uncertainties present in the analysis of these decays. These theoretical uncertainties have been considerably reduced through the inclusion of the next-to-leading QCD corrections [64,65]. The present estimates in the SM read [8]

$$Br(K^+ \to \pi^+ \nu \bar{\nu}) = (7.9 \pm 3.1) \cdot 10^{-11}, \qquad Br(K_L \to \pi^0 \nu \bar{\nu}) = (2.8 \pm 1.1) \cdot 10^{-11} \tag{35}$$

where the errors come dominantly from the uncertainties in the CKM parameters. On the experimental side we have

$$Br(K^+ \to \pi^+ \nu \bar{\nu}) = (1.5^{+3.4}_{-1.2}) \cdot 10^{-10}, \qquad Br(K_L \to \pi^0 \nu \bar{\nu}) < 5.9 \cdot 10^{-7}. \tag{36}$$

from the BNL787 collaboration at Brookhaven [66] and from FNAL experiment KTeV–E799 [67] respectively. The latter bound is substantially weaker than the *model independent* bound [68] from isospin symmetry:

$$Br(K_L \to \pi^0 \nu \bar{\nu}) < 4.4 \cdot Br(K^+ \to \pi^+ \nu \bar{\nu}) < 2 \cdot 10^{-9}. \tag{37}$$

The measurement of $Br(K^+ \to \pi^+ \nu \bar{\nu})$ and $Br(K_L \to \pi^0 \nu \bar{\nu})$ can determine the unitarity triangle completely provided m_t and V_{cb} are known [69]. $Br(K^+ \to \pi^+ \nu \bar{\nu})$ and $Br(K_L \to \pi^0 \nu \bar{\nu})$ determine R_t and $\bar{\eta}$ respectively. Using these two branching ratios simultaneously allows to eliminate $|V_{ub}/V_{cb}|$ from the analysis which removes a considerable uncertainty. The most interesting are the determinations of $\text{Im}\lambda_t$ and $\sin 2\beta$ [69]:

$$\text{Im}\lambda_t = \lambda^5 \frac{\sqrt{B_2}}{X(x_t)}, \qquad \sin 2\beta = \frac{2r_s}{1+r_s^2} \tag{38}$$

with

$$r_s(B_1, B_2) = \sqrt{\sigma} \frac{\sqrt{\sigma(B_1 - B_2)} - P_0(X)}{\sqrt{B_2}}. \tag{39}$$

Here we have defined the "reduced" branching ratios

$$B_1 = \frac{Br(K^+ \to \pi^+ \nu \bar{\nu})}{4.11 \cdot 10^{-11}} \qquad B_2 = \frac{Br(K_L \to \pi^0 \nu \bar{\nu})}{1.80 \cdot 10^{-10}}, \tag{40}$$

$\sigma = 1/(1-\lambda^2/2)^2$ and $X(x_t) = 1.51$ for $m_t(m_t) = 165$ GeV.

It should be stressed that $\sin 2\beta$ determined this way depends only on two measurable branching ratios and on the function $P_0(X) = 0.42 \pm 0.06$ which is completely calculable in perturbation theory [64]. Consequently this determination is free from any hadronic uncertainties. $\text{Im}\lambda_t$ as seen in (38) can be obtained from $K_L \to \pi^0 \nu \bar{\nu}$ alone.

Measuring the branching ratios to within $\pm 10\%$ and m_t within ± 3 GeV allows the derminations of $\sin 2\beta$ and of $\text{Im}\lambda_t$ with the errors ± 0.05 and $\pm 5\%$ respectively [69]. $|V_{td}|$ can be determined with an error of $\pm 10\%$.

The accuracy to which $\sin 2\beta$ can be obtained from $K \to \pi \nu \bar{\nu}$ is, in the example discussed above, comparable to the one expected in determining $\sin 2\beta$ from $B_d \to \psi K_S$ prior to LHC experiments. Due to very small theoretical uncertainties involved, the comparison of these two measurements of $\sin 2\beta$ is particularly suited for tests of CP violation in the SM and offers a powerful tool to probe the physics beyond it.

The SM sensitivity for $Br(K^+ \to \pi^+ \nu \bar{\nu})$ is expected to be reached at AGS in the coming years after all data from E787 and from its future version (E949) have been analyzed. Also Fermilab with the Main Injector could measure this decay. Concerning $Br(K_L \to \pi^0 \nu \bar{\nu})$ FNAL-E799 expects to reach the accuracy $\mathcal{O}(10^{-8})$ and a very interesting new experiment at Brookhaven (BNL E926) expects to reach the single event sensitivity $2 \cdot 10^{-12}$ allowing a 10% measurement of the expected branching ratio. There are furthermore plans to measure this gold-plated decay with comparable sensitivity at Fermilab and KEK.

Other Strategies

The three classes of perspectives discussed so far will offer rather clean tests of CP violation providing simultaneously precise measurements of CP-violating phases and of the elements of the CKM matrix. On the other hand there are other avenues to explore. Inspite of the fact that they all are subject to considerable hadronic uncertainties, they all should be explored in order to gain better insight into the physics of CP violation. These are in particular:

- ε'/ε discussed already in sections 2 and 3.

- The decay $K_L \to \pi^0 e^+ e^-$. Here one has three competing contributions: CP-conserving, indirect CP-violating and direct CP-violating. Only the last one can be calculated without hadronic uncertainties. As all contributions are $\mathcal{O}(10^{-12})$ it will be a challange to separate them. KTeV-E799 has the best chance to measure this decay. At present they find $Br(K_L \to \pi^0 e^+ e^-) < 5.6 \cdot 10^{-10}$. On the other hand, DAPHNE could contribute to the estimate of the indirect CP violation in this decay by measuring $K_S \to \pi^0 e^+ e^-$.

- Electric dipole moments. Any measurement of this type would be a discovery of new physics and of flavour-diagonal CP violation.

- $D^0 - \bar{D}^0$ mixing and CP violation in D-decays. As the Standard Model CP-violating effects are expected to be very small in this sector, this is a very good place to look for new physics.

- Search for CP violation in hyperon decays.

- Last but not least one should search for CP violation in neutrino oscillations.

The field of CP violation had already an exciting time over the last 36 years since the discovery of this intriguing and important phenomenon. This decade should provide new insights in the origin of CP violation and it should teach us more about the flavour dynamics. Hopefully it will also give us new hints for the physics beyond the Standard Model.

I would like to thank Tony Thomas for inviting me to this nice Symposium and his crew for hospitality. Without the financial help from Max-Planck Institute for Physics my attandence would not be possible. Last but certainly not least I would like to thank my secretary, Elke Krüger, for such a wonderful help in preparing my transparencies.

REFERENCES

1. Kobayashi M., Maskawa K., *Prog. Theor. Phys.* **49** 652 (1973).
2. Cabibbo N., *Phys. Rev. Lett.* **10** 531 (1963).
3. Particle Data Group, Caso C. et al., Eur. Phys. J. **C3** 1 (1998).
4. Wolfenstein L., *Phys. Rev. Lett.* **51** 1945 (1983).
5. Buras A.J., Lautenbacher M.E. and Ostermaier G. , *Phys. Rev.* **D50** 3433 (1994).
6. Christenson J.H., Cronin J.W., Fitch V.L. and Turlay R., *Phys. Rev. Lett.* **13** 128 (1964).
7. Buchalla G., Buras A.J. and Lautenbacher M.E., *Rev. Mod. Phys.* **68** 1125 (1996).
8. Buras A.J., hep-ph/9806471, hep-ph/9905437.
9. This remark originated in recent discussions with Kleinknecht K. and Wahl H..
10. Alavi-Harati A. et al., *Phys. Rev. Lett.* **83** 22 (1999).
11. Fanti V. et al., *Phys. Lett.* **B465** 335 (1999).
12. Burkhardt H. et al., *Phys. Lett.* **B206** 169 (1988); Barr G.D. et al., *Phys. Lett.* **B317** 233 (1993).
13. Gibbons L.K. et al., *Phys. Rev. Lett.* **70** 1203 (1993).
14. Bosch S., Buras A.J., Gorbahn M., Jäger S., Jamin M., Lautenbacher M.E. and Silvestrini L., *Nucl. Phys.* **B565** 3 (2000).
15. Ecker G., Müller G., Neufeld H. and Pich A., *Phys. Lett.* **B477** 88 (2000); Gardner S. and Valencia G., *Phys. Lett.* **B466** 355 (1999).
16. Ciuchini M., Franco E., Giusti L., Lubicz V. and Martinelli G., hep-ph/9910237.
17. Ciuchini M. and Martinelli G., hep-ph/0006056.
18. Bertolini S., Fabbrichesi M. and Eeg J.O., *Rev. Mod. Phys* **72** 65 (2000); hep-ph/0002234.

19. Hamby T., Köhler G.O., Paschos E.A. and Soldan P.H., *Nucl. Phys.* **B564** 391 (2000).
20. Narison S., hep-ph/0004247.
21. Bijnens J. and Prades J., hep-ph/0005189.
22. Cheng H.-Y., hep-ph/9911202.
23. Belkov A.A., Bohm G., Lanyov A.V. and Moshkin A.A., hep-ph/9907335.
24. Buras A.J., hep-ph/9908395; Jamin M., hep-ph/9911390.
25. Pallante E. and Pich A., *Phys. Rev. Lett.* **84** 2568 (2000).
26. Paschos E.A., hep-ph/9912230.
27. Buras A.J., Ciuchini M., Franco E., Isidori G., Martinelli G. and Silvestrini L., *Phys. Lett.* **B480** 80 (2000).
28. Lellouch L. and Lüscher M., hep-lat/0003023.
29. Keum Y.-Y., Nierste U. and Sanda A.I., *Phys. Lett.* **B457** 157 (1999).
30. Masiero A. and Murayama H., *Phys. Rev. Lett.* **83** 907 (1999).
31. Buras A.J., Colangelo G., Isidori G., Romanino A. and Silvestrini L., *Nucl. Phys.* B **566** 3 (2000).
32. Kagan A. and Neubert M., *Phys. Rev. Lett.* **83** 4929 (1999).
33. Colangelo G. and Isidori G., JHEP 09 009 (1998).
34. Buras A.J. and Silvestrini L., *Nucl. Phys.* B **546** 299 (1999).
35. Chanowitz M.S., hep-ph/9905478(v2).
36. Buras A.J., Gambino P., Gorbahn M., Jäger S., and Silvestrini L., in preparation.
37. Gabrielli E. and Giudice G.F. *Nucl. Phys.* **B433** 3 (1995); Erratum *Nucl. Phys.* **B507** 549 (1997).
38. Nir Y., hep-ph/9911321.
39. The BaBar Physics Book, eds. Harrison P.F. and Quinn H.R., SLAC-R-504 (1998).
40. B Decays at LHC, Ball P. et al., hep-ph/0003238.
41. Fleischer R., *Int. J. of Mod. Phys.* **A12** 2459 (1997).
42. Buras A.J. and Fleischer R., hep-ph/9704376.
43. Bigi I.I.Y. and Sanda A.I., *Nucl. Phys.* **B193** 85 (1981).
44. Gronau M. and London D., *Phys. Rev. Lett.* **65** 3381 (1990).
45. Quinn H.R. and Silva J.P., hep-ph/0001290.
46. Aleksan R., Dunietz I. and Kayser B., *Z.Phys.* **C54** 653 (1992); Fleischer R. and Dunietz I., *Phys. Lett.* **B387** 361 (1996).
47. Gronau M. and Wyler D., *Phys. Lett.* **B265** 172 (1991).
48. Gronau M. and London D., *Phys. Lett.* **B253** 483 (1991). Dunietz I., *Phys. Lett.* **B270** 75 (1991).
49. Atwood D., Dunietz I. and Soni A., *Phys. Rev. Lett.* **B78** 3257 (1997).
50. Cronin-Hennessy D. et al., (CLEO), hep-ex/0001010 and references therein.
51. Fleischer R., *Phys. Lett.* **B365** 399 (1996); Fleischer R. and Mannel T., *Phys. Rev.* **D57** 2752 (1998);
52. Gronau M. and Rosner J.L., *Phys. Rev.* **D57** 6843 (1998);
53. Fleischer R., *Eur. Phys. J.* **C6** 451 (1999);
54. Buras A.J. and Fleischer R., *Eur. Phys. J.* **C11** 93 (1999); hep-ph/0003323.
55. Neubert M., JHEP 9902 014 (1998).
56. Beneke M., Buchalla G., Neubert M. and Sachrajda C.T., *Phys. Rev. Lett.* **83** 1914

(1999).
57. Neubert M. and Rosner J.L., *Phys. Lett.* **B441** 403 (1998); *Phys. Rev. Lett.* **81** 5076 (1998);
58. Fleischer R., hep-ph/9904313.
59. Neubert M., hep-ph/9904321.
60. Hou W-S. and Yang K-Ch., *Phys. Rev.* **D61** 073014 (2000); Hou W-S., Smith J.G. and Würthwein F., hep-ex/9910014.
61. Fleischer R., *Eur. Phys. J.* **C10** 299 (1999).
62. Fleischer R., *Phys. Lett.* **B459** 306 (1999).
63. Littenberg L., *Phys. Rev.* **D39** 3322 (1989).
64. Buchalla G. and Buras A.J., *Nucl. Phys.* **B400** 225 (1993); *Nucl. Phys.* **B412** 106 (1994); *Nucl. Phys.* **B548** 309 (1999).
65. Misiak M. and Urban J., *Phys. Lett.* **B541** 161 (1999).
66. Adler S. et al., *Phys. Rev. Lett.* **79** 2204 (1997); hep-ex/0002015.
67. Alavi-Harati A. et al., *Phys. Rev.* **D61** 072006 (2000).
68. Grossman Y., Nir Y. and Rattazzi R., hep-ph/9701231, in Heavy Flavours II, eds. Buras A.J. and Lindner M., *World Scientific*, page 755 (1998). Nir Y., hep-ph/9904271.
69. Buchalla G. and Buras A.J., *Phys. Lett.* **B333** 221 (1994); *Phys. Rev.* **D54** 6782 (1996).

Latest Results from Super-Kamiokande

Takaaki Kajita
for the Super-Kamiokande collaboration

Research Center for Cosmic Neutrinos, Institute for Cosmic Ray Research, Univ. of Tokyo, Kashiwa-no-ha 5-1-5, Kashiwa, Chiba 277-8582, Japan
E-mail: kajita@icrr.u-tokyo.ac.jp

Abstract. Recent data from Super-Kamiokande on solar and atmospheric neutrinos and their constraints on the neutrino mass and mixing are presented. Also results on proton decay searches are presented.

I SUPER-KAMIOKANDE

Super-Kamiokande is a 50 k·ton water Cherenkov detector which is located at the depth of 2700 meters water equivalent underground. The detector is consisted of two parts, the inner detector and the outer detector. In the inner(outer) detector, Cherenkov photons are detected by 11146(1885) 50(20) cm diameter PMTs. The charge and timing information recorded by the inner detector PMTs are used to reconstruct kinematics of neutrino events. The outer detector is useful to identify incoming cosmic ray muons and exiting particles in neutrino events which occurred in the inner detector. The fiducial volume for various physics analyses is 22.5 kton.

II SOLAR NEUTRINOS

Five solar neutrino experiments [1] [2] [3] [4] [5] confirmed the basic mechanism of the energy generation in the Sun. However, the observed fluxes from these experiments were lower than the Standard Solar Model (SSM) prediction [6]. Neutrino oscillations [7] can explain the existing solar neutrino data naturally. For the definite confirmation of solar neutrino oscillations, experimental results which can not be predicted by any solar models are highly desirable. Furthermore, the parameter region of neutrino oscillations should be uniquely determined. There are 3 (or 4) different parameter regions (solutions) which can explain the existing solar neutrino data. It is predicted that the spectrum of the ^{8}B solar neutrinos should be distorted significantly for the cases of the "small angle MSW solution"

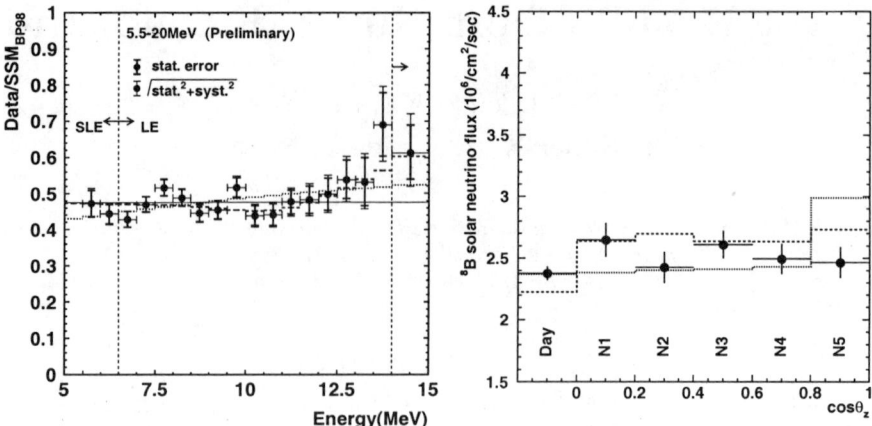

FIGURE 1. Left: Plot of Data/SSM as a function of electron energy. The dotted and dashed histograms show the expected energy spectrum for a typical small-mixing solution ($\sin^2 2\theta = 5 \times 10^{-3}$, $\Delta m^2 = 5 \times 10^{-6} \mathrm{eV}^2$) and the best-fit vacuum-oscillation solution (0.79, $4.3 \times 10^{-10} \mathrm{eV}^2$), respectively. Right: The day-night data. The night time data are divided into 5 bins according to the relative direction to the Sun. The dotted and dashed histograms show the expected day-night effects for a small-mixing solution (0.01, $6 \times 10^{-6} \mathrm{eV}^2$) and a large-mixing solution (0.56, $1.2 \times 10^{-5} \mathrm{eV}^2$), respectively. These parameters are excluded by the day-night data at 99% C.L..

and the "vacuum oscillation solution", or the day-night effect should be observed for a region of the "large mixing MSW solution".

Super-Kamiokande detects ^{8}B solar neutrinos through $\nu e \rightarrow \nu e$. The measured ^{8}B solar ν flux above 6.5 MeV during 825 days of the observation time was $0.475 ^{+0.008}_{-0.007}$(stat.) ± 0.013(sys. of the data) of the SSM prediction [6]. Solar neutrino events in a lower energy region (5.5 to 6.5 MeV) were also observed (524 days of data). The observed flux in this energy range was consistent with the higher energy data, see Figure 1 (left).

Super-Kamiokande provides the information for both the energy spectrum and the day-night effect. The day-night data are shown in Figure 1 (right). The day-time and night-time fluxes were compared as: $Night/Day = 1.067 \pm 0.033$(stat.)$\pm 0.013$(sys.). The possible excess of the flux in the night time was about 2 σ and was not significant.

Figure 1 (left) shows the energy spectrum of the recoil-electrons divided by the SSM prediction. The Data/SSM values in the end-point region were higher than the average. The statistical significance of the possible distortion of the energy spectrum was tested by a χ^2 method. The absolute flux value was assumed to be a free parameter and only the shape of the distributions of the data and the

		Data	MC
Sub-GeV	e-like	2185	2081.8
	μ-like	2178	3137.4
		$R = 0.66 \pm 0.02 \pm 0.05$	
Multi-GeV	e-like	492	481.3
	μ-like(FC+PC)	984(421+563)	1458.9(640.0+818.9)
		$R = 0.66 \pm 0.04 \pm 0.08$	

TABLE 1. Summary of the atmospheric neutrino data and the $(\mu/e)_{data}/(\mu/e)_{MC}$ ratio measurements in Super-Kamiokande (61 kton·yr).

expectations was compared. The χ^2 values were obtained to be 24.3, 25.0 and 17.4 (DOF=17) for the no distortion case (i.e., no oscillation effect), the small-mixing MSW case and the best fit in the vacuum oscillation solution, respectively. The energy spectrum measurement was not conclusive. More data (and probably more experiments) are needed for further understanding of the solar neutrino problem.

III ATMOSPHERIC NEUTRINOS

Cosmic ray interactions in the atmosphere produce neutrinos. The measured values of the (μ/e) ratio (which is closely related to the (ν_μ/ν_e) flux ratio) by Kamiokande were significantly smaller than the Monte Carlo (MC) prediction [8] [9]. Also a zenith-angle dependent deficit of μ-like events was measured by Kamiokande [9] at high energies. These observations strongly suggested neutrino oscillations, and therefore Kamiokande estimated the allowed parameter regions of neutrino oscillations. Because of the relatively small statistics, both $\nu_\mu \to \nu_e$ and $\nu_\mu \to \nu_\tau$ oscillations were allowed. Recently, long baseline reactor experiments [12] [13] excluded the $\nu_\mu \to \nu_e$ solution of the atmospheric neutrino problem. In 1998, Super-Kamiokande reported that the atmospheric neutrino data gave evidence for neutrino oscillations [14]. After that, the statistics of the Super-Kamiokande data were increased by about a factor of two: a total of 8145 fully-contained (FC) events and 563 partially-contained (PC) events were observed in a 61 kton·year exposure.

For the analysis of FC events, only single-ring events were used. Single-ring events were identified as e-like or μ-like. The FC events were separated into "sub-GeV" (E_{vis} < 1330 MeV) and "multi-GeV" (E_{vis} > 1330 MeV) samples, where E_{vis} was defined to be the energy of an electron that would produce the observed amount of Cherenkov light.

Table 1 summarizes the atmospheric neutrino data and the $(\mu/e)_{data}/(\mu/e)_{MC}$ measurements. The observed values of this ratio were significantly smaller than unity and were consistent with the Kamiokande [8] [9], IMB (sub-GeV) [10] and Soudan-2 [11] results.

The zenith angle distributions from Super-Kamiokande are shown in Figure 2.

The μ-like data exhibited a strong up-down asymmetry in zenith angle (Θ) while no significant asymmetry was observed in the e-like data.

Energetic atmospheric ν_μ's passing through the Earth interact with rock surrounding a detector and produce muons via charged current (CC) interactions. These neutrino events are observed as upward going muons. Super-Kamiokande [15] observed 1196 (285) upward through-going (stopping) muon events (including 8.6±0.8 (20.2±8.0)estimated background events) during 1074 (1053) detector live days. Figure 3 shows the zenith-angle distributions of the upward-going muon fluxes. The prediction for the through-going muons had a flatter zenith-angle distribution than the data. Similar data were obtained in Kamiokande [16] and MACRO [17]. The observed flux of upward stopping muons by Super-Kamiokande was significantly smaller than the prediction. These data are explained by neutrino oscillations.

Since the contained events and upward-going muon events consistently suggested neutrino oscillations, an allowed region of the neutrino oscillation parameters assuming pure $\nu_\mu \to \nu_\tau$ oscillations was obtained by using all the atmospheric neutrino data from Super-Kamiokande. The allowed region [18] is shown in Figure 4(left) together with the allowed regions from Kamiokande [16], Soudan-2 [11] and MACRO [17]. The best fit point was found at $(2.8\times10^{-3}\mathrm{eV}^2, 1.0)$. Even if the analysis was extended to the unphysical region of $\sin^2 2\theta > 1$, the best fit point was the same. The 90% C.L. allowed region of the oscillation parameters was; $2 \times 10^{-3} < \Delta m^2 < 5 \times 10^{-3} \mathrm{eV}^2$ and $\sin^2 2\theta > 0.88$.

As an extension of this analysis, Super-Kamiokande carried out a 3 flavor neutrino oscillation analysis. It was assumed that the mass difference between the lightest and the second lightest neutrinos is too small and therefore the effect of the oscillation between these two neutrinos is invisible in atmospheric neutrinos. The allowed region was obtained on the 3 dimensional space of $\sin^2\theta_{13}$, $\sin^2\theta_{23}$ and $\Delta m^2 (\equiv \Delta m_{13}^2 = \Delta m_{23}^2)$. An allowed region on the $\sin^2\theta_{13}$ and $\sin^2\theta_{23}$ plane is shown in Figure 4(right). (Note that the axes are $\sin^2\theta_{ij}$, not $\sin^2 2\theta_{ij}$.) Super-Kamiokande found no evidence for non-zero $\sin^2\theta_{13}$. The constraint from the CHOOZ experiment [12] on $\sin^2\theta_{13}$ in the Δm^2 range between 2 and 5 $\times 10^{-3}\mathrm{eV}^2$ is much stronger than that from Super-Kamiokande. The Super-Kamiokande result on $\sin^2\theta_{13}$ is consistent with the CHOOZ result.

As another extension of the oscillation analysis, Super-Kamiokande studied if the energy dependence of the neutrino oscillations is really as expected by the standard neutrino oscillations generated by neutrino mass and mixing. For this purpose, the atmospheric neutrino data (including upward going muons) were fitted with the form of $\mathrm{P}(\nu_\mu \to \nu_\mu) = 1 - \sin^2\alpha \cdot \sin^2(\beta \cdot L \cdot E^n)$, where α, β and n are fitted parameters. The energy dependence expected by the standard neutrino oscillations generated by neutrino mass and mixing ($n = -1$) was favored by the data (the fitted value of n was about -1.0 ± 0.1), and models which predict other energy dependence of the oscillations were strongly disfavored.

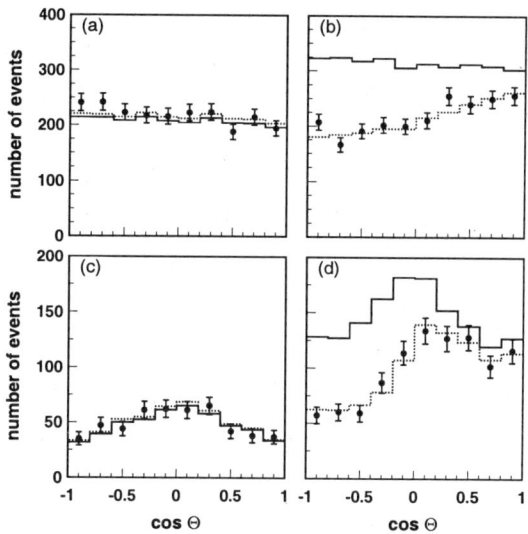

FIGURE 2. Zenith angle distributions for (a) sub-GeV e-like, (b) sub-GeV μ-like, (c) multi-GeV e-like and (d) multi-GeV (FC+PC) μ-like events observed in Super-Kamiokande. $\cos\Theta = 1$ means down-going particles. The solid histograms show the MC prediction without neutrino oscillations. The dashed histograms show the MC prediction for $\nu_\mu \rightarrow \nu_\tau$ oscillations with $\sin^2 2\theta = 1$ and $\Delta m^2 = 2.8 \times 10^{-3}$ eV^2. In the oscillation histograms, the absolute normalization was adjusted to get the minimum χ^2.

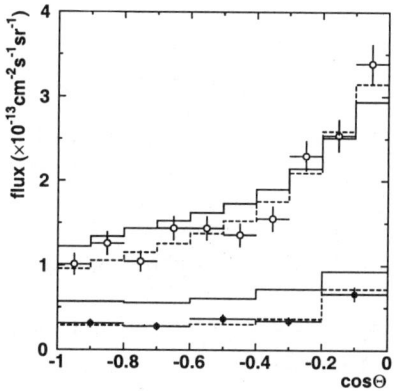

FIGURE 3. Zenith angle distributions for upward-going muon fluxes observed in Super-Kamiokande. Blank (filled) circles with error bars show the upward through-going (stopping) muon data and the histograms show the corresponding predictions. Error bars show statistical + experimental systematic errors. Estimated background events are subtracted. The solid histograms show the expected fluxes for the null neutrino oscillation case. The dashed histograms show the expected fluxes for the $\nu_\mu \rightarrow \nu_\tau$ oscillation case with $\sin^2 2\theta = 1.0$ and $\Delta m^2 = 2.8 \times 10^{-3}$ eV^2.

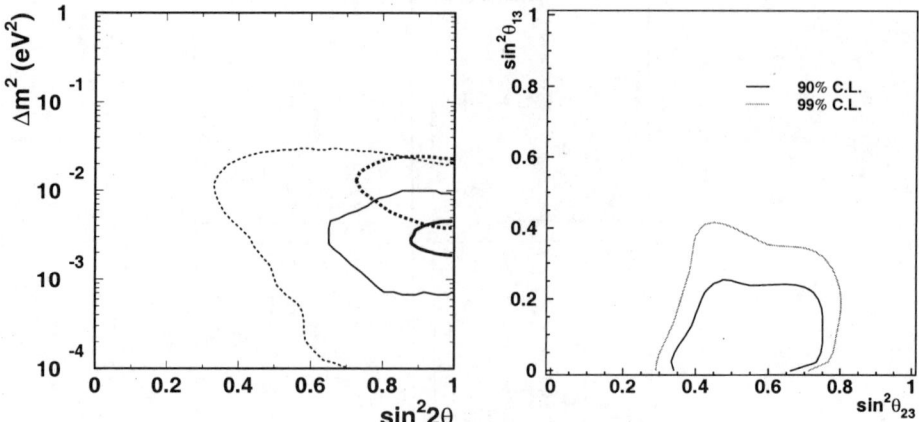

FIGURE 4. Left: The 90% C.L. allowed parameter region of $\nu_\mu \to \nu_\tau$ neutrino oscillations by a combined analysis of FC, PC, upward stopping muon and upward through-going muon events from Super-Kamiokande (thick solid line, preliminary). Results from Kamiokande (thick dashed line), Soudan-2 (thin dashed line) and MACRO (thin solid line) are also shown. Right: Allowed region on the $\sin^2\theta_{13}$ and $\sin^2\theta_{23}$ plane obtained by a 3 flavor analysis of the Super-Kamiokande FC+PC data.

FC data from Super-Kamiokande could also be explained by the $\nu_\mu \to \nu_{sterile}$ neutrino oscillations. $\nu_{sterile}$ is a hypothetical neutrino-like particle which does not interact with matter. Indeed, an oscillation analysis using FC events suggests an equally good fit to the $\nu_\mu \to \nu_{sterile}$ hypothesis. In order to distinguish these two possibilities, Super-Kamiokande studied zenith angle distribution of neutral current (NC) enriched sample. Since a $\nu_{sterile}$ does not interact with matter via NC, there should be a deficit of upward going NC events for $\nu_\mu \to \nu_{sterile}$ oscillations. In addition, it is possible to discriminate these two possibilities by using the matter effect [19]. In the case of $\nu_\mu \to \nu_\tau$ oscillations, the matter effect does not change the oscillation probability, however in the case of $\nu_\mu \to \nu_{sterile}$ oscillations, matter effect may change the oscillation probability significantly. The difference of the neutrino oscillation probability is large only for high energy (>20 GeV) atmospheric neutrinos traveling through the Earth.

NC enriched event sample was made by the following criteria; multi-ring events, most energetic ring must be e-like and E_{vis} >400 MeV. The fraction of NC events in this sample is estimated to be 29% for no neutrino oscillations. Figure 5(a) shows the zenith angle distribution for these events together with predictions of neutrino oscillations. The up-down ratio, $Up(-1 < \cos\Theta < -0.4)/Down(0.4 < \cos\Theta < 1)$, of the data was $0.99 \pm 0.07 \pm 0.005$ while the predictions at ($\sin^2 2\theta = 1$, $\Delta m^2 = 3 \times 10^{-3}$) were 0.93 ± 0.03(sys.) for $\nu_\mu \to \nu_\tau$ and 0.81 ± 0.02(sys.) for $\nu_\mu \to \nu_{sterile}$. The expected up-down ratio for $\nu_\mu \to \nu_{sterile}$ neutrino oscillation hypothesis did not agree with the data at 2 σ level.

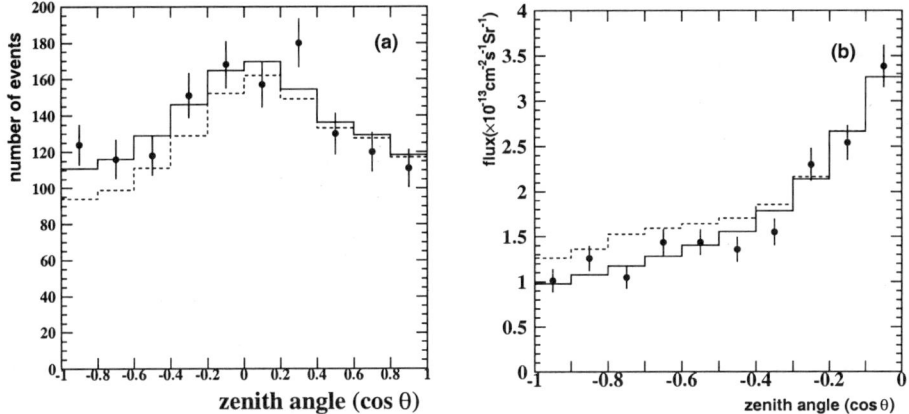

FIGURE 5. Zenith angle distributions for; (a)NC enriched multi-ring events and (b)upward through-going muon events. In (a) and (b), solid (dashed) histograms show the expectations for $\nu_\mu \to \nu_\tau$ ($\nu_\mu \to \nu_{sterile}$). In these figures, $\sin^2 2\theta = 1$ and $\Delta m^2 = 3 \times 10^{-3} \text{eV}^2$ are assumed for the expectations.

To separate the two neutrino oscillation hypotheses using the matter effect, Super-Kamiokande studied the upward through-going muons and high energy PC events. Figure 5(b) shows the zenith angle distribution of the upward through-going muons together with predictions of neutrino oscillations. The vertical-horizontal ratio, $Vertical(-1 < \cos\Theta < -0.4)/Horizontal(-0.4 < \cos\Theta < 0)$, of the data was $0.77 \pm 0.045 \pm 0.007$ while the predictions at $(\sin^2 2\theta = 1, \Delta m^2 = 3 \times 10^{-3})$ were $0.76 \pm 0.024(sys.)$ for $\nu_\mu \to \nu_\tau$ and $0.91 \pm 0.029(sys.)$ for $\nu_\mu \to \nu_{sterile}$. The expected vertical-horizontal ratio for $\nu_\mu \to \nu_{sterile}$ neutrino oscillation hypothesis did not agree with the data at 2.5 σ level. A similar (but less significant) result was obtained by the high-energy PC events.

Finally, the two oscillation hypotheses were tested by combining these studies by a χ^2 method. Most of the parameter region suggested by the analysis of FC events for $\nu_\mu \to \nu_{sterile}$ was disfavored at about 99% C.L.(preliminary) [18]. On the other hand, $\nu_\mu \to \nu_\tau$ oscillation hypothesis did not contradict with the data in this analysis.

IV SEARCH FOR NUCLEON DECAY

One of the most unique predictions of Grand Unified Theories (GUTs) is the baryon number violation. In the past two decades, several large underground experiments have searched for nucleon decays. However no clear evidence has been

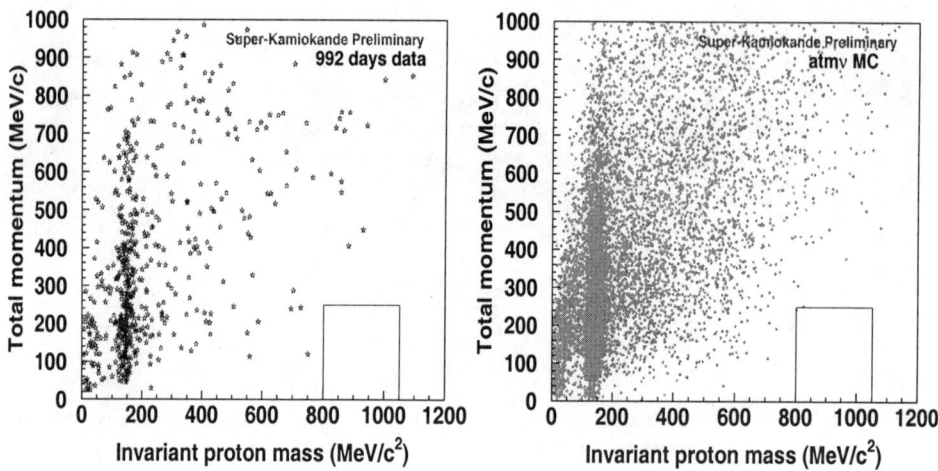

FIGURE 6. Invariant mass and total momentum plot in the search for $p \to e^+\pi^0$ for the data (left) and for the atmospheric neutrino MC events of 900 kton·yr exposure equivalent (right). Events in this plots satisfies the other criteria mentioned in the text.

observed. In many GUT models, either $p \to e^+\pi^0$ or $p \to \nu K^+$ is the dominant decay mode. Accordingly, we have searched for nucleon decays for these modes.

$p \to e^+\pi^0$: This decay mode has a characteristic event signature, in which the electromagnetic shower caused by the positron is balanced against the two showers caused by the gamma rays from the decay of the π^0.

Proton decay events were searched for in the fully-contained atmospheric neutrino sample. The essential selection criteria [20] were: (1) the number of Cherenkov rings should be 2 or 3, (2) all rings have a showering particle type (e-like), (3) no electron from muon decay, (4) π^0 invariant mass cut for 3 ring events, and (5) $800 <$ total invariant mass < 1050 MeV/c^2 and total momentum < 250 MeV/c. The detection efficiency was estimated to be 43%. The expected number of background events was 0.1. Figure 6 shows a two dimensional plot of the total invariant mass versus the total momentum for the data and atmospheric neutrino MC events. No event was observed in the signal region. The lower limit on the partial lifetime for $p \to e^+\pi^0$ was obtained to be $\tau/B_{p \to e^+\pi^0} > 3.8 \times 10^{33}$ years at 90% C.L.

$p \to \nu K^+$: The momentum of the K^+ from $p \to \nu K^+$ is 340 MeV/c and is below the threshold momentum for producing Cherenkov light in water. Candidate events for this decay mode are therefore identified through the decay products of the K^+: $K^+ \to \mu^+\nu$ and $K^+ \to \pi^+\pi^0$ [21].

Two separate decay methods were used to search for $K^+ \to \mu^+\nu$. The first method searched for mono-energetic muons from the stopped K^+. The signal can be found as an excess of single-ring μ-like events at 236 MeV/c above the continuous background due to atmospheric neutrino interactions. No significant excess was observed.

If a proton in ^{16}O decays, the remaining ^{15}N emits a gamma ray of 6.3 MeV (or slightly higher energy) for about a half of the cases. Since K^+ has a lifetime of 12 nsec, the decay signal of K^+ should be delayed relative to the gamma ray. The delayed coincidence of the gamma and the muon (from the K decay) is not expected in most of the atmospheric neutrino events which do not involve K mesons in the final state. Therefore, the expected number of background events was 1.0 in the 61 kton·yr data. There was no candidate event.

The signals of $K^+ \to \pi^+\pi^0$ were searched for by asking (1)mono-energetic π^0 and (b)weak Cherenkov radiation by the π^+ in the opposite direction of the π^0. The expected number of background events was 1.7 in the 61 kton·yr data. No candidate event was observed.

By combining all the three results, no evidence for $p \to \nu K^+$ was observed and the lower limit of the partial lifetime for this mode was obtained to be 1.6×10^{33} yr at 90%C.L..

V SUMMARY

To distinguish the solutions of the solar neutrino problem, the day-night effect and the energy spectrum of ^8B solar neutrinos were studied. The present statistics of the Super-Kamiokande data is not enough to distinguish these solutions.

With the increased statistics of the atmospheric neutrino data, the main feature of the data was essentially unchanged: Both the zenith angle distribution of μ-like events and the (μ/e) values were significantly different from the predictions in the absence of neutrino oscillations. The data were in good agreement with $\nu_\mu \to \nu_\tau$ oscillations. This conclusion was supported by the upward-going muon data. By using all the atmospheric neutrino data from Super-Kamiokande, the 90% C.L. allowed region on $\nu_\mu \to \nu_\tau$ neutrino oscillation parameters was: $2 \times 10^{-3} < \Delta m^2 < 5 \times 10^{-3} eV^2$ and $\sin^2 2\theta > 0.88$. The matter effect and the NC events were used to discriminate $\nu_\mu \to \nu_\tau$ and $\nu_\mu \to \nu_{sterile}$ oscillations. The preliminary result disfavored the $\nu_\mu \to \nu_{sterile}$ oscillation hypothesis at about 99% C.L..

Proton decay events were searched for. No evidence was found. The lower limit on the partial lifetime of proton was; $3.8 \times 10^{33} yr$ for the $e^+\pi^o$ mode, and $1.6 \times 10^{33} yr$ for the νK^+ mode.

This work was partly supported by the Japanese Ministry of Education, Science, Sports and Culture.

REFERENCES

1. B.T. Cleveland *et al.*, *Ap. J.* **496**, 505 (1998).

2. Y. Fukuda et al., *Phys. Rev. Lett.* **77**, 1683 (1996).
3. J.N. Abdurashitov et al., *Phys. Rev. Lett.* **83**, 4686 (1999).
4. W. Hampel et al., *Phys. Lett. B* **447**, 127 (1999).
5. Y. Fukuda et al., *Phys. Rev. Lett.* **81**, 1158 (1998); *Phys. Rev. Lett.* **82**, 1810 (1999); *Phys. Rev. Lett.* **82**, 2430 (1999).
6. J.N. Bahcall, S. Basu, and M.H. Pinsonneault, *Phys. Lett. B* **433**, 1 (1998).
7. L. Wolfenstein, *Phys. Rev. D* **17**, 2369 (1978); S.P. Mikheyev and A.Yu. Smirnov, *Sov. J. Nucl. Phys.* **42**, 1441 (1985); S.P. Mikheyev and A.Yu. Smirnov, *Nuovo Cim.* **C9**, 17 (1986).
8. K.S. Hirata et al., *Phys. Lett. B* **205**, 416 (1988), *Phys. Lett. B* **280**, 146 (1992).
9. Y. Fukuda et al., *Phys. Lett. B* **335**, 237 (1994).
10. D. Casper et al., *Phys. Rev. Lett.* **66**, 2561 (1991); R. Becker-Szendy et al., *Phys. Rev. D* **46**, 3720 (1992).
11. W.W.M. Allison et al., *Phys. Lett. B* **391**, 491 (1997); *Phys. Lett. B* **449**, 137 (1999); W.A. Mann, hep-ex/9912007.
12. M. Apollonio et al., *Phys. Lett. B* **420**, 397 (1998); *Phys. Lett. B* **466**, 415 (1999).
13. F. Boehm et al., hep-ex/9912050; hep-ex/0003022.
14. Y. Fukuda et al., *Phys. Lett. B* **433**, 9 (1998); *Phys. Lett. B* **436**, 33 (1998); *Phys. Rev. Lett.* **81**, 1562 (1998).
15. Y. Fukuda et al., *Phys. Rev. Lett.* **82**, 2644 (1999); Y. Fukuda et al., *Phys. Lett. B* **467**, 185 (1999).
16. S. Hatakeyama et al., *Phys. Rev. Lett.* **81**, 2016 (1998).
17. M. Ambrosio et al., *Phys. Lett. B* **434**, 451 (1998); M. Spurio, for the MACRO collaboration, hep-ex/9908066.
18. The Super-Kamiokande collaboration, draft in preparation.
19. Q.Y. Liu, S.P. Mikheyev, and A.Yu. Smirnov, *Phys. Lett. B* **440**, 319 (1998); P. Lipari and M. Lusignoli, *Phys. Rev. D* **58**, 073005 (1998).
20. M. Shiozawa et al., *Phys. Rev. Lett.* **81**, 3319 (1998).
21. Y. Hayato et al., *Phys. Rev. Lett.* **83**, 1529 (1999).

Additional Isospin-Breaking Effects in ϵ'/ϵ

S. Gardner*,[1]

* *Department of Physics and Astronomy, University of Kentucky, Lexington, KY 40506-0055 USA*

Abstract. Isospin-breaking effects, in particular those associated with electroweak-penguin contributions and π^0-η, η' mixing, have long been known to affect the Standard Model prediction of ϵ'/ϵ in a significant manner. We have found an heretofore unconsidered isospin-violating effect of importance; namely, the u-d quark mass difference can spawn $|\Delta I| = 3/2$ components in the matrix elements of the gluonic penguin operators. Using chiral perturbation theory and the factorization approximation for the hadronic matrix elements, we find within a specific model for the low-energy constants that we can readily accommodate an increase in ϵ'/ϵ by a factor of two.

INTRODUCTION

The recent measurement of a non-zero value of $\mathrm{Re}\,(\epsilon'/\epsilon)$ [1] establishes the existence of CP violation in direct decay and thus provides an important first check of the mechanism of CP violation in the Standard Model (SM). Nevertheless, the world average which emerges is $\mathrm{Re}\,(\epsilon'/\epsilon) = (19.3 \pm 2.4) \cdot 10^{-4}$ [2], which is larger than the "central" SM prediction of $7.0 \cdot 10^{-4}$ [3,4] by nearly a factor of three. This compels us to scrutinize the SM prediction in further detail: we study isospin-violating effects arising from the u-d quark mass difference.

Isospin violation plays an important role in the analysis of ϵ'/ϵ, for the latter is predicated by the difference of the imaginary to real part ratios in the $|\Delta I| = 1/2$ and $|\Delta I| = 3/2$ $K \to \pi\pi$ amplitudes. The differing charges of the u and d quarks engender $|\Delta I| = 3/2$ electroweak penguin contributions, whereas π^0-η, η' mixing, driven by the u-d quark mass difference, modifies the relative contribution of the $|\Delta I| = 1/2$ and $|\Delta I| = 3/2$ amplitudes in a significant way.

Here we describe isospin-breaking effects in the matrix elements of the gluonic penguin operators [5], such as Q_6. These operators have always been thought to induce exclusively $|\Delta I| = 1/2$ transitions, but this is true only in the limit of isospin symmetry. The difference in the up and down quark masses effectively distinguishes

---

[1] Presenter. Based on work performed in collaboration with G. Valencia.

the interaction of gluons with up and down quarks, so that the $\langle\pi\pi|Q_6|K\rangle$ matrix element possesses a $|\Delta I| = 3/2$ component as well [6].

Let us begin by showing why the numerical prediction of ϵ'/ϵ is sensitive to the presence of isospin violation. The value of ϵ'/ϵ is inferred from a ratio of ratios, namely

$$\mathrm{Re}\left(\frac{\epsilon'}{\epsilon}\right) = \frac{1}{6}\left[\left|\frac{\eta_{+-}}{\eta_{00}}\right|^2 - 1\right], \tag{1}$$

where

$$\eta_{+-} \equiv \frac{\mathcal{A}(K_L \to \pi^+\pi^-)}{\mathcal{A}(K_S \to \pi^+\pi^-)} \approx \epsilon + \epsilon' \ ; \ \eta_{00} \equiv \frac{\mathcal{A}(K_L \to \pi^0\pi^0)}{\mathcal{A}(K_S \to \pi^0\pi^0)} \approx \epsilon - 2\epsilon' . \tag{2}$$

In the isospin-perfect limit, the two independent amplitudes present in $K \to \pi\pi$ decay are distinguished by the isospin of the final-state pions, namely $A_I \equiv \mathcal{A}(K \to (\pi\pi)_I)$ with $I = 0, 2$. ϵ'/ϵ can thus be written

$$\frac{\epsilon'}{\epsilon} = -\frac{\omega}{\sqrt{2}|\epsilon|}\xi(1 - \Omega) , \tag{3}$$

with

$$\omega \equiv \frac{\mathrm{Re}\,A_2}{\mathrm{Re}\,A_0} \ ; \ \xi \equiv \frac{\mathrm{Im}\,A_0}{\mathrm{Re}\,A_0} \ ; \ \Omega \equiv \frac{\mathrm{Im}\,A_2}{\omega\mathrm{Im}\,A_0} . \tag{4}$$

In standard practice, $\omega \approx 1/22$ and $\mathrm{Re}\,A_0$ are taken from experiment whereas $\mathrm{Im}\,A_I$ is computed using the operator-product expansion [3,4], that is, via

$$\mathcal{H}_{\mathrm{eff}}(|\Delta S| = 1) = 4\frac{G_F}{\sqrt{2}}V^*_{us}V_{ud}\sum_{i=1}^{10}C_i(\mu)Q_i(\mu) + \mathrm{h.c.} \tag{5}$$

The numerical value of ϵ'/ϵ is driven by the matrix elements of the QCD penguin operator Q_6 and the electroweak penguin operator Q_8 [7]. Writing $\langle Q_i \rangle_I$ as $\langle(\pi\pi)_I|Q_i|K\rangle \equiv B_i^{((I+1)/2)}\langle(\pi\pi)_I|Q_i|K\rangle^{(\mathrm{vac})}$, where "vac" indicates the use of the vacuum saturation approximation, one recovers the schematic formula [3]

$$\frac{\epsilon'}{\epsilon} = 13\,\mathrm{Im}\lambda_t\left[B_6^{(1/2)}(1 - \Omega_{\eta+\eta'}) - 0.4\,B_8^{(3/2)}\right] . \tag{6}$$

Using $B_6^{(1/2)} = 1.0$, $B_8^{(3/2)} = 0.8$, and $\Omega_{\eta+\eta'} = 0.25$ yields the "central" SM value of $\epsilon'/\epsilon \sim 7.0 \cdot 10^{-4}$ [3], roughly a factor of three smaller than the measured value. Larger estimates of $B_6^{(1/2)}$, and hence of ϵ'/ϵ, exist [8–10]; we investigate sources of $\Omega_{\eta+\eta'}$. Note that under $\Omega_{\eta+\eta'} \to -\Omega_{\eta+\eta'}$, $\mathrm{Re}\,(\epsilon'/\epsilon) \to 2.2\,\mathrm{Re}\,(\epsilon'/\epsilon)$. Were $|B_6^{(1/2)}| \gg |B_8^{(3/2)}|$, flipping the sign of $\Omega_{\eta+\eta'}$ would increase ϵ'/ϵ by a factor of 1.7.

Let us consider possible sources of $\Omega_{\eta+\eta'}$. We replace $\Omega_{\eta+\eta'}$ by Ω_{IB}, where

$$\Omega_{\text{IB}} = \left(\frac{\sqrt{2}}{3\omega}\right)\frac{\text{Im}(A_{\text{P}}(K^0 \to \pi^+\pi^-) - A_{\text{P}}(K^0 \to \pi^0\pi^0))}{\text{Im}\,A_{\text{P}}(K^0 \to \pi\pi)} \quad (7)$$

and $\text{Im}\,A_{\text{P}}(K^0 \to \pi\pi) = (\text{Im}\,A_{\text{P}}(K^0 \to \pi^+\pi^-) + \text{Im}\,A_{\text{P}}(K^0 \to \pi^0\pi^0))/2$. "$A_{\text{P}}$" denotes an amplitude induced by $(8_L, 1_R)$ (e.g., Q_6) operators — the empirical $|\Delta I| = 1/2$ rule suggests such operators dominate the isospin-violating effects. Ω_{IB} vanishes in the absence of isospin violation, i.e., if $m_u = m_d$, $e_u = e_d$. It can be generated by both strong-interaction and electromagnetic effects, mediated by $m_d \neq m_u$ and $e_u \neq e_d$ [11], respectively. We focus on $m_d \neq m_u$ effects. The latter include π^0-η, η' mixing [12–14]; in $\mathcal{O}(p^2, 1/N_c)$ this yields $\Omega_{\eta+\eta'} = 0.25 \pm 0.05$ [13,14], used in the analysis of Ref. [3]. However, $m_u \neq m_d$ effects can also spawn a $|\Delta I| = 3/2$ component in the matrix elements of the gluonic penguin operators [6], as illustrated in Fig. 1. We turn to a chiral Lagrangian analysis in order to estimate the size of this effect [5].

CHIRAL LAGRANGIAN ANALYSIS

The weak chiral Lagrangian for $K \to \pi\pi$ decay is written in terms of the unitary matrix $U = \exp(i\phi/f)$ and the function χ, both of which transform as $U \to RUL^\dagger$ under the chiral group $SU(3)_L \times SU(3)_R$. The function ϕ represents the octet of pseudo-Goldstone bosons, i.e., $\phi = \sum_{a=1,...,8} \lambda_a \phi_a$. In the absence of external fields, $\chi = 2B_0 M$ with $M = \text{diag}(m_u, m_d, m_s)$ and $B_0 \propto \langle \bar{q}q \rangle$. The leading-order, $\mathcal{O}(p^2)$, weak chiral Lagrangian contains no mass-dependent terms [15], so that $m_d \neq m_u$ effects in the hadronization of the gluonic penguin operators first appear in $\mathcal{O}(p^4)$. This is illustrated in Fig. 2.

Let us enumerate the possible isospin-violating effects which occur in $\mathcal{O}(m_d - m_u)$ and $\mathcal{O}(p^4)$:

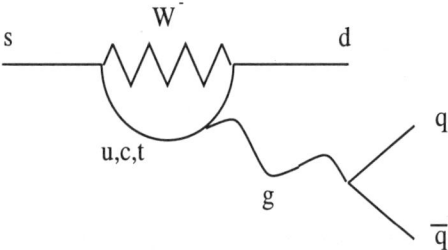

FIGURE 1. Quark line diagram illustrating the "strong penguin" $s \to d\bar{q}q$ transition in the Standard Model. Note that $\bar{q}q \in \bar{u}u, \bar{d}d$; in the isospin-perfect limit, $m_u = m_d$ and only $|\Delta I| = 1/2$ transitions are generated. If $m_u \neq m_d$, the $K \to \pi\pi$ matrix element associated with this operator contains a $|\Delta I| = 3/2$ component as well.

i) π^0-η mixing realized from the $\mathcal{O}(p^2)$ strong chiral Lagrangian, in concert with the $\mathcal{O}(p^2)$ weak chiral Lagrangian, computed to one-loop order.

ii) π^0-η mixing, realized from the $\mathcal{O}(p^2)$ strong chiral Lagrangian, combined with the isospin-conserving vertices of the $\mathcal{O}(p^4)$ weak chiral Lagrangian.

iii) π^0-η mixing as realized from the strong chiral Lagrangian in $\mathcal{O}(p^4)$, combined with the $\mathcal{O}(p^2)$ weak chiral Lagrangian. The π^0-η' mixing effects included in Refs. [13,14] ape this effect.

iv) Isospin violation in the vertices of the $\mathcal{O}(p^4)$ weak chiral Lagrangian. This serves as our focus here, for it contains the qualitatively new effects we argue.

We use the octet terms in the $\mathcal{O}(p^4)$, CP-odd weak chiral Lagrangian of Ref. [16]. Collecting the χ-dependent terms as per iv), working to $\mathcal{O}(m_d - m_u)$, and dropping terms suppressed by M_π^2/M_K^2, we find

$$\Omega_P = \frac{2\sqrt{2}}{3\omega} \frac{M_{K^0}^2}{M_{K^0}^2 - M_\pi^2} \frac{B_0(m_d - m_u)}{c_2^-} \tilde{E}^- \approx \frac{0.12\,\text{GeV}^2}{c_2^-} \tilde{E}^- \qquad (8)$$

with $\tilde{E}^- = 2E_1^- - 2E_3^- - 4E_4^- - E_{10}^- - E_{11}^- - 4E_{12}^- - E_{15}^-$. Note that c_2^- is the low-energy constant associated with the $\mathcal{O}(p^2)$, $(8_L, 1_R)$ weak chiral Lagrangian [16]. As per ii), π^0-η mixing in $\mathcal{O}(p^2)$ also enters when combined with the isospin-conserving vertices of the $\mathcal{O}(p^4)$ weak chiral Lagrangian. Including the χ-dependent octet terms, we find

$$\Omega_{\eta+\eta'}^{(4)} = \frac{2\sqrt{2}}{3\omega} \frac{M_{K^0}^2}{M_{K^0}^2 - M_\pi^2} \frac{B_0(m_d - m_u)}{c_2^-} E_{\eta+\eta'}^- \approx \frac{0.12\,\text{GeV}^2}{c_2^-} E_{\eta+\eta'}^- \qquad (9)$$

with $E_{\eta+\eta'}^- = -2(E_3^- + E_4^- - E_5^-) + (E_{10}^- - E_{11}^-)/2 - 2E_{12}^- + E_{14}^- + 3E_{15}^-/2$ — so that no manifest cancellation with the terms of Eq. (8) occurs.

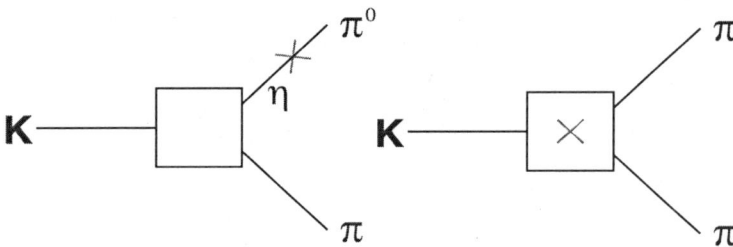

FIGURE 2. Isospin violation in $K \to \pi\pi$ decays. The square box represents the weak $|\Delta S| = 1$ transition at low energies, whereas the "×" represents the presence of $m_d \neq m_u$ effects. Mass effects do not occur in the weak transition in leading order in chiral perturbation theory, so that only the left-hand diagram occurs in $\mathcal{O}(p^2)$ and in $\mathcal{O}(m_d - m_u)$ — π^0-η mixing, mediated by $m_d - m_u$ effects in the strong chiral Lagrangian, can occur. In $\mathcal{O}(p^4)$ both diagrams are present; the new effect we discuss is associated with the right-hand diagram.

The low-energy constants E_i^- are unknown, so that we turn to the factorization approximation to proceed. The construction relevant to $(8_L, 1_R)$ transitions in $K^0 \to \pi\pi$ decay is [17]

$$\mathcal{L}_P = -\frac{G_F}{\sqrt{2}} V_{us}^\star V_{ud} C_6 \left(-8(\bar{s}_L q_R)(\bar{q}_R d_L)\right) + \text{h.c.} \tag{10}$$

$$\to \frac{G_F}{\sqrt{2}} V_{us}^\star V_{ud} C_6 \, 32 B_0^2 \frac{\delta\mathcal{L}_{\text{str}}}{\delta\chi_{3i}^\dagger} \frac{\delta\mathcal{L}_{\text{str}}}{\delta\chi_{i2}} + \text{h.c.}, \tag{11}$$

where $\mathcal{L}_{\text{str}}$ is the strong chiral Lagrangian. To generate terms of $\mathcal{O}(p^4)$ in $\mathcal{L}_P$ requires terms of both $\mathcal{O}(p^4)$ [18] and $\mathcal{O}(p^6)$ [19] in $\mathcal{L}_{\text{str}}$. Unfortunately, the low-energy constants of the latter are also unknown; the use of "resonance saturation" allows us to estimate some of them. We explicitly consider the scalar nonet of resonances as per Ref. [20]. An example of the manner in which the scalar resonances can generate contributions to the E_i^- is illustrated in Fig. 3. Integrating out the scalar resonances for $p^2 \ll M_S^2$, we find two terms which contribute to the scalar densities in the bosonization of Q_6 [20],

$$\mathcal{L}_S^{(6)} = \frac{d_m c_m^2}{2 M_S^4} \langle \chi_+^3 \rangle + \frac{c_d c_m d_m}{M_S^4} \langle \chi_+^2 L^2 \rangle, \tag{12}$$

yielding contributions to E_1^- and E_{10}^- in terms of d_m, c_m, c_d, and M_S. The parameter d_m is ill-known; we find $d_m \sim -2.4\,(-0.76)$. The sign of d_m and thus of Ω_P in our model results from the mass of the lowest-lying strange scalar being greater than that of the lowest-lying isovector scalar. As per our earlier classification, $\Omega_{\text{IB}}^{(4)} = \Omega_{\text{IB}}^{(4),i} + \Omega_{\text{IB}}^{(4),ii} + \Omega_{\text{IB}}^{(4),iii} + \Omega_{\text{IB}}^{(4),iv}$, so that with $d_m = -2.4\,(-0.76)$, we have $\Omega_{\text{IB}}^{(4),iv} = -0.79\,(-0.21)$. Estimating $\Omega_{\text{IB}}^{(4),ii}$ using the χ-dependent E_i^- yields $\Omega_{\text{IB}}^{(4),ii} = -0.12\,(-0.03)$. $\Omega_{\text{IB}}^{(4),iii}$ has been partially determined through the inclusion

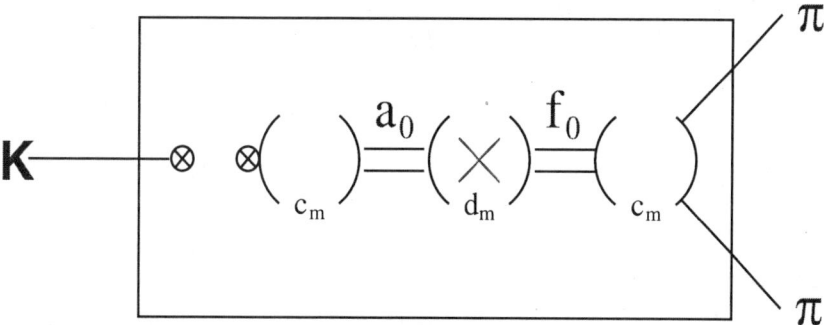

FIGURE 3. A contribution to the right-hand diagram of Fig. 2 in $\mathcal{O}(m_d - m_u)$, estimated in the factorization approximation with explicit, scalar-resonance degrees of freedom. The "⊗" represents a bosonized current; the open parentheses denote contributions from the vertices of the $\mathcal{O}(p^6)$ model Lagrangian of Ref. [20]. In this case the isospin-violating contribution is driven by a_0-f_0 mixing.

of π^0-η' mixing in $\Omega_{\eta+\eta'} = 0.25 \pm 0.05$ [13,14]. Using the result $\Omega_{IB}^{(2)} + \Omega_{IB}^{(4),iii} = 0.16 \pm 0.03$ [21] and neglecting $\Omega_{IB}^{(4),i}$, as the ill-known E_i^- do not warrant such a calculation, we estimate, finally, that $\Omega_{IB} = \Omega_{IB}^{(2)} + \Omega_{IB}^{(4)} \sim -0.05 \rightarrow -0.78$. For reference, note that $\Omega_{IB}^{(2)} \sim 0.13$. The large value of $\Omega_{IB}^{(4)}$ is driven by the numerical prefactor of Eqs. (8,9) — the contributions in $\Omega_{IB}^{(4)}$ are "naturally" of the same size as $\Omega_{IB}^{(2)}$. Thus we find a very large correction to the value of $\Omega_{\eta+\eta'} = 0.25 \pm 0.05$, used in "central value" of ϵ'/ϵ. The large negative change in Ω_{IB} found in $\mathcal{O}(p^4)$ generates a substantial increase in ϵ'/ϵ.

The Ω_{IB} we calculate impacts ϵ'/ϵ in a significant manner. Our estimate of Ω_{IB} from the specific $m_d \neq m_u$ effects we consider ranges from $-0.05 \rightarrow -0.78$; this range exceeds the central value, $\Omega_{\eta+\eta'} = 0.25 \pm 0.05$, used in earlier analyses and reflects a variation in ϵ'/ϵ of more than a factor of two. The presence of unknown low-energy constants implies that we lack a reliable way to calculate the effects we consider. Such limitations, however, underscore the need for a larger uncertainty in the Standard Model prediction of ϵ'/ϵ.

The collaboration of G. Valencia is gratefully acknowledged. The work of S.G. is supported in part by the U.S. DOE under contract number DE-FG02-96ER40989.

REFERENCES

1. A. Alavi-Harati et al. (KTeV Collaboration), *Phys. Rev. Lett.* **83**, 22 (1999); V. Fanti et al. (NA48 Collaboration), *Phys. Lett.* **B465**, 335 (1999).
2. A. Ceccucci, CERN seminar, February 29, 2000, http://www.cern.ch/NA48/ .
3. S. Bosch et al., *Nucl. Phys.* **B565**, 3 (2000).
4. A. Buras, hep-ph/9806471, to appear in *Probing the Standard Model of Particle Physics*, F. David and R. Gupta, eds. (Elsevier Science B. V., Amsterdam, 1998).
5. S. Gardner and G. Valencia, *Phys. Lett.* **B466**, 355 (1999).
6. S. Gardner, *Phys. Rev.* **D59**, 077502 (1999); hep-ph/9906269.
7. A. J. Buras, M. Jamin, and M. E. Lautenbacher, *Nucl. Phys.* **B408**, 209 (1993).
8. S. Bertolini, M. Fabbrichesi, and J. Eeg, hep-ph/0002234, and references therein.
9. T. Hambye, G. O. Kohler, E. A. Paschos, and P. H. Soldan, *Nucl. Phys.* **B564**, 391 (2000).
10. J. Bijnens and J. Prades, hep-ph/0005189.
11. V. Cirigliano, J. F. Donoghue, and E. Golowich, *Phys.Rev.* **D61**, 093001 (2000); *Phys.Rev.* **D61**, 093002 (2000).
12. J. Bijnens and M. Wise, *Phys. Lett.* **137B**, 245 (1984).
13. J. F. Donoghue et al., *Phys. Lett.* **179B**, 361 (1986).
14. A. J. Buras and J. M. Gerard, *Phys. Lett.* **192B**, 156 (1987).
15. J. A. Cronin, *Phys. Rev.* **161**, 1483 (1967).
16. J. Kambor, J. Missimer, and D. Wyler, *Nucl. Phys.* **B346**, 17 (1990).
17. S. Chivukula, J. Flynn, and H. Georgi, *Phys. Lett.* **B171**, 453 (1986).
18. J. Gasser and H. Leutwyler, *Nucl. Phys.* **B250**, 465 (1985).

19. H. W. Fearing and S. Scherer, *Phys. Rev.* **D53**, 315 (1996); J. Bijnens, G. Colangelo, and G. Ecker, *JHEP* **9902**, 020 (1999).
20. G. Amorós, J. Bijnens, and P. Talavera, *Nucl.Phys.* **B568**, 319 (2000).
21. G. Ecker, G. Müller, H. Neufeld, and A. Pich, *Phys. Lett.* **B477**, 88 (2000).

Isospin Breaking in the Extraction of Isovector and Isoscalar Spectral Functions from $e^+e^- \to hadrons$

K. Maltman[*] and C.E. Wolfe[†]

[*]Dept. Mathematics and Statistics, York Univ., 4700 Keele St., Toronto, ON Canada, and CSSM, Univ. of Adelaide, Adelaide, SA Australia[1]
[†]Nuclear Theory Center, Indiana University, Bloomington, IN, USA

Abstract. A finite energy sum rule (FESR) analysis of the isospin-breaking vector current correlator $\langle 0|T\left(V_\mu^3 V_\nu^8\right)|0\rangle$ is used to determine the isospin-breaking electromagnetic (EM) decay constants of the low-lying vector mesons. These results are used to evaluate the corrections required to extract the flavor diagonal 33 and 88 resonance contributions from the full resonance EM contributions to the EM spectral function. A large ($\sim 15\%$) correction is found in the case of the ω contribution to the isoscalar spectral function. The implications of these results for sum rules based on the isovector-isoscalar spectral difference are considered.

INTRODUCTION

If we define the flavor ab $(a, b = 3, 8)$ vector current correlators via

$$i\int d^4x\, e^{iqx} \langle 0|T(V_\mu^a(x)V_\nu^b(0)^\dagger)|0\rangle \equiv (-g_{\mu\nu}q^2 + q_\mu q_\nu)\,\Pi^{ab}(q^2)\,, \tag{1}$$

the corresponding EM current-current correlator is given by

$$\Pi^{EM} = \Pi^{33} + \frac{2}{\sqrt{3}}\Pi^{38} + \frac{1}{3}\Pi^{88}\,, \tag{2}$$

which reduces to $\Pi^{33} + \Pi^{88}/3$ in the isospin limit. Π^{EM} is related to the hadronic electroproduction cross-section by

$$\sigma(e^+e^- \to hadrons) = \frac{16\pi^2\alpha^2}{s}\,\mathrm{Im}\Pi^{EM}(s)\,. \tag{3}$$

In the isospin limit, the 33 and 88 contributions associated with $n\pi$ states can be classified by G-parity, making an experimental determination of the separate

---

[1] Supported by the Natural Sciences and Research Engineering Council of Canada

isovector (33) and isoscalar (88) components of the EM spectral function, $\rho^{EM}(s)$, possible. Isospin breaking (IB), however, complicates this separation because of the presence of a non-zero 38 spectral component. One contribution to the 38 component, associated with the process $e^+e^- \to \omega \to \pi^+\pi^-$, is conventionally removed by hand in determining the 33 spectral function, but other 38 contributions also exist which are not so subtracted. If we define, in obvious notation, the isovector (3) and isoscalar (8) components of the EM decay constant of vector meson, V, via

$$F_V^{EM} = F_V^3 + \frac{1}{\sqrt{3}} F_V^8 \qquad (4)$$

then, for example, the intermediate ρ contribution to $\rho^{EM}(s)$ contains not only a 33 component, proportional to $[F_\rho^3]^2$, but also, to leading order in IB, a 38 component proportional to $F_\rho^3 F_\rho^8$. The ω contribution to $\rho^{EM}(s)$ similarly contains a 38 component proportional to $F_\omega^3 F_\omega^8$.

The presence of these 38 "contaminations" in the conventionally extracted 33 and 88 spectral functions creates potential problems for

- CVC tests
- the determination of m_s using Narison's $33 - 88$ "tau-decay-like" sum rule [1]
- The inverse weighted chiral $33 - 88$ sum rule for the 6th order ChPT LEC, Q_V, which governs flavor/isospin breaking in vector current correlators [2].

In the latter two cases, the IB effect can be particularly important (1) because the cancellation between the nominal isovector and isoscalar integrals is typically rather strong (to the $\sim 10\%$ level) and, (2) because the 38 component of the ρ and ω EM decay constants induced by ρ-ω mixing has the opposite sign for the ρ and ω, the 38 corrections add constructively when one forms the nominal isovector-isoscalar difference.

It is easy to see that the relative importance of this effect should be much larger for the isovector "contamination" of the ω decay constant than for the isoscalar "contamination" of the ρ decay constant. Indeed, taking $SU(3)_F$ for the vector current vacuum-to-vector-meson matrix elements, and assuming ideal mixing of the vector meson sector, $F_\rho^{(0)} \simeq 3 F_\omega^{(0)}$, where the superscript (0) indicates the isospin-unmixed states. With ϵ the ρ-ω mixing angle, one then has

$$F_\rho^{EM} = F_\rho^{(0)} - \epsilon F_\omega^{(0)} \simeq F_\rho^{(0)} \left(1 - \frac{\epsilon}{3}\right)$$
$$F_\omega^{EM} = F_\omega^{(0)} + \epsilon F_\rho^{(0)} \simeq F_\omega^{(0)} (1 + 3\epsilon) \ . \qquad (5)$$

The fractional contribution of the "wrong-isospin" current to the EM decay constant of the ω should thus be 9 times that of the corresponding IB contribution to F_ρ^{EM}, in the limit that the effect is dominated by ρ-ω mixing.

In this paper, we employ a FESR analysis of the IB vector current correlator, Π^{38}, in order to extract the product, $F_V^3 F_V^8$, of IB and isospin-conserving (IC) decay constants for the low-lying vector mesons. This allows us to determine the 38 contamination of the resonance contributions to the nominally extracted 33 and 88 spectral functions, and hence to make the corrections required to extract the actual versions thereof from data.

FESR'S AND THE DETERMINATION OF THE IB VECTOR MESON DECAY CONSTANTS

The analyticity structure of Π^{38} is such that, for any function $w(s)$ analytic in the region of the contour, one has, using Cauchy's theorem, the FESR relation

$$\int_{s_{th}}^{s_0} ds\, w(s) \rho^{38}(s) = \frac{-1}{2\pi i} \oint_{|s|=s_0} ds\, w(s) \Pi^{38}(s) , \qquad (6)$$

where $\rho^{38}(s) = \frac{1}{\pi} \mathrm{Im}\, \Pi^{38}(s)$. Choosing s_0 large enough that Π^{38} can be represented by the OPE on the RHS, but small enough that ρ^{38} on the LHS is still dominated by the known isovector and isoscalar vector mesons, one obtains a relation between the products $f_V = F_V^3 F_V^8$ which govern the sizes of resonance contributions to ρ^{38} and the parameters (α_s and vacuum condensates) entering the OPE. Since the OPE is well known this allows us to constrain the resonance parameters.

In order to make the analysis suggested above reliable, we will employ a form of FESR tested in the isovector vector channel and shown to be very accurately satisfied there [3]. These "pinch-weighted" FESR's are those corresponding to weights satisfying $w(s_0) = 0$. In the isovector vector channel, where they can be tested, they are known to be very accurately satisfied down to rather low scales (~ 2 GeV^2, even when more general FESR's, which do not suppress contributions from the region near the timelike real axis, are poorly satisfied [3]). It is convenient to work with two families of pinched weights, $w_s(s) = [1 - s/s_0][1 + As/s_0]$ and $w_d(s) = [1 - s/s_0]^2[1 + As/s_0]$, where in each case A is a free parameter, used to vary the weight profile. It is worth noting that, using the FESR's resulting from these two weight families to fit the decay constants of the first three ρ resonances, one obtains a determination of the $\rho(770)$ decay constant in terms of OPE parameters which is accurate to within experimental errors [3,4].

On the hadronic side of our FESR's we assume a sum of Breit-Wigner resonance contributions for the ρ, ω, and ϕ with PDG values of the masses and widths. In the ρ'-ω' region, a single effective resonance contribution with the average values of the masses and widths is employed since the individual ρ' and ω' masses and widths are very similar, making it impossible for our sum rule to discriminate between them. The necessity of including a possible ϕ term has been discussed in Ref. [5], while the importance of including the resonance widths has been stressed in Ref. [6]). The only remaining unknowns in the spectral ansatz are the products $f_V = F_V^3 F_V^8$, which determine the heights of the resonance peaks.

The $D = 0, 2, 4, 6$ terms in the OPE representation of Π^{38} are given by

$$\left[\Pi^{38}_{1\gamma E}\right]_{D=0} = -\frac{\alpha}{16\pi^3}\frac{1}{4\sqrt{3}}ln(Q^2)$$

$$\left[\Pi^{38}(Q^2)\right]_{D=2} = \frac{3}{2\pi^2 Q^2}\frac{1}{4\sqrt{3}}\left[(m_d^2 - m_u^2)(Q^2)\right]\left[1 + \frac{8}{3}a(Q^2)\right.$$
$$\left. + \left(\frac{17981}{432} + \frac{62}{27}\zeta(3) - \frac{1045}{54}\zeta(5)\right)a^2(Q^2)\right]$$

$$\left[\Pi^{38}(Q^2)\right]_{D=4} = \frac{2\left(\langle m_u \bar{u}u\rangle - \langle m_d \bar{d}d\rangle\right)}{4\sqrt{3}Q^4}\left[1 + \frac{1}{3}a(Q^2) + \frac{11}{2}a^2(Q^2)\right]$$

$$\left[\Pi^{38}(Q^2)\right]_{D=6} = \frac{112\pi}{81\sqrt{3}Q^6}\rho_{red}\gamma(\rho\alpha_s\langle\bar{q}q\rangle^2) , \tag{7}$$

where α is the usual EM coupling, $a(Q^2) = \alpha_s(Q^2)/\pi$, $\zeta(n)$ is the Riemann zeta function, $\gamma = [<\bar{d}d>/<\bar{u}u>] - 1$, and ρ_{red} is the ratio of the violation of the vacuum saturation approximation (VSA) estimate for the $D = 6$ condensate combination in the 38 channel to that in the flavor-diagonal 33 vector channel (see Ref. [8] for the latter value). Numerically, the $D = 4$ term is the largest, owing to the smallness of α and $m_{u,d}$. It can be rewritten in terms of the IB mass ratio $r \equiv (m_d - m_u)/(m_d + m_u)$ (for which we use the value from Leutwyler's ChPT analysis [7]) and $<(m_u + m_d)\bar{u}u> \simeq -f_\pi^2 m_\pi^2$ (where we have used the GMOR relation). We employ the 4-loop versions of the running mass and coupling.

Lack of knowledge of ρ_{red} would normally limit the accuracy of the extraction of the f_V. Using both the w_s and w_d families of FESR's, however, it turns out to be possible to determine ρ_{red} self-consistently. In Figure 1 we illustrate the dependence of f_ρ on ρ_{red} for both the w_s and w_d weight families. Obviously only one value of ρ_{red} provides a consistent determination of f_ρ. Checking analogous figures for the other f_V, one finds that the values of ρ_{red} for which they become consistent agree with that in Figure 1 to better than 1%. This clearly demonstrates that this "self-consistency" determination of ρ_{red} is physically meaningful. It is interesting to note that the degree of VSA violation is very similar for the flavor-diagonal isovector and IB flavor 38 correlators. Further details of the calculation, including OPE input values, uncertainties, and a description of the contour-improved implementation of the OPE integrals [9], may be found in Ref. [10].

Having extracted f_V, and knowing the EM decay constants, F_V^{EM}, one may then separately determine F_V^3 and F_V^8 for $V = \rho, \omega, \phi$. To understand the impact of the IB decay constants on the extraction of the 33 and 88 spectral functions, it is convenient to quote the results in terms of the squared ratios, r_V, shown below. In each case, the numerator in the ratio of decay constants represents the IC decay constant of the meson in question (whose square is relevant to the contribution to either the 33 or 88 spectral function), while the denominator represents the nominal value of the numerator, obtained from EM data neglecting the presence of the "wrong-isospin" IB contributions discussed above. The deviations of the r_V

from 1 reflect the presence of the IB F_ρ^8, F_ω^3 and F_ϕ^3 contributions to $F_{\rho,\omega,\phi}^{EM}$. The ratios have been defined in such a way that the true resonance contributions to the 33 (or 88) spectral functions are obtained by multiplying the nominal results, obtained in the usual analysis, by r_V. The results of the analysis outlined above are then

$$r_\rho = \left[\frac{F_\rho^3}{F_\rho^{EM}}\right]^2 = 0.982 \pm 0.0021$$

$$r_\omega = \left[\frac{F_\omega^8}{\sqrt{3}\, F_\omega^{EM}}\right]^2 = 1.154 \pm 0.017$$

$$r_\phi = \left[\frac{F_\phi^8}{\sqrt{3}\, F_\phi^{EM}}\right]^2 = 1.009 \pm 0.001 , \qquad (8)$$

where the errors are dominated by the uncertainty in the quark mass ratio, r, and hence are strongly correlated.

We note the following features of the above results:

- The effect on the ρ contribution to the 33 spectral function is small, and hence has no impact on CVC tests, given the current experimental errors on the EM cross-sections.

- As expected from the argument above, the effect is much larger for the ω contribution to the 88 spectral function. Our result corresponds to a $\sim 7.5\%$ isovector contribution to the physical ω decay constant.

- The ratio of ρ and ω corrections is $8.6 \simeq 9$, suggesting dominance by ρ-ω mixing.

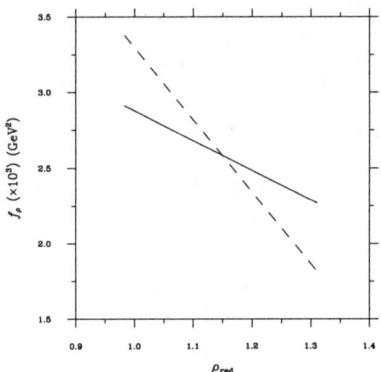

FIGURE 1. Dependence of the ρ spectral parameter, f_ρ, on the reduced VSA-violation parameter, ρ_{red}, for the two weight families, w_s and w_d. The solid line corresponds to w_s, the dashed line to w_d.

The impact of these corrections on the 33-88 sum rules discussed above is as follows.

- Using τ decay data for the isovector part, and the above corrections for the EM isoscalar data, the Narison m_s sum rule implies

$$m_s(1 \text{ GeV}^2) = 154(\pm \sim 50) \text{ MeV} , \qquad (9)$$

compatible with value $157 \pm 16 \pm 15 \pm 16$ obtained in a recent analysis based on flavor breaking in hadronic τ decays [11], and to be compared to the central value 197 MeV obtained neglecting the presence of IB corrections [1].

- The correction to the inverse chiral $33 - 88$ sum rule extraction of the 6th order ChPT low-energy constant Q_V is also large. The FESR solution above implies, via an inverse moment 38 sum rule, the value [10]

$$Q_V(m_\rho^2) = (3.3 \pm 0.4) \times 10^{-5} , \qquad (10)$$

to be compared to the value from a recent analysis [12] of the vector current Π^{ud}-Π^{us} difference, employing τ decay spectral data (which does not need IB corrections)

$$Q_V(m_\rho^2) = (2.8 \pm 1.3) \times 10^{-5} . \qquad (11)$$

The agreement of the results of Eq. (10) with the independent determination of Eq. (11) provides further strong support for the values of the IB vector meson decay constants obtained in the present analysis.

REFERENCES

1. Narison, S., *Phys. Lett.* **B358**, 113 (1995).
2. Golowich, E. and Kambor, J., *Nucl. Phys.* **B447**, 373 (1995); *Phys. Rev.* **D53**, 2651 (1996).
3. Maltman, K., *Phys. Lett.* **B440**, 367 (1998).
4. Maltman, K., *Phys. Lett.* **B467**, 14 (1999).
5. Maltman, K., *Phys. Rev.* **D53**, 2563 (1996).
6. Iqbal, M.J., Jin, X-M. and Leinweber, D.B., *Phys. Lett.* **B367**, 45 (1996).
7. Leutwyler, H., *Phys. Lett.* **B374**, 163 (1996); *Phys. Lett.* **B378**, 313 (1996).
8. Narison, S. *Phys. Lett.* **B361**, 121 (1995).
9. Le Diberder, F. and Pich, A., *Phys. Lett.* **B286**, 147 (1992); **B289**, 165 (1992).
10. Maltman, K. and Wolfe, C.E., *Phys. Rev.* **D59**, 096003 (1999).
11. Kambor, J. and Maltman, K., "The Strange Quark Mass From Flavor Breaking in Hadronic τ Decays", in preparation.
12. Dürr, S. and Kambor, J., hep-ph/9907539.

Neutrino Mixing and CP Violation in Matter

Zhi-zhong Xing

Sektion Physik, Universität München, Theresienstrasse 37A, 80333 Munich, Germany

Abstract. Within the framework of three lepton families I present a transparent analytical relationship between the neutrino mixing and CP-violating parameters in vacuum and thos in matter. Such a model- and parametrization-independent result will be particularly useful to recast the fundamental lepton flavor mixing matrix from the future long-baseline neutrino experiments.

Today strong evidence, that neutrinos are massive and lepton flavors are mixed, has been accumulated from a variety of neutrino experiments. The mixing matrix of three different lepton families may in general consist of non-removable complex phases, leading to CP or T violation. Leptonic CP violation can manifest itself in neutrino oscillations. The best way to observe CP- and T-violating effects is to carry out the long-baseline neutrino experiments. In such experiments the earth-induced matter effects, which are likely to deform the neutrino oscillation behaviors in vacuum and fake the genuine CP-violating signals, must be taken into account. To single out the "true" theory of lepton mass generation and CP violation depends crucially upon how accurately the fundamental parameters of lepton flavor mixing can be measured and disentangled from the matter effects. It is therefore desirable to explore the most transparent analytical relationship between the genuine flavor mixing matrix and the matter-corrected one.

In this talk I present an exact and compact formula to describe the matter effect on lepton flavor mixing and CP violation within the framework of three lepton families [1]. The result is completely independent of the specific models of neutrino masses and the specific parametrizations of neutrino mixing. Therefore it will be particularly useful, in the long run, to recast the fundamental flavor mixing matrix from the precise measurements of neutrino oscillations in a variety of long-baseline neutrino experiments.

In vacuum the 3×3 lepton flavor mixing matrix V links the neutrino mass eigenstates (ν_1, ν_2, ν_3) to the neutrino flavor eigenstates $(\nu_e, \nu_\mu, \nu_\tau)$. If neutrinos are massive Dirac fermions, V can be parametrized in terms of three rotation angles and one CP-violating phase. If neutrinos are Majorana fermions, however, two additional CP-violating phases are in general needed to fully parametrize V. The strength of CP violation in neutrino oscillations, no matter whether neutrinos are of the Dirac or Majorana type, depends only upon a universal parameter $\mathcal{J}$ [2]:

$$\text{Im}\left(V_{\alpha i}V_{\beta j}V_{\alpha j}^{*}V_{\beta i}^{*}\right) = \mathcal{J}\sum_{\gamma,k}\epsilon_{\alpha\beta\gamma}\epsilon_{ijk}, \quad (1)$$

where (α,β,γ) and (i,j,k) run over (e,μ,τ) and $(1,2,3)$, respectively. In the specific models of fermion mass generation V can be derived from the mass matrices of charged leptons and neutrinos [3]. To test such theoretical models one has to compare their predictions for V with the experimental data of neutrino oscillations. The latter may in most cases be involved in the potential matter effects and must be carefully handled.

In the flavor basis where the charged lepton mass matrix M_l is diagonal (i.e., $M_l = \text{Diag}\{m_e, m_\mu, m_\tau\}$), the effective Hamiltonian responsible for neutrinos propagating in matter can be written as $\mathcal{H}_\nu = \Phi_\nu^m/(2E)$ with $\Phi_\nu^m = \Phi_\nu + \Phi_A$ and

$$\Phi_\nu = V\begin{pmatrix} m_1^2 & 0 & 0 \\ 0 & m_2^2 & 0 \\ 0 & 0 & m_3^2 \end{pmatrix}V^\dagger, \quad \Phi_A = \begin{pmatrix} A & 0 & 0 \\ 0 & 0 & 0 \\ 0 & 0 & 0 \end{pmatrix}. \quad (2)$$

Here m_i (for $i = 1, 2, 3$) denote neutrino masses, $A = 2\sqrt{2}G_F N_e E$ describes the charged-current contribution to the $\nu_e e^-$ forward scattering, N_e is the background density of electrons, and E stands for the neutrino beam energy. The neutral-current contributions, which are universal for ν_e, ν_μ and ν_τ neutrinos, lead only to an overall unobservable phase and have been neglected. One can diagonalize Φ_ν^m with a unitary transformation: $V^{m\dagger}\Phi_\nu^m V^m = \text{Diag}\{\lambda_1, \lambda_2, \lambda_3\}$, where λ_i denote the effective mass-squared eigenvalues of three neutrinos in matter. Explicitly we have

$$\lambda_1 = m_1^2 + \frac{1}{3}x - \frac{1}{3}\sqrt{x^2 - 3y}\left[z + \sqrt{3(1-z^2)}\right],$$

$$\lambda_2 = m_1^2 + \frac{1}{3}x - \frac{1}{3}\sqrt{x^2 - 3y}\left[z - \sqrt{3(1-z^2)}\right],$$

$$\lambda_3 = m_1^2 + \frac{1}{3}x + \frac{2}{3}z\sqrt{x^2 - 3y}, \quad (3)$$

where x, y and z are given by [4]

$$x = \Delta m_{21}^2 + \Delta m_{31}^2 + A,$$

$$y = \Delta m_{21}^2 \Delta m_{31}^2 + A\left[\Delta m_{21}^2\left(1 - |V_{e2}|^2\right) + \Delta m_{31}^2\left(1 - |V_{e3}|^2\right)\right], \quad (4)$$

$$z = \cos\left[\frac{1}{3}\arccos\frac{2x^3 - 9xy + 27A\Delta m_{21}^2\Delta m_{31}^2|V_{e1}|^2}{2\left(x^2 - 3y\right)^{3/2}}\right]$$

with $\Delta m_{21}^2 \equiv m_2^2 - m_1^2$ and $\Delta m_{31}^2 \equiv m_3^2 - m_1^2$. Note that the unitary matrix V^m is just the lepton flavor mixing matrix in matter. After a lengthy calculation we arrive at the elements of V^m as [1]

$$V_{\alpha i}^m = \frac{N_i}{D_i} V_{\alpha i} + \frac{A}{D_i} V_{ei} \left[\left(\lambda_i - m_j^2 \right) V_{ek}^* V_{\alpha k} + \left(\lambda_i - m_k^2 \right) V_{ej}^* V_{\alpha j} \right], \tag{5}$$

where α runs over (e, μ, τ) and (i, j, k) over $(1, 2, 3)$ with $i \neq j \neq k$, and

$$N_i = \left(\lambda_i - m_j^2 \right) \left(\lambda_i - m_k^2 \right) - A \left[\left(\lambda_i - m_j^2 \right) |V_{ek}|^2 + \left(\lambda_i - m_k^2 \right) |V_{ej}|^2 \right],$$
$$D_i^2 = N_i^2 + A^2 |V_{ei}|^2 \left[\left(\lambda_i - m_j^2 \right)^2 |V_{ek}|^2 + \left(\lambda_i - m_k^2 \right)^2 |V_{ej}|^2 \right]. \tag{6}$$

Obviously $A = 0$ leads to $V_{\alpha i}^m = V_{\alpha i}$. This exact and compact formula shows clearly how the flavor mixing matrix in vacuum is corrected by the matter effects. Instructive analytical approximations can be made for Eq. (5), once the hierarchy of neutrino masses is experimentally known or theoretically assumed [5].

If leptonic CP were an exact symmetry in vacuum, the determinant of the commutator $[\Phi_\nu, M_l^2]$ would vanish. As both M_l^2 and Φ_A are real diagonal matrices in the chosen flavor basis, we find that $[\Phi_\nu^m, M_l^2] = [\Phi_\nu, M_l^2]$ holds. Then one can derive the relationship between the CP-violating parameter $\mathcal{J}$ and its counterpart in matter $\mathcal{J}_m$ from the equality $\text{Det}[\Phi_\nu^m, M_l^2] = \text{Det}[\Phi_\nu, M_l^2]$. Following the calculations in Ref. [2], we arrive at

$$\mathcal{J}_m (\lambda_1 - \lambda_2)(\lambda_2 - \lambda_3)(\lambda_3 - \lambda_1) = \mathcal{J} \Delta m_{21}^2 \Delta m_{31}^2 \Delta m_{32}^2 . \tag{7}$$

Such an elegant relation has already been observed in Ref. [6]. It indicates that the matter contamination to CP- and T-violating observables, which must be dependent upon $\mathcal{J}_m$, is in general unavoidable. Note that both $\mathcal{J}_m$ and λ_i are complicated functions of the matter parameter A.

The results obtained above are valid for neutrinos interacting with matter. As for antineutrinos, the matter effects arise from the charged-current contribution to the "$\bar{\nu}_e e^+$ forward scattering". The corresponding formulas for antineutrino mixing can straightforwardly be obtained from Eqs. (5) and (7) through the replacements $V \Longrightarrow V^*$ and $A \Longrightarrow -A$.

The matter-corrected flavor mixing and CP-violating parameters can be determined from a variety of long-baseline neutrino experiments. We calculate the conversion probabilities of ν_α (or $\bar{\nu}_\alpha$) to ν_β (or $\bar{\nu}_\beta$) neutrinos in matter and obtain

$$P_m(\nu_\alpha \to \nu_\beta) = -4 \sum_{i<j} [\text{Re}(V_{\alpha i}^m V_{\beta j}^m V_{\alpha j}^{m*} V_{\beta i}^{m*})\, \sin^2 \Delta_{ij}] + 8 \mathcal{J}_m \prod_{i<j} \sin \Delta_{ij} ,$$
$$P_m(\bar{\nu}_\alpha \to \bar{\nu}_\beta) = -4 \sum_{i<j} [\text{Re}(\tilde{V}_{\alpha i}^m \tilde{V}_{\beta j}^m \tilde{V}_{\alpha j}^{m*} \tilde{V}_{\beta i}^{m*})\, \sin^2 \tilde{\Delta}_{ij}] - 8 \tilde{\mathcal{J}}_m \prod_{i<j} \sin \tilde{\Delta}_{ij} , \tag{8}$$

where (α, β) run over (e, μ), (μ, τ) or (τ, e); $\tilde{V}_{\alpha i}(A) \equiv V_{\alpha i}(-A)$, $\tilde{\Delta}_{ij}(A) \equiv \Delta_{ij}(-A)$, and $\tilde{\mathcal{J}}_m(A) \equiv \mathcal{J}_m(-A)$; and $\Delta_{ij} \equiv 1.27(\lambda_i - \lambda_j)L/E$ with L the distance between the production and interaction points of ν_α (in unit of km) and E the neutrino beam energy (in unit of GeV). $P_m(\nu_\beta \to \nu_\alpha)$ and $P_m(\bar{\nu}_\beta \to \bar{\nu}_\alpha)$ can be read off from Eq. (8) with the replacements $\mathcal{J}_m \Longrightarrow -\mathcal{J}_m$ and $\tilde{\mathcal{J}}_m \Longrightarrow -\tilde{\mathcal{J}}_m$, respectively.

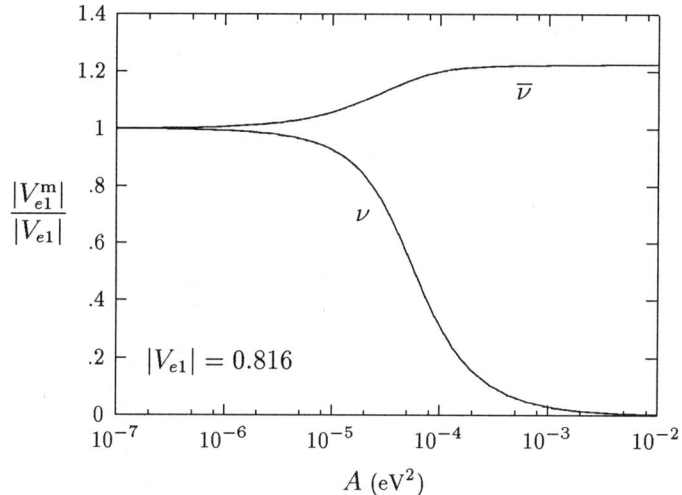

FIGURE 1. Illustrative plot for matter effects on $|V_{e1}|$ associated with neutrinos ($+A$) and antineutrinos ($-A$), where $\Delta m_{21}^2 = 5 \cdot 10^{-5}$ eV^2 and $\Delta m_{31}^2 = 3 \cdot 10^{-3}$ eV^2 have been input.

Let me give a numerical illustration of the matter-induced corrections to the lepton flavor mixing matrix in vacuum. The elements of V^m, except the unknown Majorana phases, can be completely determined by four rephasing-invariant quantities (e.g., four independent $|V_{\alpha i}^m|$ or three independent $|V_{\alpha i}^m|$ plus $\mathcal{J}_m$). As the solar and atmospheric neutrino oscillations in vacuum are essentially associated with the elements in the first row and the third column of V, it is favored to choose $|V_{e1}|$, $|V_{e2}|$, $|V_{\mu 3}|$ and $\mathcal{J}$ as the four basic parameters. To be specific we take $|V_{e1}| = 0.816$, $|V_{e2}| = 0.571$, $|V_{\mu 3}| = 0.640$, and $\mathcal{J} = \pm 0.020$ for neutrinos and antineutrinos. Such a choice is consistent with the CHOOZ experiment [7], the large-angle MSW solution to the solar neutrino problem, and a nearly maximal mixing in the atmospheric neutrino oscillation [8]. The relevant neutrino mass-squared differences are typically taken to be $\Delta m_{21}^2 = 5 \cdot 10^{-5}$ eV^2 and $\Delta m_{31}^2 = 3 \cdot 10^{-3}$ eV^2. Then we compute the ratios $|V_{\alpha i}^m|/|V_{\alpha i}|$ and $\mathcal{J}_m/\mathcal{J}$ as functions of the matter parameter A in the range 10^{-7} eV$^2 \leq A \leq 10^{-2}$ eV^2. The numerical results are shown in Figs. 1 to 4.

We observe that matter effects can be significant for the elements in the first and the second columns of V, if $A \geq 10^{-5}$ eV^2. In comparison, the magnitudes of $|V_{e3}|$, $|V_{\mu 3}|$ and $|V_{\tau 3}|$ may be drastically enhanced or suppressed only for $A > 10^{-3}$ eV^2. The neutrinos are relatively more sensitive to the matter effects than the antineutrinos.

The magnitude of $\mathcal{J}_m$ decreases, when the matter effect becomes significant (e.g., $A \geq 10^{-4}$ eV^2). However, this does not imply that the CP- or T-violating asymmetries in realistic long-baseline neutrino oscillations would be smaller than their values in vacuum. Large matter effects can significantly modify the frequencies of

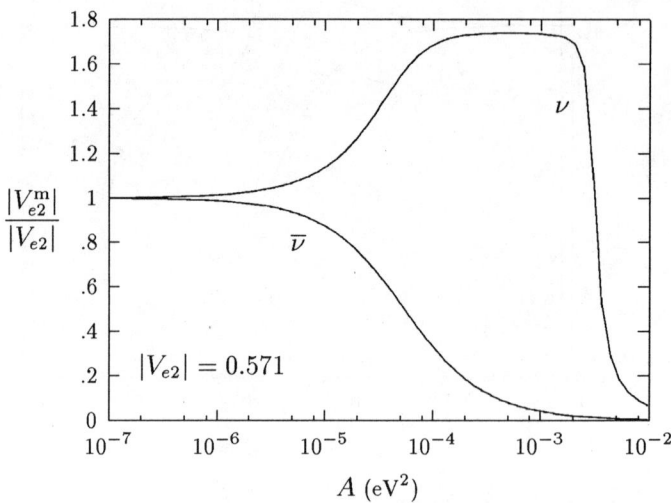

FIGURE 2. Illustrative plot for matter effects on $|V_{e2}|$ associated with neutrinos ($+A$) and antineutrinos ($-A$), where $\Delta m_{21}^2 = 5 \cdot 10^{-5}$ eV^2 and $\Delta m_{31}^2 = 3 \cdot 10^{-3}$ eV^2 have been input.

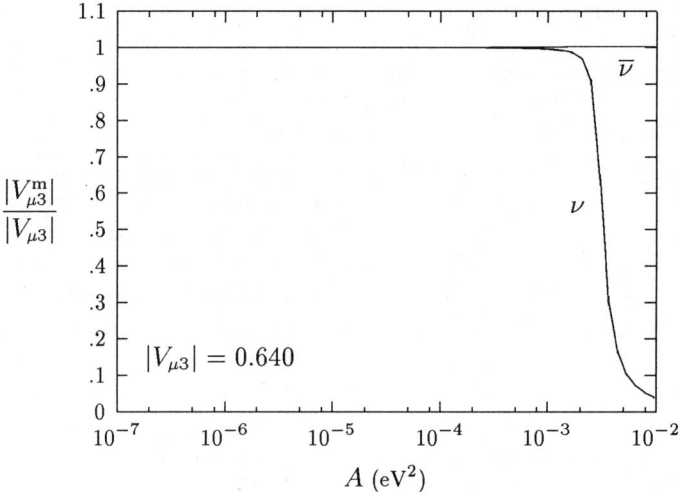

FIGURE 3. Illustrative plot for matter effects on $|V_{\mu 3}|$ associated with neutrinos ($+A$) and antineutrinos ($-A$), where $\Delta m_{21}^2 = 5 \cdot 10^{-5}$ eV^2 and $\Delta m_{31}^2 = 3 \cdot 10^{-3}$ eV^2 have been input.

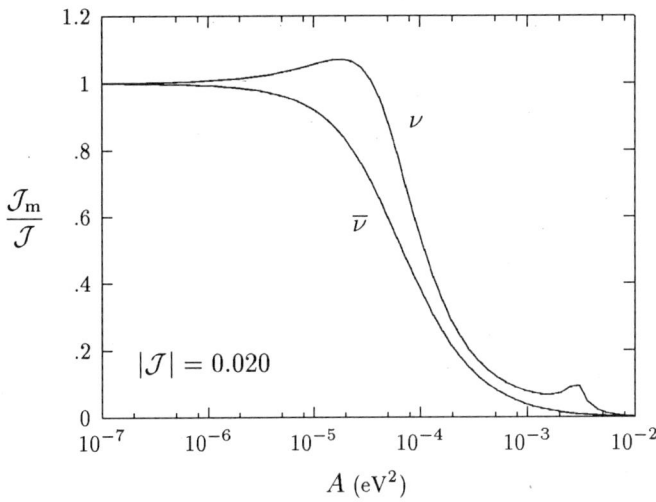

FIGURE 4. Illustrative plot for matter effects on $\mathcal{J}$ associated with neutrinos ($+A$) and antineutrinos ($-A$), where $\Delta m_{21}^2 = 5 \cdot 10^{-5}$ eV^2 and $\Delta m_{31}^2 = 3 \cdot 10^{-3}$ eV^2 have been input.

neutrino oscillations and thus enhance or suppress the genuine signals of CP or T violation.

If the earth-induced matter effects can well be controlled, it is possible to recast the fundamental flavor mixing matrix V from a variety of measurements of neutrino oscillations. Such a goal is expected to be reached in the neutrino factories [9,10].

Acknowledgment: I would like to thank H. Fritzsch and A. Thomas for supporting my participation in this interesting conference.

REFERENCES

1. Z.Z. Xing, hep-ph/0002246.
2. C. Jarlskog, in *CP Violation* (World Scientific, Singapore, 1989), pp.3.
3. For a recent review with extensive references, see: H. Fritzsch and Z.Z. Xing, hep-ph/9912358; S.M. Barr and I. Dorsner, hep-ph/0003058.
4. V. Barger, S. Pakvasa, R.J.N. Phillips, and K. Whisnant, Phys. Rev. D **22**, 2718 (1980); H.W. Zaglauer and K.H. Schwarzer, Z. Phys. C **40**, 273 (1988);
5. See, e.g., H. Fritzsch and Z.Z. Xing, Phys. Lett. B **372**, 265 (1996); Phys. Lett. B **440**, 313 (1998); Phys. Rev. D **61**, 073016 (2000).
6. P.F. Harrison and W.G. Scott, Phys. Lett. B **476**, 349 (2000).
7. M. Apollonio *et al.*, Phys. Lett. B **420**, 397 (1998).
8. Y. Fukuda *et al.*, Phys. Rev. Lett. **81**, 1562 (1998); *ibid.* **81**, 4279 (1998).
9. B. Autin *et al.*, CERN 99-02 (1999); D. Ayres *et al.*, physics/9911009.
10. For a recent review with extensive references, see: V. Barger, hep-ph/0003212.

Tritium Decay, Neutrino Mixing and Neutrino Interactions.

Bruce H. J. McKellar*, G. J. Stephenson, Jr.† and T. Goldman‡

*University of Melbourne, Parkville, Victoria 3052, Australia
†University of New Mexico, Albuquerque, NM 87131, USA
‡Los Alamos National Laboratory, Los Alamos, NM 87545, USA

Abstract. Recent Tritium beta decay experiments continue to exhibit an excess of counts near the end point. The representation of the data in terms of the extracted mass squared can be reproduced by allowing interference between the Standard Model V-A current and other Lorentz currents allowed by existing experimental limits. This is true only if the electron antineutrino is nearly, but not exactly, a single mass eigenstate with a mixing probability to other mass eigenstates of the order of a few parts per thousand, and if the mass of that eigenstate is a few electron volts. In conjunction with double beta decay constraints, this precludes a solitary Majorana neutrino in the electron family. Combining this with solar neutrino results leads to pseudo-Dirac neutrinos and implies that neutral current events in SNO will suffer the same suppression as charge current events.

INTRODUCTION

The 1998 Particle Data Group report shows that all data taken in the period 1990-1995 with errors of less than about 100eV^2 give negative values of $m^2_{\nu_e}$. The best result of this period (from Troitsk) is

$$m^2_{\nu_e} = -22 \pm 4.8 \quad \text{eV}^2 \tag{1}$$

One obvious interpretation of the anomalous value of $m^2_{\nu_e}$ is that there is an unexpected and unexplained excess of counts near the endpoint of the electron spectrum.

Then in 1999 new data were reported by the Troitsk [1] and Mainz [2] groups which, although analyzed differently, confirm the existence of anomalous results. The Troitsk data show an enhancement about 15eV below the endpoint, with evidence that the position of the bump is varying with time. In 4 of the 5 runs the Mainz group report values of $m^2_{\nu_e}$ which "are significantly negative".

It was once well known that the effect of a finite mass on the beta decay spectrum depends on the interactions [3], but with the dominance of $V - A$ theory and its elevation to the standard model this observation seems to have passed out of the general consciousness. We explore its consequences, emphasizing that

1. new interactions can distort the spectrum near the end point, especially through interference with the dominant standard model interaction,

2. mixing in the lepton sector plays an important role in the structure of the interference [4],

3. time dependence can be introduced by a background density of neutrinos.

Perhaps the anomalous results are giving us a clue about physics beyond the Standard Model.

NEW INTERACTIONS

For low energy physics, like nuclear β-decay in general and Tritium β-decay in particular, any new interactions can only appear as effective currents in the four fermion formulation of the theory with the usual space time structures of S, P, T, V or A. Given the dominance of the SM, it is reasonable to recast this as S_L, S_R, T, R or L, where $R \sim (V+A)$ and $L \sim (V-A)$, with a similar construction for S_L and S_R from S and P. The effect on the spectrum displayed below only occurs if there are additional currents in the lepton sector that are different from L. Corresponding changes in the hadron currents affect only the scale of each new contribution. To emphasize this fact, we employ an unconventional notation, writing the effective low energy Hamiltonian for semi-leptonic decays as

$$H_I = \sum_{\alpha,\beta=S_L,S_R,R,L,T} G^{\alpha\beta} \sum_f \left(J_{h\alpha}^\dagger \cdot J_{f\beta} + h.c.\right) \tag{2}$$

where, for example,

$$J_{f\alpha} = \overline{\psi_{\nu_f}} \Gamma_\alpha \psi_f \tag{3}$$

with ψ_f representing a charged lepton of a given flavor, ψ_{ν_f} a neutral lepton associated with that lepton through the particular interaction (see the discussion below for more detail). A similar construction can be made on the hadron side. Note that, in this convention, the first Greek index in Eq.(1) refers to the hadron current. Explicitly,

$$\begin{aligned}\Gamma_{S_L} &= (1-\gamma^5) \\ \Gamma_{S_R} &= (1+\gamma^5) \\ \Gamma_R &= \gamma^\mu(1+\gamma^5) \\ \Gamma_L &= \gamma^\mu(1-\gamma^5) \\ \Gamma_T &= [\gamma^\mu, \gamma^\nu]/2.\end{aligned} \tag{4}$$

Most off diagonal combinations ($\alpha \neq \beta$) in the sum vanish, the exceptions being for (S_R, S_L) and (R, L). For nuclear beta decay in the SM, only $\beta = L$ and $\alpha = L, R$ survive.

We need to evaluate the effect on the Tritium beta spectrum of possible interferences between the SM Left-chiral vector current the various possible new currents, and well as the contribution of the new currents by themselves. Taking the no-recoil approximation for the nuclei, the leptonic currents which survive after contraction with the hadronic currents give the following contributions to the beta spectrum:

$$
\begin{array}{ll}
LL(SM) & E_\nu E_\beta \\
RR & E_\nu E_\beta \\
LR + RL & -2m_\nu m_e \\
S_L S_L & E_\nu E_\beta \\
S_R S_R & E_\nu E_\beta \\
S_L S_R + S_R S_L & -2m_\nu m_e \\
TT & [E_\nu E_\beta - m_\nu m_e] \\
LS_L + S_L L & -2m_\nu E_\beta \\
LS_R + S_R L & 2E_\nu m_e \\
LT + TL & 2[E_\nu m_e - m_\nu E_\beta]
\end{array}
\tag{5}
$$

The negative sign of the terms with a factor of m_ν arises from the fact that the neutral lepton in Tritium beta decay is an anti-neutrino.

It is important to emphasize that most scenarios for "Physics beyond the standard model" include some of these interactions, eg

1. Exchange of a charge-changing scalar, arising, for example, in supersymmetric models

2. Exchange of a vector boson coupled to Right-chiral fermions, as occurs for example in left-right symmetric models. The new W_R may or may not mix with the W_L bosons of the SM.

Thus the question which should be asked is not "do new interactions exist?" — almost every attempt to generalize physics beyond the standard model will introduce such interactions. The question should instead be "how strong are the new interactions?"

LEPTON MIXING

Interference of different currents has been considered before in terms of trying to limit the strength of these non-standard currents. However, the previous analyses have omitted consideration of lepton mixing. The Superkamiokande data demonstrate the phenomenon of neutrino oscillation, at least between the muon neutrino and at least one other neutrino species. The solar neutrino problem strongly suggests oscillation between the electron neutrino and at least one other species. Oscillation thus implies that there is mixing of lepton families, just as there is mixing of the hadron families.

For simplicity we assume an orthogonal mixing matrix, neglecting CP violation, and restrict the present discussion to Dirac neutrinos. The generalization to Majorana neutrinos is straightforward, and is discussed in detail elsewhere [5]. In particular, we assume the existence of three massive Dirac neutrinos, ν_i^D, with Dirac masses m_i^D. We allow for the possibility that, as in the quark sector, current eigenstates and mass eigenstates are not necessarily identical. In particular, we shall refer to the ν_f, $f = (e, \mu, \tau)$ as those linear combinations of mass eigenstates which are coupled to the corresponding charged leptons by the $SU(2)_L$ currents of the SM. These current eigenstates are expressed in terms of the mass eigenstates as

$$\nu_f = \sum_i \cos \theta_f^i \nu_i \tag{6}$$

where the $\cos \theta_f^i$ are the direction cosines in the coordinate system spanned by the mass eigenstates.

If there are additional charged bosons, besides the known massive weak vector bosons, that couple to both the quarks and the leptons, we may define neutral leptons which are current eigenstates of this new interaction in exactly the same way, denoted by $\hat{\nu}_f$, $f = (e, \mu, \tau)$, where now

$$\hat{\nu}_f = \sum_i \cos \hat{\theta}_f^i \nu_i \tag{7}$$

and one would naturally expect that, in general,

$$\hat{\theta}_f^i \neq \theta_f^i. \tag{8}$$

The intrinsic strength of the interaction X relative to the $L-L$ interaction of the SM, $\hat{\rho}_X'$, is determined by the mass of the non-SM boson, M_X and its coupling g_X

$$\hat{\rho}_X' = \frac{g_X^2}{g^2} \frac{M_W^2}{M_X^2} \tag{9}$$

The practical coupling will be modified by the mixing and the relative strength of the "new interaction" now includes the ratio of mixing matrix elements:

$$\hat{\rho}_X = \hat{\rho}_X' \frac{\tilde{\Lambda}_{e1}^X}{\Lambda_{e1}} \tag{10}$$

It is important to realize that the sign of $\hat{\rho}_X$ is not fixed, although $\hat{\rho}_X'$ is positive.

EFFECTS ON THE TRITIUM SPECTRUM

In general we may write for the beta decay spectrum of tritium[1], in the standard model, but with massive neutrinos and allowing for mixing

---

[1] We simplify this discussion by omitting the discussion of the effect of excited molecular states on the spectrum, which must of course be included in a realistic analysis [5]

$$\left(\frac{dN}{dE_\beta}\right)_{SM} = KF(E_\beta)q_\beta E_\beta E_\nu \sum_i \cos^2\theta_e^i q_\nu^{(i)} \Theta(E_\nu - m_i) \tag{11}$$

Here the sum goes over the neutrino mass eigenstates of mass m_i, E_ν is the neutrino energy $E_\nu = \mathcal{E}_0^0 - E_\beta$, $\mathcal{E}_0^0$ is the beta spectrum endpoint if the neutrinos have zero mass, and $q_\nu^{(i)} = \sqrt{E_\nu^2 - m_i^2}$ is the momentum of the neutrino of mass m_i. The factor K includes various normalization factors which are unimportant in the analysis near the end point, and $F(E)$ is the Fermi function.

Taking as an example the standard model interaction, with an additional S_L interaction, the beta spectrum has the form

$$\left(\frac{dN}{dE_\beta}\right) = KF(E_\beta)q_\beta \sum_i \cos^2\theta_e^i q_\nu^{(i)} \Theta(E_\nu - m_i) \left\{ E_\nu E_\beta + \hat{\rho}_{S_L}^{(i)}(-m_i E_\beta) \right.$$
$$\left. + \hat{\rho}_{S_L}^{(i)\,2} E_\nu E_\beta \right\} \tag{12}$$
$$= KF(E_\beta)q_\beta \sum_i \cos^2\theta_e^i q_\nu^{(i)} \Theta(E_\nu - m_i) \left(1 + \hat{\rho}_{S_L}^{(i)\,2}\right) E_\nu E_\beta$$
$$\times \left\{ 1 + \epsilon_{S_L}^{(i)} \left(-\frac{m_i}{E_\nu}\right) \right\} \tag{13}$$

The parameter $\epsilon_{S_L}^{(i)}$ is given by

$$\epsilon_{S_L}^{(i)} = \frac{\hat{\rho}_{S_L}^{(i)}}{1 + \hat{\rho}_{S_L}^{(i)\,2}} \tag{14}$$

We are interested only in the shape of the spectrum closer than 100eV to the end point. In this region, the dependence on E_β is very gentle. The product $F(E_\beta)q_\beta$ is nearly constant, and the ratio m_e/E_β varies by less than 2 parts in 10^3. However, the dependence on $E_\nu = (\mathcal{E}_0^i - E_\beta)$ is important. Ignoring the E_β variation in this region, we can absorb the functions of E_β into a new 'constant', K', and we find that all of the modifications to the spectrum from new interactions can be expressed in the form

$$\frac{dN}{dE_\beta} \cong K' \sum_i \cos^2\theta_e^i E_\nu^2 \left[1 + \varepsilon_i + \frac{\phi_i}{E_\nu}\right] \sqrt{1 - \frac{m_i^2}{E_\nu^2}} \Theta(E_\nu - m_i) \tag{15}$$

After making these approximations it is straightforward to do the integrals necessary to express the integral spectrum from some energy $E_{\beta,\min}$ to the endpoint, which corresponds to the way the data is taken. We are then able to show, if the electron antineutrino is nearly, but not exactly, a single mass eigenstate with a mixing probability to other mass eigenstates of the order of a few parts per thousand, and if the mass of that eigenstate is a few electron volts, that one can obtain

1. a fitted value of m_ν^2 which is negative when just one neutrino mass and a standard model interaction is assumed

2. this negative value of m_ν^2 varies with the value of $E_{\beta,\mathrm{min}}$

It is just these properties which summarize the effects observed by the Mainz and Troitsk groups.

In conjunction with double beta decay constraints, this precludes a solitary Majorana neutrino in the electron family. Combining this with solar neutrino results leads to pseudo-Dirac neutrinos and implies that neutral current events in SNO will suffer the same suppression as charge current events [6].

We [7] have previously studied the consequences that may arise if there is a very light scalar particle coupled only to neutrinos. The primary effects are that it is energetically favourable for neutrinos to condense into clouds, and that in these regions the effective mass (which depends on the density of the neutrinos) which will govern the interference effects discussed above. Thus this scenario introduces the possibility of describing the time dependence of the excess counts.

These qualitative results are encouraging. An analysis of the data should be carried out using the formula of equation (15) — or its more complicated exact form — to test whether there is indeed evidence here for physics beyond the standard model.

ACKNOWLEDGMENTS

We are happy to acknowledge valuable conversations on these topics with Maurice Goldhaber, Bill Louis and Peter Herczeg. This research is partially supported by the Department of Energy under contract W-7405-ENG-36, by the National Science Foundation and by the Australian Research Council. One of us (GJS) acknowledges the hospitality of the Institute for Nuclear Theory at the University of Washington on several occasions, where a portion of this work was carried out.

REFERENCES

1. V.M. Lobashev et al. , *Phys. Lett.* **B460**, 227 (1999).
2. Ch. Weinheimer et al. , *Phys. Lett.* **B460**, 219 (1999).
3. O. Koefed-Hansen, *Phys. Rev.* **71**, 451 (1947); *Phys. Rev.* **74**, 1785 (1948); *Physica* **XVIII**, 1287 (1952).
4. G.J. Stephenson, Jr. and T. Goldman, *Phys. Lett.* **B440**, 89 (1998).
5. G. J. Stephenson, Jr, T. Goldman and B. H. J. McKellar, to be published.
6. T. Goldman, G. J. Stephenson, Jr., and B. H. J. McKellar, *Modern Physics Letters***A15**, 439 (2000).
7. G.J. Stephenson, Jr. , T. Goldman and B.H.J. McKellar, *Int. J. Mod. Phys.***A13**, 2765 (1998).

Constraints on a Parity-even/Time-Reversal-odd Interaction [1]

Willem T.H. van Oers

Department of Physics, University of Manitoba, Winnipeg, MB, Canada R3T 2N2
and
TRIUMF, 4004 Wesbrook Mall, Vancouver, BC Canada V6T 2A3

Abstract. Time-Reversal-Invariance non-conservation has for the first time been unequivocally demonstrated in a direct measurement, one of the results of the CPLEAR experiment. What is the situation then with regard to time-reversal-invariance non-conservation in systems other than the neutral kaon system? Two classes of tests of time-reversal-invariance need to be distinguished: the first one deals with parity violating (P-odd)/time-reversal-invariance non-conserving (T-odd) interactions, while the second one deals with P-even/T-odd interactions (assuming CPT conservation this implies C-conjugation non-conservation). Limits on a P-odd/T-odd interaction follow from measurements of the electric dipole moment of the neutron. This in turn provides a limit on a P-odd/T-odd pion-nucleon coupling constant which is 10^{-4} times the weak interaction strength. Limits on a P-even/T-odd interaction are much less stringent. The better constraint stems also from the measurement of the electric dipole moment of the neutron. Of all the other tests, measurements of charge-symmetry breaking in neutron-proton elastic scattering provide the next better constraint. The latter experiments were performed at TRIUMF (at 477 and 347 MeV) and at IUCF (at 183 MeV). Weak decay experiments (the transverse polarization of the muon in $K^+ \to \pi^0 \mu^+ \nu_\mu$ and the transverse polarization of the positrons in polarized muon decay) have the potential to provide comparable or possibly better constraints.

INTRODUCTION

Time-reversal-invariance non-conservation has for the first time been unequivocally demonstrated in a direct measurement in the CPLEAR experiment. [1] The experiment measured the difference in the transition probabilities $P(\overline{K}^0 \to K^0)$ and $P(K^0 \to \overline{K}^0)$. Assuming CPT conservation but allowing for a possible breaking of the $\Delta S = \Delta Q$ rule, the result obtained for A_T

---

[1] Work supported in part by the Natural Sciences and Engineering Research Council of Canada.

$$A_T = \frac{R(\overline{K^0} \to K^0) - R(K^0 \to \overline{K^0})}{R(\overline{K^0} \to K^0) + R(K^0 \to \overline{K^0})} = [6.6 \pm 1.3(\text{stat.}) \pm 1.0(\text{syst.})] \times 10^{-3} \quad (1)$$

is in good agreement with the measure of CP violation in neutral kaon decay. A more recent reported result is a large asymmetry in the distribution of $K_L \to \pi^+\pi^-e^+e^-$ events in the CP-odd/T-odd angle ϕ between the decay planes of the $\pi^+\pi^-$ and e^+e^- pairs in the K_L centre of mass system. The overall asymmetry found was $[13.6 \pm 2.5(\text{stat.}) \pm 1.2(\text{syst.})]\%$. [2] The question then is: what is the situation with regard to time-reversal-invariance in systems other than the kaon system?

Tests of time-reversal-invariance can be distinguished as belonging to two classes: the first one deals with P-odd/T-odd interactions, while the second one deals with P-even/T-odd interactions (assuming CPT conservation this implies C-conjugation non-conservation). But it should be noted that constraints on these two classes of interactions are not independent since the effects due to P-odd/T-odd interactions may also be produced by P-even/T-odd interactions in conjunction with Standard Model parity violating radiative corrections. The latter can occur at the 10^{-7} level and consequently could present a limit on the constraint of a P-even/T-odd interaction, derived from experiment. Limits on a P-odd/T-odd interaction follow from measurements of the electric dipole moment of the neutron (which currently stands at $< 6 \times 10^{-26}$ e.cm [95% C.L.]). This provides a limit on a P-odd/T-odd pion-nucleon coupling constant which is less than 10^{-4} times the weak interaction strength. Measurements of ^{129}Xe and ^{199}Hg ($< 8 \times 10^{-28}$ e.cm [95% C.L.]) give similar constraints. [see Ref. 3]

Experimental limits on a P-even/T-odd interaction are much less stringent. Following the standard approach of describing the nucleon-nucleon interaction in terms of meson exchanges, it can be shown that only charged rho-meson exchange and A_1-meson exchange can lead to a P-even/T-odd interaction. [4] The better constraints stem from measurements of the electric dipole moment of the neutron and next from measurements of charge symmetry breaking in neutron-proton (n-p) elastic scattering. All other experiments, like gamma decay experiments [5], detailed balance experiments [6], polarization - analyzing power difference measurements [7], and five-fold correlation experiments with polarized incident nucleons and aligned nuclear targets [8], have been shown to be at least an order of magnitude less sensitive. Haxton, Hoering, and Musolf [3] have deduced constraints on a P-even/T-odd interaction from nucleon, nuclear, and atomic electric dipole moments with the better constraint coming from the electric dipole moment of the neutron. In terms of a ratio to the strong rho-meson nucleon coupling constant, they deduced for the P-even/T-odd rho-meson nucleon coupling: $|\bar{g}_\rho| < 0.53 \times 10^{-3} \times |f_\pi^{\text{DDH}}/f_\pi^{\text{meas.}}|$. But the ratio of the theoretical to the experimental value of f_π may be as large as 15! [9] However, constraints derived from one-loop contributions to the electric dipole moment of the neutron exceed the two-loop limits by more than an order of magnitude and are much more stringent. [10] However, a translation in terms of coupling strengths in the hadronic sector still needs to be made.

It is very difficult to accommodate a P-even/T-odd interaction in the Standard Model. It requires C-conjugation non-conservation, which cannot be introduced at the first generation quark level. It can neither be introduced into the gluon self-interaction. Consequently, one needs to consider C-conjugation non-conservation between quarks of different generations and/or between interacting fields. [11]

CHARGE SYMMETRY BREAKING IN NEUTRON PROTON ELASTIC SCATTERING

Charge symmetry breaking (CSB) in neutron-proton elastic scattering manifests itself as a non-zero difference of the neutron (A_n) and proton (A_p) analyzing powers, $\Delta A = A_n - A_p = 2 \times [\text{Re}(b^*f) + \text{Im}(c^*h)]/\sigma_0$. Here the complex amplitude f is charge symmetry breaking, while the complex amplitude h is both charge symmetry breaking and time-reversal-invariance non-conserving. The complex amplitudes b and c belong to the usual five n-p scattering amplitudes and σ_0 is the unpolarized differential cross section. The three precision experiments performed (at TRIUMF at 477 MeV [12] and at 347 MeV [13], and at IUCF at 183 MeV [14]) have unambiguously shown that charge symmetry is broken and that the results for ΔA at the zero-crossing angle of the average analyzing power are very well reproduced by meson exchange model calculations (see Fig. 1). A P-even/T-odd interaction produces a term in the scattering amplitude which is simultaneously charge symmetry breaking (the complex amplitude h in the expression above). Thus, Simonius [15] deduced an upper limit on a P-even/T-odd CSB interaction from a comparison of the experimental results with the theoretical predictions for the three n-p CSB experiments. The upper limit so derived is $|\bar{g}_\rho| < 6.7 \times 10^{-3}$ [95% C.L.]. This is therefore comparable to the upper limit deduced from the electric dipole moment of the neutron, taking present experimental limits of f_π, and is considerably lower than the limits inferred from direct tests of a P-even/T-odd interaction. For instance the detailed balance experiments give a limit of $|\bar{g}_\rho| < 2.5 \times 10^{-1}$ [see Ref. 8], while measurements of the five-fold correlation parameter $A_{y,xz}$ in polarized neutron transmission through nuclear spin-aligned ^{165}Ho give a limit of $|\bar{g}_\rho| < 5.9 \times 10^{-2}$ even though the measured value of $A_{y,xz}$ was $(8.6 \pm 7.7) \times 10^{-6}$. [8] It is effectively only the valence proton in ^{165}Ho which contributes to $A_{y,xz}$. Even though it is inconceivable in the Standard Model to account for a P-even/T-odd interaction, there is a need to clarify the experimental situation by providing a better experimental result.

Such an experimental constraint may be provided by an improved upper limit on the electric dipole moment of the neutron. In fact a new measurement with a sensitivity of 4×10^{-28} e.cm has been proposed at the Los Alamos Neutron Science Center. [16] Performing an improved n-p elastic scattering CSB experiment also appears to be a very attractive possibility. One can calculate with a great deal of confidence the contributions to CSB due to one-photon exchange and due to the n-p mass difference affecting charge one-pion and rho-meson exchange. Furthermore,

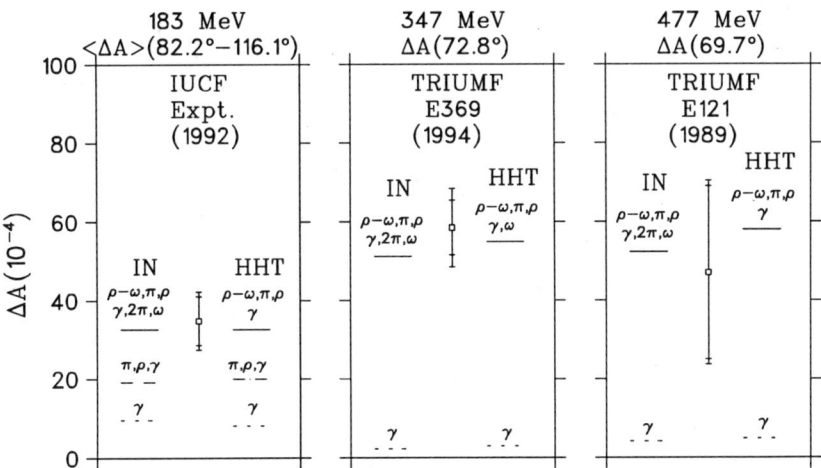

FIGURE 1. Experimental results of ΔA at the zero-crossing angle at incident neutron energies of 183, 347, and 477 MeV compared with theoretical predictions of Iqbal and Niskanen, and Holzenkamp, Holinde, and Thomas. The inner error bars present the statistical uncertainties; the outer error bars have the systematic uncertainties included (added in quadrature). For further details see Ref. 13.

one can select an energy where the $\rho^\circ - \omega$ meson mixing contribution changes sign at the same angle where the average of the analyzing powers A_n and A_p changes sign and therefore does not contribute to ΔA. This occurs at a neutron energy of 320 MeV and is caused by the particular interplay of the n-p phase shifts and the form of the spin/isospin operator connected with the $\rho^\circ - \omega$ mixing term. But also the one-photon exchange term changes sign at about the same angle at 320 MeV. The contribution due to two-pion exchange with an intermediate Δ is expected to be less than one tenth of the overall CSB effect, essentially determining an upper limit on the theoretical uncertainty (see Fig. 2). [17] It has been shown that simultaneous γ-π exchanges can only contribute to ΔA through second order processes and can therefore be neglected. [18] Also the effects of inelasticity are negligibly small at 320 MeV. It appears therefore well within reach to reduce the theoretical uncertainty in the comparison of experiment and theory. Subtracting the calculated ΔA from the measured ΔA permits establishing an upper limit on a P-even/T-odd/CSB interaction.

In the TRIUMF CSB experiments polarized neutrons were scattered from unpolarized protons and vice versa. The polarized (or unpolarized) neutron beam was obtained using the (p,n) reaction with a 369 (and 497) MeV polarized (or unpolarized) proton beam incident on a 0.20 m long LD_2 target. At these energies one makes use of the large sideways-to-sideways polarization transfer coefficient r_t at $9°$ in the lab. The only difference in obtaining the polarized and unpolarized 347 MeV neutron beams was turning off the pumping laser light in the optically pumped polarized ion source (OPPIS). The polarized proton target was of the frozen spin type

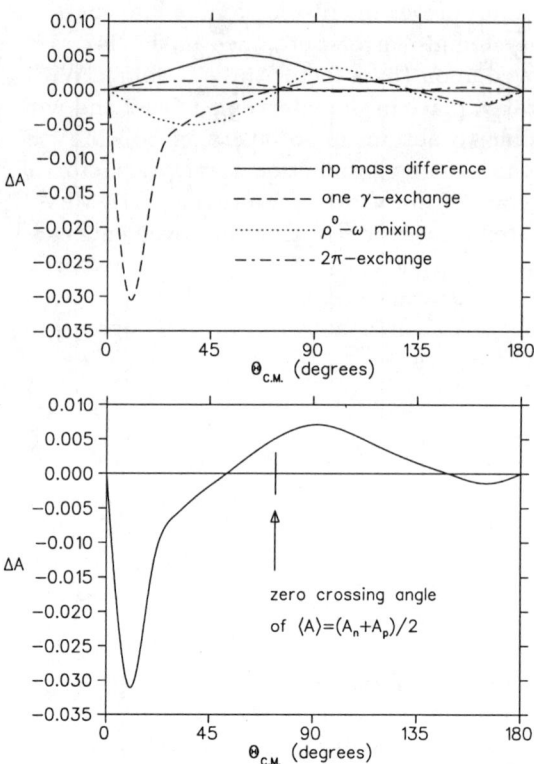

FIGURE 2. Angular distributions of the different contributions to ΔA at an incident neutron energy of 320 MeV. (Ref. 17) Note that the ρ^0 - ω mixing contribution passes through zero at the same angle as the average of A_n and A_p (vertical bar). The lower part of the figure gives the total ΔA angular distribution.

with butanol beads as target material. The same target after depolarization was used as the unpolarized proton target. Great care was taken that the two interleaved phases of the experiments were performed with identical beam and target parameters except for the polarization states. At 347 MeV scattered neutrons and recoiling protons were detected in coincidence in the c.m. angular range 53.4 to 86.9 degrees in two left-right symmetric detector systems. Rather than measuring A_n and A_p directly (which would limit the accuracy attainable by not having polarization calibration standards of the required precision), the zero-crossings of A_n and A_p were determined by fitting the partial angular distributions with polynomials, deduced/. from $n-p$ phase shift analyses. The difference ΔA followed then by multiplying the difference in the zero-crossing angles by the average slope of the analyzing powers (the experiment measured the slope of A_p at the zero-crossing angle, which is a good approximation for the average slope at the zero-crossing angle and introduces a negligible error). The execution of the experiments depended on a great deal of simultaneous monitoring and online control measurements. Both the statistical and systematic errors, obtained in the 347 MeV experiment, can be considerably improved upon (by a factor three to four). With the OPPIS developments which have taken place in the intervening years and with a biased Na-ionizer cell it will be possible to obtain up to 50 μA of 342 MeV 80% polarized proton beam incident on the neutron production target (a factor of 50 increase in neutron beam intensity at 320 MeV over the previous 347 MeV CSB experiment). In addition various systematic error reducing improvements can be introduced in the experimental arrangements and procedures. Such an experiment would constitute a measurement of CSB in n-p elastic scattering of unprecedented precision of great value on its own and would simultaneously provide a greatly improved upper limit on a P-even/T-odd interaction.

PARTICLE DECAYS

Searches for P-even/T-odd interactions are also made in particle decays, e.g., in the decay $\mu^+ \to e^+ \nu_e \bar{\nu}_\mu$ and in the decay $K^+ \to \mu^+ \pi^0 \nu_\mu$. A non-zero value of the muon polarization transverse to the decay plane would be an indication of time-reversal-invariance non-conservation. An experiment to measure the first decay is presently being executed at PSI. [19] Several experiments have been performed to measure the transverse muon polarization in both neutral and charged kaon decay. There is a unique feature to the transverse muon polarization in that it does not have contributions from the Standard Model at tree level and that higher order effects are of order 10^{-6}. When only one charged particle is present in the final state, a final state interaction, which can mimic a time-reversal-invariance breaking effect, is greatly reduced and is estimated to occur only at the same level of 10^{-6}. The more recent effort of measuring the time-reversal-invariance non-conserving transverse muon polarization is being done at KEK using a stopped K^+ beam. The experiment reported a result for P_T = -0.0042 $\pm$ 0.0049(stat.) $\pm$ 0.0009(syst.), based on the data taken in 1996 and 1997, which translates into a value of Imξ = -0.013 $\pm$ 0.016(stat.) $\pm$ 0.003(syst.). [20] The quantity ξ is defined as the ratio of two form factors, $f_+(q^2)$ and $f_-(q^2)$, in the $K_{\mu 3}$ decay. Imξ must be equal to zero for time-reversal-invariance to hold. With the data already in hand and with the approved data taking time, it is anticipated to arrive at a statistical error of $\pm$0.008 in Imξ. The best previous experimental limits were obtained with both neutral and charged kaons at the BNL-AGS. [21] A combination of both experimental results provided a limit on the imaginary part of the hadron form factor, Imξ = -0.010 $\pm$ 0.019. A new search for the time-reversal-invariance non-conserving transverse muon polarization with in-flight decays of $K^+ \to \mu^+ \pi^0 \nu_\mu$ was proposed at the BNL-AGS. [22] It was intended to obtain a sensitivity to the transverse muon polarization of $\pm$0.00013, corresponding to a sensitivity to Imξ of $\pm$0.0007. Similar searches for the time-reversal-invariance non-conserving transverse τ polarization in B semileptonic decays, $B \to M \tau \nu_\tau$ are under consideration. Significant transverse τ

lepton polarizations have been predicted. Clearly, a non-zero value of the transverse muon polarization in $K_{\mu 3}$ decay, or of the τ lepton in semi-leptonic B decays would constitute evidence for new physics.

SUMMARY

The searches made so far for a P-even/T-odd interaction have resulted in only very modest constraints on such an interaction. Most promising are the continued efforts to measure the electric dipole moment of the neutron and secondly charge symmetry breaking in neutron-proton elastic scattering at around 320 MeV. But also measurements of transverse lepton polarizations in μ, K, and B decays have the potential to set better experimental limits on a P-even/T-odd interaction.

REFERENCES

1. Angelopoulos, A. *et al.*, (CPLEAR Collaboration), Phys. Lett. B444, 43 (1998).
2. Alavi-Harati, A. *et al.*, (KTeV Collaboration), Phys. Rev. Lett. 84, 408 (2000).
3. Haxton, W.C., Hoering, A. and Musolf, M.J., Phys. Rev. D50, 3422 (1994).
4. Simonius, M., Phys. Lett. 58B, 147 (1975).
5. Boehm, F. in Symmetries and Fundamental Interactions in Nuclei, ed. Haxton, W.C. and Henley, E.M. (World Scientific, Singapore, 1997), p. 67.
6. Blanke, E. *et al.*, Phys. Rev. Lett. 51, 355 (1983).
7. Conzett, H.E. in Polarization Phenomena in Nuclear Physics - 1980, ed. Ohlsen, G.G., Brown, R.E., Jarmie, N., McNaughton, W.W., and Hale, G.M. AIP Conference Proceedings No 69, p. 1452.
8. Huffman, P.R. *et al.*, Phys. Rev. C55, 2684 (1997).
9. Page, S.A. *et al.*, Phys. Rev. C35, 1119 (1987); Bini, M. *et al.*, Phys. Rev. C38, 1195 (1988).
10. Ramsey-Musolf, M.J., Phys. Rev. Lett. 83, 3997 (1999).
11. Simonius, M. in Intersections between Particle and Nuclear Physics - 1984, ed. Mischke, R.E. AIP Conference Proceedings No 123, p. 1115.
12. Abegg, R. *et al.*, Phys. Rev. D39, 2464 (1989).
13. Zhao, J. *et al.*, Phys. Rev. C57, 2126 (1998).
14. Vigdor, S.E. *et al.*, Phys. Rev. C46, 410 (1992).
15. Simonius, M. Phys. Rev. Lett. 78, 4161 (1997).
16. LANSCE Proposal, spokespersons Cooper, M.J. and Lamoreaux, S.K.
17. Iqbal, J. private communication (1999).
18. Friar, J.L. and Coon, S.A. Phys. Rev. C53, 588 (1996).
19. Fetscher, W. these Proceedings.
20. Abe, M. *et al.*, Phys. Rev. Lett. 83, 4253 (1999).
21. Morse, W.M. *et al.*, Phys. Rev. D21, 1750 (1980); Blatt, S.R. *et al.*, Phys. Rev. D27, 1056 (1983).
22. BNL-AGS proposal, spokespersons Divan, M.V., Ma, Hong and Adair, R.

What Can We Learn from QED at Large Couplings?

A. W. Schreiber[a], R. Rosenfelder[b] and C. Alexandrou[c]

[a] Department of Physics and Mathematical Physics and Research
Centre for the Subatomic Structure of Matter
University of Adelaide, Adelaide, S. A. 5005, Australia

[b] Paul Scherrer Institut, CH-5232 Villigen PSI, Switzerland

[c] Department of Physics, University of Cyprus
CY-1678 Nicosia, Cyprus

Abstract. In order to understand QCD at the energies relevant to hadronic physics one requires analytical methods for dealing with relativistic gauge field theories at large couplings. Strongly coupled quenched QED provides an ideal laboratory for the development of such techniques, in particular as many calculations suggest that – like QCD – this theory has a phase with broken chiral symmetry. In this talk we report on a nonperturbative variational calculation of the electron propagator within quenched QED and compare results to those obtained in other approaches. We find surprising differences among these results.

INTRODUCTION

It is well known that the content of a relativistic field theory (let us take QCD as an example) may be expressed in terms of functional averages of operators of the type

$$\int \mathcal{D}[\overline{\Psi}, \Psi, A] \, \mathcal{O}[\overline{\Psi}, \Psi, A] \, e^{-S_{\text{QCD}}[\overline{\Psi}, \Psi, A]} \quad . \tag{1}$$

Even though it is not possible to perform these integrals exactly, we have learnt an awful lot over the years by studying Eq. (1) in various approximations. For example, if there is a small parameter (e.g. g, $1/N_c$...) one may develop well-ordered expansions of Eq. (1). Alternatively one can use stationary phase methods to approximate the integral – relevant, for example, if one is interested in elucidating the importance of classical configurations. Also, these methods are used in order to gain understanding of the behaviour of very high orders of perturbation theory

(HOPT). One may study the symmetries of the theory (e.g. chiral perturbation theory) in order to relate observables, or one can study its equations of motion (e.g. Dyson-Schwinger equation (DSE) studies). Finally, much progress has been made in recent years by actually evaluating Eq. (1) directly by discretizing it, as is done in Lattice QCD.

We have reported elsewhere on yet another technique, the so-called "worldline variational approach", in which the path integral is approximated, in a rigorous and systematically correctable way, via a variational principle. This approach was initially applied to relativistic field theory in a scalar model [1] and more recently to quenched QED [2]. The common property that QED shares with QCD is that they are both renormalizable gauge field theories, in contrast to the scalar model which is super-renormalizable. Of course QCD, even after quenching (i.e. neglecting pair creation), differs from quenched QED by the fact that it is asymptotically free whereas quenched QED has a constant coupling. However, many of the above-mentioned approximative schemes for dealing with Eq. (1) should be applicable within either theory. Quenched QED therefore serves, and is often used, as a test-ground for these methods. In this talk we compare results for a particular quantity characterising quenched QED, namely its anomalous mass dimension, obtained via a) the variational approach, b) via perturbation theory, c) via Dyson-Schwinger equation studies and d) via HOPT. Surprisingly, we find that these results are not entirely consistent with each other, which indicates that our present understanding of these methods may be incomplete.

In Ref. [2] we derived from the variational approach an analytic, implicit, result for the anomalous mass dimension $\gamma_m(\alpha)$ of quenched, dimensionally regularized, QED within the MS scheme:

$$\alpha = \frac{4}{3}(1 + \gamma_m^{\text{var}}) \cot \frac{\pi/2}{1 + \gamma_m^{\text{var}}}. \quad (2)$$

This is a remarkable result. At small couplings it should (approximately) agree with the known result of perturbation theory [1]. Indeed, as may be seen in Fig. 1 (taken from Ref. [2]), the variational result agrees with perturbation theory to within $\approx$ 20 % for couplings where 4th order perturbation theory appears applicable, i.e. for α less than about 1. We have also plotted an estimate of perturbation theory up to 5th order term of the perturbative expansion [2].

It is interesting to note that for a value of α around 1 the various orders of perturbation theory appear to 'fan out'. It is tempting to speculate that this effect arises because of a finite radius of convergence of the perturbation expansion. This would be unexpected: the general wisdom from HOPT studies is that, as a first rough estimate, the nth order contribution to a perturbative expansion should

---

[1] $\gamma_m(\alpha)$ has been calculated to fourth order in α within SU(N) [3]; by suitable choice of Casimirs (i.e. $C_A = 0$, $C_F = 1$ and, for a quenched theory, $N_f = 0$) the result for U(1) may be extracted from this. Recently, Broadhurst [4] has also calculated $\gamma_m(\alpha)$ to fourth order directly within QED
[2] The 5th order estimate has been derived using the Padé approximation methods of Ref. [5]

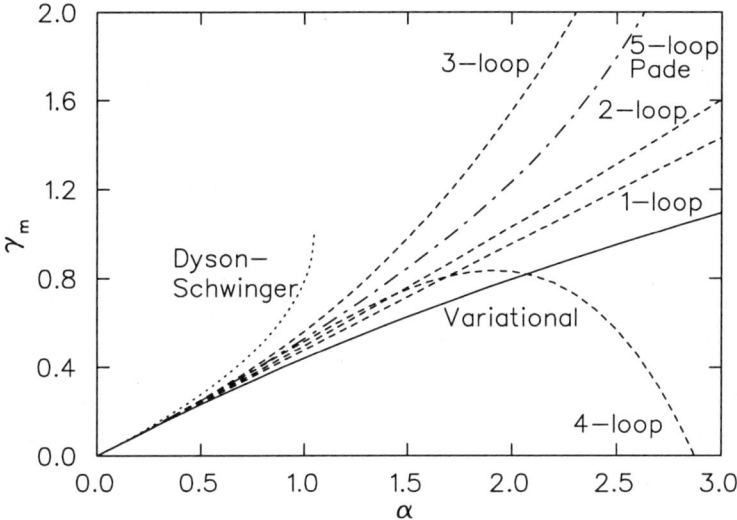

FIGURE 1. Anomalous mass dimension γ_m as function of the coupling constant α in quenched QED. The variational result (Eq. 2) is shown as a solid curve. The curves labeled "n-loop" show the result up to n-loop perturbation theory. The Padé estimation of the 5-loop result is shown as a dot-dashed line and finally, the solution from the Dyson-Schwinger equations in rainbow approximation is indicated as a dotted curve.

scale roughly like the number of diagrams at that order. For the quenched QED propagator, the number of diagrams at order n is given by

$$N_n = \frac{(2n)!}{2^n \, n!} = (2n-1)!! \stackrel{n \to \infty}{\longrightarrow} 2^{n+1/2} \, e^{-n} \, n^n \qquad (3)$$

and hence the radius of convergence ($= \lim_{n \to \infty} |N_n|^{-1/n}$) of the perturbative expansion vanishes. More sophisticated analyses [6] confirm this general result (although, apparently, the high order behaviour of γ_m does not itself appear to ever have been calculated explicitly within quenched QED).

The final curve plotted in Fig. 1 is the prediction of a Dyson-Schwinger equation calculation within the rainbow approximation. This was obtained, for the dimensionally regularized theory within the MS scheme, in Ref. [2,7] and was found to be identical with the well known result for γ_m in the theory regularized with a hard cut-off, i.e. $\gamma_m^{DS} = 1 - \sqrt{1 - \frac{3}{\pi}\alpha}$. At small couplings this agrees with perturbation theory but then diverges from the latter in a region where (4th order) perturbation theory still appears to be applicable. Above $\alpha = \pi/3$, γ_m^{DS} becomes complex, which is also the value α_{cr} of the coupling at which the famous chiral symmetry breaking of quenched QED occurs. Note also that the perturbative expansion of γ_m^{DS} has a finite radius of convergence given by $\alpha_{con}^{DS} = \alpha_{cr}$. It is, however, not a great surprise that α_{con}^{DS} is finite in this calculation as the rainbow approximation contains exactly one diagram at each order and hence does not reproduce the factorial growth (Eq. 3) in the number of diagrams. Nevertheless, the coincidence of α_{con}^{DS} and α_{cr}^{DS} is interesting and one wonders if it will persist in Dyson-Schwinger calculations when going beyond the rainbow approximation. Also, if in the exact theory α_{con} actually vanishes one wonders what will happen to α_{cr} in that case.

The variational result plotted in Fig. 1 appears to show no sign of chiral symmetry breaking. Indeed it is easy to derive that at large couplings $\gamma_m^{var} \to \sqrt{3\pi/8}\sqrt{\alpha}$. However, it turns out that a perturbative expansion of γ_m^{var} also has a finite radius of convergence, not too different from the one mentioned above. We shall discuss this, as well as the large order behaviour of the perturbative expansion of γ_m^{var}, below. It is important to note that in this case a finite radius of convergence is not a trivial result of the approximation, as it was for rainbow QED: it can be shown [1] that each diagram, at any order in perturbation theory, is represented in some approximate way within the variational approximation; i.e. a perturbative expansion of γ_m^{var} would receive contributions from $(2n-1)!!$ diagrams at order n.

THE RADIUS OF CONVERGENCE OF γ_M^{VAR}

Equation (2) is an implicit equation for γ_m^{var} which can easily be solved numerically for arbitrary α. Importantly it can also be solved for complex α, so the determination of α_{con} amounts to a search for nontrivial analytic structure in the complex α plane. In Fig. 2 we show a plot of the (negative) real part of γ_m as a function of the (complex) coupling. Branch cuts, limiting the region of convergence of the perturbation expansion of γ_m^{var}, are clearly visible for negative $\mathcal{R}e\,\alpha$. Note that, as opposed to the Dyson-Schwinger equation result γ_m^{DS}, there are no cuts on the real axis (also for positive $\mathcal{R}e\,\alpha$, which is not shown in Fig. 2). Hence, in Fig. 1, γ_m^{var} is real for all α while γ_m^{DS} terminates at α_{con} [8].

The position of the cuts in Fig. 2 can be obtained straightforwardly by searching for the value of γ_m at which Eq. (2) has two distinct solutions infinitesimally close to each other; i.e. the value of γ_m^{var} at which the derivative with respect to γ_m^{var} of Eq. (2) vanishes. One obtains that the branchpoints are located at $\alpha = -0.496127 \pm 0.619172\,i$, which yields a radius of convergence for the perturbation expansion of γ_m^{var} of $\alpha_{con}^{var} = 0.79342$.

THE BEHAVIOUR OF γ_M^{VAR} AT LARGE ORDERS IN PERTURBATION THEORY

The radius of convergence of the perturbative expansion of the anomalous mass dimension can also be obtained, of course, by directly examining the behaviour of its expansion coefficients at large orders, i.e. if we expand

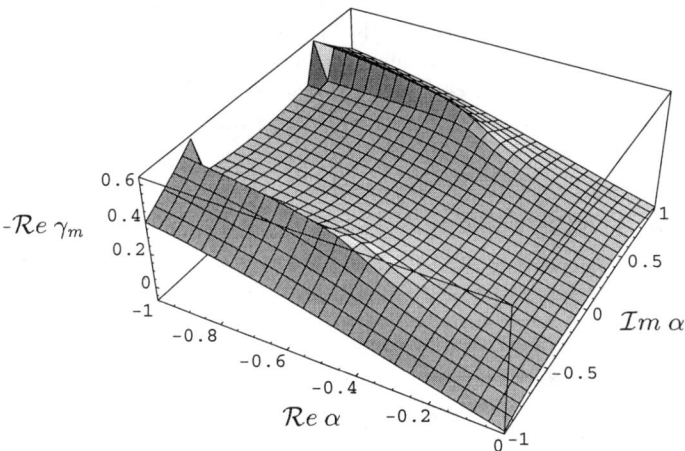

FIGURE 2. Solution of Eq. (2) for complex α. The figure was obtained by using Newton's method, with an initial seed value of $\gamma_m = 0$. No interesting analytic structure is found at positive $\mathcal{R}e\,\alpha$, hence this region is not shown.

$$\gamma_m = \sum_{n=1}^{\infty} c_n\, \alpha^n, \qquad (4)$$

then radius of convergence is given by

$$\alpha_{con} = \lim_{n\to\infty} |c_n|^{-1/n}. \qquad (5)$$

For n up to ≈ 30, it is possible to obtain these expansion coefficients by direct substitution of Eq. (4) into Eq. (2). The results are tabulated in Table 1, where it is clearly seen that although $|c_n|^{-1/n}$ does perhaps tend to a finite limit, the rate of convergence is rather slow (and non-uniform).

In order to find the c_n's for higher values of n, it proves to be advantageous to convert Eq. (2) into a differential equation in order to eliminate the cotangent. We obtain

TABLE 1. Expansion coefficients of the perturbative expansion of γ_m^{var}.

n	5	10	15	20	25	30
$|c_n|^{-1/n}$	2.00	1.31	1.16	1.11	1.02	1.16

$$1 + \gamma_m^{\text{var}} = \left(\alpha + \frac{2\pi}{3}\right) \gamma_m^{\text{var}\,\prime} - \frac{3\pi}{8}\alpha^2 \left(\frac{1}{1+\gamma_m^{\text{var}}}\right)'. \tag{6}$$

If we now substitute Eq. (4) for γ_m^{var}, and $\sum_{n=0}^{\infty} a_n \alpha^n$ for $1/(1+\gamma_m^{\text{var}})$ (hence $a_0 = 1$ and $a_n = -\sum_{k=1}^{n-1} c_{n-k} a_n$) we may solve for the coefficients c_n and a_n in an iterative manner. The results are shown in Fig. 3.

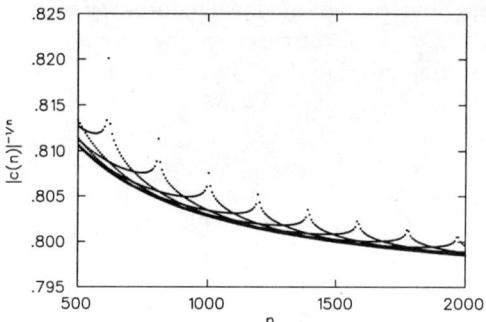

FIGURE 3. The coefficients c_n of the perturbative expansion of γ_m^{var}.

The slow rate of convergence, as well as its non-uniformity, is clearly visible in this figure. Indeed, numerically one finds that for large n the points in Fig. 3 are fitted exceedingly well by the functional form

$$c_n \approx (\alpha_{\text{con}}^{\text{var}})^{-n} \frac{e^{-\beta}}{n^{3/2}} \sin\left[\left(a + \frac{5\pi}{7}\right)n - \frac{3\pi}{7} + b\right], \tag{7}$$

with $\beta \approx 1.376$, $a \approx 2.32 \times 10^{-3}$ and $b \approx -8.268 \times 10^{-2}$. These values of c_n may be compared to the equivalent expansion coefficients obtained from the Dyson-Schwinger equation result:

$$c_n^{\text{DS}} \approx \left(\alpha_{\text{con}}^{\text{DS}}\right)^{-n} \frac{e^{-1.27}}{n^{3/2}}. \tag{8}$$

In fact, the variational result and the rainbow Dyson-Schwinger result for the c_n's are surprisingly similar, not only in their functional form but even in the numerical coefficients. The main difference is the occurance of the sine function in the former. It is because of this sine that the branchcut, which for γ_m^{DS} lies on the real axis, has moved into the complex plane for γ_m^{var}.

CONCLUSION

In this contribution we have compared predictions for the anomalous mass dimension of quenched QED obtained through the use of a variety of techniques: the worldline variational approach, perturbation theory up to $O[\alpha^4]$ (as well as a Padé estimate for the $O[\alpha^5]$ term), rainbow Dyson-Schwinger equation studies and general expectations from studies of high order perturbation theory. Both the rainbow DSE's and the variational approach yield a cut in γ_m as a function of the coupling. In the former this cut is on the real axis, and is associated with chiral symmetry breaking, while in the latter it has moved (a long way) off the real axis, so that in that calculation there is no obvious sign of chiral symmetry breaking. In either case, the cuts necessitate a finite radius of convergence for the perturbative expansion of $\gamma_m(\alpha)$, and one can argue that circumstantial evidence for this may be seen in the perturbative result as well. General expectations from HOPT, on the other hand, suggest that the radius of convergence should be zero. Clearly, as quenched QED is the prototype gauge theory for the investigation of chiral symmetry breaking, efforts should be made to clarify these disagreements. It may even be the case that the eventual resolution of these issues will teach us something about the limitations of one or more of these approximate techniques for dealing with nonperturbative gauge field theories.

REFERENCES

1. Rosenfelder, R. and Schreiber, A. W., *Phys. Rev.* D **53**, 3337–3353 (1996); *ibid* 3354–3365 (1996); Schreiber, A. W., Rosenfelder, R. and Alexandrou, C., *Int. Journ. Mod. Phys.* E **5**, 681–716 (1996); Schreiber, A. W. and Rosenfelder, R., *Nucl. Phys.* A **601**, 397–424 (1996); Alexandrou, C., Rosenfelder, R. and Schreiber, A. W., *Nucl. Phys.* A **628**, 427–457 (1998).
2. Alexandrou, C., Rosenfelder, R. and Schreiber, A. W., submitted to *Phys. Rev.* D, hep-th/0003253; see also Schreiber, A. W., Rosenfelder, R. and Alexandrou, C., contribution to appear in "Proceedings of the Workshop on Light-Cone QCD and Nonperturbative Hadron Physics", Adelaide, Dec. 1999, ADP-00-16/T400.
3. Chetyrkin, K. G., *Phys. Lett.* B **404**, 161–165 (1997); Vermaseren, J. A. M., Larin, S. A. and van Ritbergen, T., *Phys. Lett.* B **405**, 327–333 (1997).
4. Broadhurst, D. J., *Phys. Lett.* B **466**, 319–325 (1999).
5. Elias, V. *et al*, *Phys. Rev.* D **58**, 116007-1 –116007-15 (1998).
6. For a review of HOPT see, for example, Bogomol'nyi, E., Fateev, V. A. and Lipatov, L. N., *Physics Reviews* **2**, 247–393 (1980); a useful collection of papers on the subject may be found in "Large-order behaviour of perturbation theory", eds. J.C. Le Guillou and J. Zinn-Justin, North-Holland (1990); an explicit calculation of the electron's anomalous magnetic moment calculated within quenched QED has been carried out in Itzykson, C., Parisi, G. and Zuber, J-B., *Phys. Rev.* D **16**, 996–1013 (1977) (paper 43 of the above volume).
7. Alexandrou, C., Rosenfelder, R. and Schreiber, A. W., to be published.
8. A more detailed discussion of the analytic structure of Eq. (2) can be found in Markushin, V. E., Rosenfelder R. and Schreiber, A. W., to be published.

CP Violation in $\Lambda \to p\pi^-$: SM vs New Physics

G. Valencia

Department of Physics, Iowa State University, Ames, IA 50011[1]

Abstract.
I discuss CP violation in $\Lambda \to p\pi^-$ comparing the standard model expectations with what could happen in new physics scenarios. I point out that Fermilab experiment E871 is sensitive to some of these scenarios.

Introduction

In non-leptonic hyperon decays such as $\Lambda \to p\pi^-$ it is possible to search for CP violation by comparing the decay with the corresponding anti-hyperon decay [1]. The Fermilab experiment E871 is currently searching for CP violation in such a decay and is sensitive to certain types of physics beyond the standard model. The observable provides information that is complementary to that obtained from the measurement of ϵ'/ϵ.

The reaction of interest is the decay of a polarized Λ, with known polarization $\vec{w}$, into a proton (whose polarization is not measured) and a π^- with momentum q. The final $p\pi^-$ state can be in an S-wave or a P-wave, and in an $I = 1/2$ or $I = 3/2$ state. The observables are the total decay rate and a correlation in the decay distribution of the form

$$\frac{d\Gamma}{d\Omega} \sim 1 + \alpha \vec{w} \cdot \vec{q} \qquad (1)$$

The branching ratio for this mode is 63.9% and the parameter α has been measured to be $\alpha = 0.64$ [2]. The CP violation in question involves a comparison of the parameter α with the corresponding parameter $\bar{\alpha}$ for the reaction $\bar{\Lambda} \to \bar{p}\pi^+$.

It is standard to write the amplitudes in terms of their isospin components in the form

$$S = S_1 e^{i\delta_1^S} + S_3 e^{i\delta_3^S}$$
$$P = P_1 e^{i\delta_1^P} + P_3 e^{i\delta_3^P} \qquad (2)$$

---

[1] Supported in part by DOE under contract number DE-FG02-92ER40730.

A $\Delta I = 1/2$ rule is observed experimentally, $S_3/S_1 \approx 0.026$ and $P_3/P_1 = 0.03\pm0.03$ [3]. The strong πN scattering phases have been measured for the $I = 1/2$ channel, $\delta_1^S \sim 6°$ and $\delta_1^P \sim -1°$ [4]. The $I = 3/2$ scattering phases have been measured with large errors but are not needed here.

To discuss CP violation, we allow the amplitudes in Eq. 2 to have a CP violating weak phase, $S_i \to S_i \exp(i\phi_i^S)$ and $P_i \to P_i \exp(i\phi_i^P)$ and compare the pair of CP conjugate reactions. CP symmetry predicts that $\Gamma = \bar{\Gamma}$ and that $\bar{\alpha} = -\alpha$. One therefore defines the CP-odd observables

$$\Delta \equiv \frac{\Gamma - \bar{\Gamma}}{\Gamma + \bar{\Gamma}} \sim \sqrt{2}\frac{S_3}{S_1}\sin(\delta_3^S - \delta_1^S)\sin(\phi_3^S - \phi_1^S)$$

$$A(\Lambda_-^0) \equiv \frac{\alpha + \bar{\alpha}}{\alpha - \bar{\alpha}} \sim -\sin(\delta_1^P - \delta_1^S)\sin(\phi_1^P - \phi_1^S) \sim 0.12\sin(\phi_1^P - \phi_1^S) \quad (3)$$

The partial rate asymmetry is very small, being suppressed by three small factors, S_3/S_1, strong phases, and weak phases. It represents an interference between amplitudes with $\Delta I = 1/2$ and $\Delta I = 3/2$. The asymmetry $A(\Lambda_-^0)$, on the other hand, is not suppressed by the $\Delta I = 1/2$ rule, as it originates in an interference of S and P-waves within the $\Delta I = 1/2$ transition. For this reason, the observable $A(\Lambda_-^0)$ is *qualitatively* different from ϵ'/ϵ.

The experiment E871 at Fermilab produces the polarized Λ from the weak decay $\Xi^- \to \Lambda \pi^-$ and for this reason what they measure is actually the combination $A(\Lambda_-^0) + A(\Xi_-^-)$. Their expected sensitivity is 10^{-4}. The weak phases in Ξ^- decay (within the standard model) have been estimated to be about two times smaller than those in Λ decay [5]. Similarly, the strong phases in Ξ^- decay are estimated to be of order $1°$ [6,7] and therefore five times smaller than the strong phase difference in Λ decay. For these two reasons we expect that the E871 measurement will be dominated by $A(\Lambda_-^0)$.

Standard Model

Within the standard model one writes the $|\Delta S| = 1$ effective weak Hamiltonian as a sum of four-quark operators multiplied by Wilson coefficients in the usual way,

$$H = \frac{G_F}{\sqrt{2}} V_{ud}^* V_{us} \sum_{i=1}^{12} c_i(\mu) Q_i(\mu) \quad (4)$$

This is, of course, the same effective Hamiltonian responsible for Kaon non-leptonic decays and is very well known. In particular the Wilson coefficients, $c_i(\mu)$ have been calculated in detail by Buras and his collaborators [8]. The remaining problem is to calculate the matrix elements of the four-quark operators between hadronic states. This problem has not been resolved yet, and there is large theoretical uncertainty in these matrix elements. The usual way to proceed (which is the same as in kaon physics) is to take the real part of the matrix element from experiment (assuming CP conservation) and to use the calculated imaginary parts.

Unlike the case of ϵ', where both $\Delta I = 1/2, 3/2$ amplitudes are important, $A(\Lambda_-^0)$ is dominated by CP violation in $\Delta I = 1/2$ amplitudes. One expects that the asymmetry will be dominated by the penguin operator with small corrections from other operators. A detailed study using vacuum saturation to estimate the matrix elements supports the view that Q_6 is dominantly responsible for $A(\Lambda_-^0)$ [9].

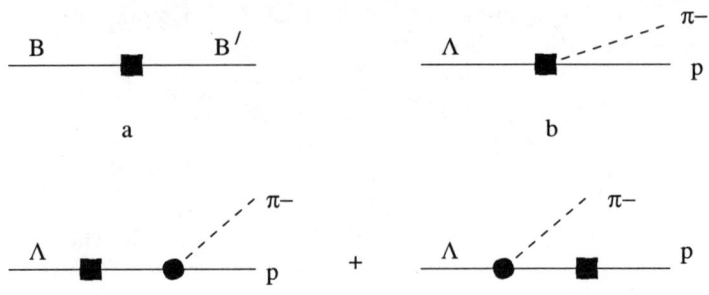

FIGURE 1. a) $B \to B'$ transition due to Q_6, solid square. b) S-wave obtained from (a) via a soft-pion theorem. c) P-wave obtained from (a) with strong pion emission (solid circle).

Once we have determined that only Q_6 is important, the strategy is to calculate the matrix elements of the form $< B'|Q_6|B >$ using a model, and then use these results to treat the non-leptonic hyperon decay at leading order in chiral perturbation theory as sketched in Figure 1. Equivalently, the S-waves are obtained with a soft-pion theorem and the P-waves with baryon poles. At present, the baryon to baryon matrix elements are taken from the MIT bag model calculation of Ref. [10].

It is difficult to quantify the theoretical error in this calculation. There are the obvious uncertainties in the short distance parameters as well as errors in the value of the strong phases. However, of greater concern is the issue of assigning an error to the hadronic matrix elements. Even if we assume that the baryon to baryon matrix elements calculated in the MIT bag model are exact, we know from the study of CP conserving amplitudes that non-leading order terms in chiral perturbation theory can be as large as the leading order amplitudes. For example, the s-wave imaginary part calculated in vacuum saturation, is a higher order correction to the bag-model plus soft pion theorem amplitude outlined above, but it is larger [9]. To get an idea for the impact of this error we assign an overall error of a factor of two to the calculated matrix elements plus an overall 30% uncorrelated error between S and P-waves. Combining all this results in,

$$A(\Lambda_-^0) = (-3.0 \pm 2.6) \times 10^{-5}. \tag{5}$$

Beyond the Standard Model

There have been several estimates of $A(\Lambda_-^0)$ beyond the standard model. For the most part these studies discuss specific models, concentrating on one or a few operators and normalizing the strength of CP violation by fitting ϵ. Some of these results (which have not been updated to incorporate current constraints on model parameters) are:

$$A(\Lambda_-^0) = \begin{cases} -2 \times 10^{-5} & \text{SM [5]} \\ -2 \times 10^{-5} & \text{3 Higgs [5]} \\ 0 & \text{Superweak} \\ 6 \times 10^{-4} & \text{LR [11]} \end{cases} \quad (6)$$

Perhaps a more interesting question is whether it is possible to have large CP violation in hyperon decays in view of what is known about ϵ and ϵ'. This question has been addressed in a model independent way by considering all the CP violating operators that can be constructed at dimension 6 that are compatible with the symmetries of the standard model [12]. With this general formalism one can compute the contributions of each new CP violating phase to ϵ, ϵ', and $A(\Lambda_-^0)$. Of course, there is the caveat that the hadronic matrix elements cannot be computed reliably. Nevertheless, one finds in general, that parity even operators generate a weak phase ϕ_1^P and do not contribute to ϵ'. Their strength can be bound from the long distance contributions to ϵ that they induce. Similarly, the parity-odd operators generate a weak phase ϕ_1^S and contribute to ϵ' (but not to ϵ).

The constraints from ϵ' turn out to be much more stringent than those from ϵ, and, therefore, the only natural way (without invoking fine cancellations between different operators) to obtain a large $A(\Lambda_-^0)$ given what we know about ϵ' is with new CP-odd, P-even interactions. Within the model independent analysis, one can identify a few new operators with the required properties, that can lead to [12],

$$A(\Lambda_-^0) \sim 5 \times 10^{-4} \quad \text{P} - \text{even}, \text{CP} - \text{odd} \quad (7)$$

This possibility has been revisited recently, motivated in part by the observation of ϵ'. The average value $\epsilon'/\epsilon = (21.2 \pm 4.6) \times 10^{-4}$ [13] appears to be larger than the standard model central prediction with simplistic models for the hadronic matrix elements. This has motivated searches for new sources of CP violation that can give large contributions to ϵ', in particular, within supersymmetric theories. One such scenario generates a large ϵ' through an enhanced gluonic dipole operator [14]. The effective Hamiltonian is of the form

$$\begin{aligned}H_{eff} &= (\delta_{12}^d)_{LR} C_g \bar{d}\sigma_{\mu\nu} t^a (1+\gamma_5) s G^{a\mu\nu} \\ &+ (\delta_{12}^d)_{RL} C_g \bar{d}\sigma_{\mu\nu} t^a (1-\gamma_5) s G^{a\mu\nu}\end{aligned} \quad (8)$$

The quantity C_g is a known loop factor, and the $(\delta_{12}^d)_{LR,RL}$ originate in the supersymmetric theory [15]. Depending on the correlation between the value of

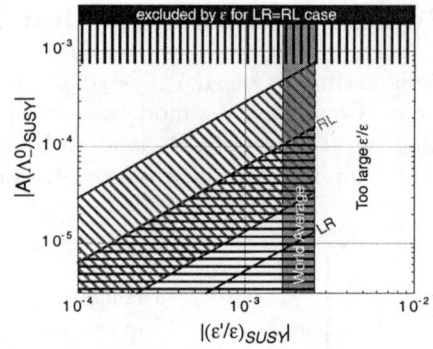

FIGURE 2. The allowed regions on ($|(\epsilon'/\epsilon)_{SUSY}|$, $|A(\Lambda_-^0)_{SUSY}|$) parameter space for three cases: a) only $\text{Im}(\delta_{12}^d)_{LR}$ contribution, which is the conservative case (hatched horizontally), b) only $\text{Im}(\delta_{12}^d)_{RL}$ contribution (hatched diagonally), and c) $\text{Im}(\delta_{12}^d)_{LR} = \text{Im}(\delta_{12}^d)_{RL}$ case which does not contribute to ϵ' and can give a large $|A(\Lambda_-^0)|$ below the shaded region (or vertically hatched region for the central values of the matrix elements). The last case is motivated by the relation $\lambda = \sqrt{m_d/m_s}$. The vertical shaded band is the world average [13] of ϵ'/ϵ. The region to the right of the band is therefore not allowed.

$(\delta_{12}^d)_{LR}$ and $(\delta_{12}^d)_{RL}$ one gets different scenarios for ϵ' and $A(\Lambda_-^0)$ as shown in Figure 2 [16]. For example, if only $(\delta_{12}^d)_{LR}$ is non-zero, there can be a large ϵ' [14], but $A(\Lambda_-^0)$ is small as in the 3-Higgs model of [5]. However, in models in which $\text{Im}(\delta_{12}^d)_{LR} = \text{Im}(\delta_{12}^d)_{RL}$ the CP violating operator is parity-even. In this case there is no contribution to ϵ' and $A(\Lambda_-^0)$ can be as large as 10^{-3} [16]. It is interesting that this type of model is not an ad-hoc model to give a large $A(\Lambda_-^0)$, but is a type of model originally designed to naturally reproduce the relation $\lambda = \sqrt{m_d/m_s}$, as in Ref. [17], for example.

Conclusion and Comments

E871 is expected to reach a sensitivity of 10^{-4} for the observable $A(\Lambda_-^0) + A(\Xi_-^-)$.

- $A(\Lambda_-^0)$ is likely to be significantly larger than $A(\Xi_-^-)$.

- $A(\Lambda_-^0) = (-3.0 \pm 2.6) \times 10^{-5}$ is our current best guess for the standard model and the theoretical uncertainty is dominated by our inability to calculate hadronic matrix elements reliably. For this reason, the error assigned to this quantity is no more than an educated guess.

- $A(\Lambda_-^0)$ can be much larger if CP violation originates in P-even new physics. A specific realization of this scenario is possible in supersymmetric theories leading to $A(\Lambda_-^0)$ as large as 10^{-3}.

I conclude that a non-zero measurement by E871 is not only possible but that it would provide valuable complementary information to what we already know from ϵ'.

Finally I would like to mention two related issues. A search for $\Delta S = 2$ hyperon non-leptonic decays is also a useful enterprise as it provides information that is complementary to what we know from $K - \bar{K}$ mixing [18]. A CP violating rate asymmetry in $\Omega \to \Xi\pi$ decay can be as large as 2×10^{-5} within the standard model (and up to ten times larger beyond), much larger than the corresponding rate asymmetries in octet-hyperon decay [19].

This work was supported by DOE under contract number DE-FG02-92ER40730. This talk summarizes work done in collaboration with John Donoghue, Xiao-Gang He, Hitoshi Murayama, Sandip Pakvasa, Herbert Steger and Jusak Tandean.

REFERENCES

1. S. Okubo, Phys. Rev. **109**, 984 (1958); A. Pais, Phys. Rev. Lett. **3**, 242 (1959).
2. C. Caso et al., Eur. Phys. J. **C3**, 1 (1998).
3. O. E. Overseth, in Review of Particle Properties, *Phys. Lett.* **111B**, 286 (1982).
4. L. D. Roper, R. M. Wright and B. Feld, Phys. Rev. **138**, 190 (1965); A. Datta and S. Pakvasa, Phys. Rev. **D56**, 4322 (1997).
5. J. Donoghue and S. Pakvasa, Phys. Rev. Lett. **55**, 162 (1985); J. Donoghue, X.-G. He and S. Pakvasa, Phys. Rev. **D34**, 833 (1986).
6. M. Lu, M. Savage and M. Wise, Phys. Lett. **B337**, 133 (1994).
7. A. Datta and S. Pakvasa, Phys. Lett. **B344**, 430 (1995); A. Kamal, Phys. Rev. **D58**, 077501 (1998).
8. G. Buchalla, A. Buras and M. Harlander, Nucl. Phys. **B337**, 313 (1990); G. Buchalla, A. J. Buras and M. E. Lautenbacher, Rev. Mod. Phys. **68**, 1125 (1996).
9. X.-G. He, H. Steger, and G. Valencia, Phys. Lett. **B272**, 411 (1991).
10. J. Donoghue et. al., Phys. Rev. **D23**, 1213 (1981).
11. D. Chang, X.-G. He and S. Pakvasa, Phys. Rev. Lett. **74**, 3927 (1995).
12. X.-G. He and G. Valencia, Phys. Rev. **D52**, 5257 (1995).
13. This is the average of E731, NA31, KTeV and NA48 with the error bar inflated to obtain $\chi^2/\text{d.o.f.} = 1$ according to the Particle Data Group prescription.
14. A. Masiero and H. Murayama, Phys. Rev. Lett. **83**, 907 (1999).
15. F. Gabbiani, et. al., Nucl. Phys. **B447**, 321 (1996).
16. Xiao-Gang He, H. Murayama, S. Pakvasa and G. Valencia, *Phys. Rev.* **D61**, 071701 (2000).
17. R. Barbieri, G. Dvali and L.J. Hall, Phys. Lett. **B377**, 76 (1996).
18. X.-G. He and G. Valencia, Phys. Lett. **B409**, 469 (1997) Erratum-ibid. **B418**, 443 (1998).
19. J. Tandean and G. Valencia, Phys. Lett. **B451**, 382 (1999).

Quark mixing angles and CP-violating phase from flavour permutational symmetry breaking [1]

A. Mondragón and E. Rodríguez-Jáuregui

Instituto de Física, Universidad Nacional Autónoma de México
Apdo. Postal 20-364, 01000 México, D. F., México.

Abstract. Different Ansätze for the breaking of flavour permutational symmetry according to $S_L(3) \otimes S_R(3) \to S_L(2) \otimes S_R(2)$ give different Hermitian mass matrices of the same modified Fritzsch type, which differ in the symmetry breaking pattern. A clear and precise indication on the preferred symmetry breaking scheme is obtained from a fit of the predicted $|V^{th}|$ to the experimentally determined absolute values of the elements of the CKM matrix. The preferred scheme leads to simple mass textures and allows us to compute the CKM mixing matrix, the Jarlskog invariant J, and the three inner angles of the unitarity triangle in terms of four quark mass ratios and only two free parameters: Z and the CP violating phase Φ. Excellent agreement with the experimentally determined absolute values of the entries in the CKM matrix is obtained for $Z^* = 2.53125$ and $\Phi^* = 90°$. The corresponding computed values of the Jarlskog invariant and the inner angles are $J = 0.000028$, $\alpha = 83°$, $\beta = 22°$ and $\gamma = 75°$ in very good agreement with current data on CP violation in the kaon-antikaon system and oscillations in the B-antiB system. It is also shown that the theoretical mixing matrix $\mathbf{V}^{th}$ derived from the breaking of the flavour permutational symmetry and the standard parametrization $\mathbf{V}^{PDG}$ advocated by the Particle Data Group are equivalent, up to a redefinition of the unobservable phases of the quark fields in the mass representation. From here, we derive exact explicit expressions for the three mixing angles θ_{12}, θ_{13}, θ_{23}, and the CP violating phase δ_{13} in terms of the quark mass ratios $(m_u/m_t, m_c/m_t, m_d/m_b, m_s/m_b)$ and the parameters $Z^{*1/2}$ and Φ^* characterizing the preferred symmetry breaking pattern. The predicted values for the CP violating phase and the mixing angles are: $\delta_{13}^* = 75°$, $\sin\theta_{12}^* = 0.2208$, $\sin\theta_{13}^* = 0.0034$, and $\sin\theta_{23}^* = 0.040$, which coincide almost exactly with the central values of the experimentally determined quantities.

---

[1] Presented by A. Mondragón

MASS MATRICES FROM THE BREAKING OF $S_L(3) \otimes S_R(3)$

Under exact $S_L(3) \otimes S_R(3)$ permutational symmetry, the mass spectrum for either up or down quark sectors consists of one massive particle in a singlet irreducible representation and a pair of massless particles in a doublet irreducible representation, the corresponding quark mass matrix is $\mathbf{M_{3q}}$. In order to generate masses for the first and second families, we add the terms $\mathbf{M_{2q}}$ and $\mathbf{M_{1q}}$ to $\mathbf{M_{3q}}$. The term $\mathbf{M_{2q}}$ breaks the permutational symmetry $S_L(3) \otimes S_R(3)$ down to $S_L(2) \otimes S_R(2)$ and mixes the singlet and doublet representation of $S(3)$. $\mathbf{M_{1q}}$ transforms as the mixed symmetry term in the doublet complex tensorial representation of $S_{diag}(2) \subset S_L(2) \otimes S_R(2)$. Putting the first family in a complex representation allows us to have a CP violating phase. Then, in a symmetry adapted basis, $\mathbf{M_q}$ takes the form

$$M_q = m_{3q}\left[\begin{pmatrix} 0 & A_q e^{-i\phi_q} & 0 \\ A_q e^{i\phi_q} & 0 & 0 \\ 0 & 0 & 0 \end{pmatrix} + \begin{pmatrix} 0 & 0 & 0 \\ 0 & -\Delta_q + \delta_q & B_q \\ 0 & B_q & \Delta_q - \delta_q \end{pmatrix}\right]$$
$$+ m_{3q}\begin{pmatrix} 0 & 0 & 0 \\ 0 & 0 & 0 \\ 0 & 0 & 1-\Delta_q \end{pmatrix} \qquad (1)$$

The entries in the mass matrix may be readily expressed in terms of the mass ratios $\tilde{m}_{1q} = m_{1q}/m_{3q}$ and $\tilde{m}_{2q} = m_{2q}/m_{3q}$: $A_q^2 = \tilde{m}_{1q}\tilde{m}_{2q}(1-\delta_q)^{-1}$, $\Delta_q = \tilde{m}_{2q} - \tilde{m}_{1q}$, $B_q = \delta_q((1-\tilde{m}_{1q}+\tilde{m}_{2q}-\delta_q) - \tilde{m}_{1q}\tilde{m}_{2q}(1-\delta_q)^{-1})$. If each possible symmetry breaking pattern (SBP) is now characterized by the ratio $Z_q^{1/2} = B_q/(-\Delta_q+\delta_q)$, the small parameter δ_q is obtained as the solution of the cubic equation

$$\delta_q[(1+\tilde{m}_{2q}-\tilde{m}_{1q}-\delta_q)(1-\delta_q) - \tilde{m}_{1q}\tilde{m}_{2q}] - Z_q(-\tilde{m}_{2q}+\tilde{m}_{1q}+\delta_q)^2 = 0 \qquad (2)$$

which vanishes when Z_q vanishes. In the symmetry adapted basis, the second term $\mathbf{M_{2q}}$ in the right hand side of Eq. (1), is decomposed as a linear combination of two linearly independent numerical matrices, $\mathbf{M_{2A}}$ and $\mathbf{M_{2S}}$, this matrices, are of the same form as $\mathbf{M_{2q}}$ with mixing parameters $Z_A = -\sqrt{8}$ and $Z_S = 1/\sqrt{8}$ respectively. There is a corresponding decomposition of the mixing parameter $Z_q^{1/2}$,

$$Z_q^{1/2} = N_{Aq}Z_A^{1/2} + N_{Sq}Z_S^{1/2} \quad \text{with} \quad 1 = N_{Aq} + N_{Sq}, \qquad (3)$$

in this way a unique linear combination of $Z_A^{1/2}$ and $Z_S^{1/2}$ is associated to the SBP. The pair of numbers (N_A, N_S) are a convenient mathematical label of the SBP. The parameter $Z_q^{1/2} = M_{2q23}/M_{2q22}$ is a measure of the amount of mixing of singlet and doublet irreducible representations of $S_L(3) \otimes S_R(3)$. It will be assumed that, the up and down mass matrices are generated following the same SBP: $Z_u^{1/2} = Z_d^{1/2} = Z^{*1/2}$.

A THE MIXING MATRIX

The Hermitian mass matrix $\mathbf{M}_q$ may be written in terms of a real matrix $\bar{\mathbf{M}}_q$ and a diagonal matrix of phases $\mathbf{P}_q$ as $\mathbf{P}_q \bar{\mathbf{M}}_q \mathbf{P}_q^\dagger$. Then, the mixing matrix $\mathbf{V}$ is given by

$$\mathbf{V} = \mathbf{O}_u{}^T \mathbf{P}^{u-d} \mathbf{O}_d \qquad (4)$$

where $\mathbf{P}^{u-d} = diag[1, e^{i(\phi_u - \phi_d)}, e^{i(\phi_u - \phi_d)}]$ is the diagonal matrix of the relative phases and $\mathbf{O}_q$ is the orthogonal matrix that diagonalizes $\bar{\mathbf{M}}_q$

$$\mathbf{O}_q = \begin{pmatrix} (\tilde{m}_{2q} f_1/D_1)^{1/2} & -(\tilde{m}_{1q} f_2/D_2)^{1/2} & (\tilde{m}_{1q}\tilde{m}_{2q} f_3/D_3)^{1/2} \\ ((1-\delta_q)\tilde{m}_{1q} f_1/D_1)^{1/2} & ((1-\delta_q)\tilde{m}_{2q} f_2/D_2)^{1/2} & ((1-\delta_q) f_3/D_3)^{1/2} \\ -(\tilde{m}_{1q} f_2 f_3/D_1)^{1/2} & -(\tilde{m}_{2q} f_1 f_3/D_2)^{1/2} & (f_1 f_2/D_3)^{1/2} \end{pmatrix} \qquad (5)$$

where $f_1 = 1 - \tilde{m}_1 - \delta_q^*$, $f_2 = 1 + \tilde{m}_2 - \delta_q^*$, $f_3 = \delta_q^*$, $D_1 = (1-\delta_q^*)(1-\tilde{m}_1)(\tilde{m}_{2q} + \tilde{m}_{1q})$, $D_2 = (1-\delta_q^*)(1+\tilde{m}_{2q})(\tilde{m}_{2q} + \tilde{m}_{1q})$ and $D_3 = (1-\delta_q^*)(1+\tilde{m}_{2q})(1-\tilde{m}_{1q})$.
From Eqs. (4) and (5), we derived closed, explicit expressions for all entries in the matrix $\mathbf{V}$ written in terms of four mass ratios ($\tilde{m}_u, \tilde{m}_c, \tilde{m}_d, \tilde{m}_s$) and two free real parameters $\Phi = \phi_u - \phi_d$ and $Z^{1/2}$ [1]. The CP violating phase Φ measures the mismatch in the $S_L(2) \otimes S_R(2)$ symmetry breaking in the u- and d-sectors.

We made a χ^2 fit of the exact expressions for the absolute values of the the entries in the mixing matrix $|V^{th}|$ and the Jarlskog invariant J to the experimentally determined values of $|V^{exp}|$ and J^{exp}. We took the values of the running quark masses evaluated at the scale of m_t from H. Fritzsch [2], and Fusaoka and Koide [3], we left the mass ratios $\tilde{m}_c$, $\tilde{m}_d$ and $\tilde{m}_s$ fixed at their central values $\tilde{m}_c = 0.0044$, $\tilde{m}_d = 0.0015$ and $\tilde{m}_s = 0.034$ but we took the value of $\tilde{m}_u = 0.000032$ close to its upper bound. We found that the $S_L(3) \otimes S_R(3)$ flavour symmetry is broken down to $S_L(2) \otimes S_R(2)$ according to a mixed symmetry breaking pattern characterized by $Z^{*1/2} = 1/2 \left(Z_S^{1/2} - Z_A^{1/2}\right) = \sqrt{81/32}$. A detailed account of the computation may be found in Mondragón and Rodríguez-Jáuregui [1]. Therefore, the theoretical expressions for the entries in the mixing matrix $\mathbf{V}$ are functions of the four mass ratios ($\tilde{m}_u, \tilde{m}_c, \tilde{m}_d, \tilde{m}_s$) with $Z^* = \sqrt{81/32}$ and the CP violating phase $\Phi = 90°$. The quark mixing matrix V^{th} computed from the theoretical expresions is

$$V^{th} = \begin{pmatrix} 0.9753 e^{i1°} & 0.221 e^{i158°} & 0.0034 e^{i84°} \\ 0.220 e^{i112°} & 0.9745 e^{i89°} & 0.040 e^{i90°} \\ 0.0085 e^{i270°} & 0.039 e^{i270°} & 0.9992 e^{i90°} \end{pmatrix} \qquad (6)$$

The Jarlskog invariant, J, may be computed directly from the commutator of the mass matrices [4]

$$J = -\frac{det\{-i[\mathbf{M}_{u,H}, \mathbf{M}_{d,H}]\}}{F} = 2.8 \times 10^{-5} \tag{7}$$

where $F = (1+\tilde{m}_c)(1-\tilde{m}_u)(\tilde{m}_c+\tilde{m}_u)(1+\tilde{m}_s)(1-\tilde{m}_d)(\tilde{m}_s+\tilde{m}_d)$. The non-vanishing of J is a necessary and sufficient condition for the violation of CP [4].

B Phase equivalence of $\mathbf{V}^{th}$ and $\mathbf{V}^{PDG}$

The standard parametrization of the mixing matrix recomended by the Particle Data Group [5] is written in terms of three mixing angles $\theta_{12}, \theta_{23}, \theta_{13}$ and one CP violating phase δ_{13},

$$\mathbf{V}^{PDG} = \begin{pmatrix} c_{12}s_{13} & s_{12}c_{13} & s_{13}e^{-i\delta_{13}} \\ -s_{12}c_{23} - c_{12}s_{23}s_{13}e^{i\delta_{13}} & c_{12}c_{23} - s_{12}s_{23}s_{13}e^{i\delta_{13}} & s_{23}c_{13} \\ s_{12}s_{23} - c_{12}c_{23}s_{13}e^{i\delta_{13}} & -c_{12}s_{23} - s_{12}c_{23}s_{13}e^{i\delta_{13}} & c_{23}c_{13} \end{pmatrix} \tag{8}$$

where $c_{ij} = \cos\theta_{ij}$ and $s_{ij} = \sin\theta_{ij}$. The range of values of the experimentally determined moduli in $|V_{ij}^{exp}|$, as given by Caso et al [3], corresponds to 90% confidence limits on the range of values of the mixing angles of: $0.217 \leq s_{12} \leq 0.222$, $0.036 \leq s_{23} \leq 0.042$, $0.0018 \leq s_{13} \leq 0.0044$. The standard parametrization $\mathbf{V}^{PDG}$ was introduced without taking the possible functional relations between the quark masses and the flavour mixing parameters into account. In contrast, these functional relations are explicitly exhibited in the theoretical expressions, V_{ij}^{th}, derived in the previous sections. Furthermore, we have seen that, when the best values of the parameters $Z^{1/2}$ and Φ are used, the mixing matrix $\mathbf{V}^{th}$ reproduces the central values of all experimentally determined quantities, that is, the moduli $|V_{ij}^{exp}|$, the Jarlskog invariant J^{exp} and the three inner angles, α, β and γ, of the unitarity triangle [1]. Since the two parametrizations reproduce the same set of experimental data equally well, we are justified in writing

$$|V_{ij}^{th}| = |V_{ij}^{PDG}| = |V_{ij}^{exp}|. \tag{9}$$

We cannot simply equate $\mathbf{V}^{th}$ and $\mathbf{V}^{PDG}$ because the arguments of corresponding matrix elements in the two parametrizations are not equal: $arg(V_{ij}^{th}) \neq arg(V_{ij}^{PDG})$. This difference is of no physical consequence, it reflects the freedom in choosing the unobservable phases of the quark fields in the mass representation. In the mass basis, the quark charged currents take the form

$$J_c^\mu = \frac{g}{\sqrt{2}} \bar{q}_{Li}^u \gamma^\mu V_{ij}^{th} q_{Lj}^d. \tag{10}$$

A redefinition of the phases of the quark fields which leaves J_c^μ invariant, will change the argument of V_{ij}^{th} but leave the moduli $|V_{ij}^{th}|$ invariant,

$$V_{ij}^{th} \rightarrow \tilde{V}_{ij}^{th} = e^{-i\chi_i^u} V_{ij}^{th} e^{i\chi_j^d}. \tag{11}$$

The phases χ_i^u and χ_j^d ocurring in Eq. (11) will be determined from the requirement that corresponding entries in $\tilde{\mathbf{V}}^{th.}$ and $\mathbf{V}^{PDG}$ be equal,

$$|V_{ij}^{th}|e^{i(w_{ij}^{th}-(\chi_i^u-\chi_j^d))} = |V_{ij}^{PDG}|e^{iw_{ij}^{PDG}}, \tag{12}$$

in this expression w_{ij}^{th} and w_{ij}^{PDG} are the arguments of V_{ij}^{th} and V_{ij}^{PDG} respectively. Since the moduli $|V_{ij}^{th}|$ and $|V_{ij}^{PDG}|$ are equal, the arguments of the entries in the two parametrizations are related by the set of nine equations

$$\chi_i^u - \chi_j^d = w_{ij}^{th} - w_{ij}^{PDG}. \tag{13}$$

The set of Eqs. (13) relate the differences of the unobservable quark field phases to the differences of the arguments of corresponding entries in $\mathbf{V}^{th}$ and $\mathbf{V}^{PDG}$. Using an elimination procedure for all possible combinations $\left(\chi_i^{(u)}-\chi_j^{(d)}\right)-\left(\chi_i^{(u)}-\chi_{j'}^{(d)}\right)$ we derive a set of nine equations, only four of which are linearly independent. Since, in $\mathbf{V}^{PDG}$ there are five entries with non-vanishing arguments, namely, $w_{13}^{PDG}=-\delta_{13}, w_{21}^{PDG}, w_{22}^{PDG}, w_{31}^{PDG}$ and w_{32}^{PDG}, we require still one more equation relating the arguments of the entries of the two parametrizations. This is obtained from the phase relations between the determinants of the two matrices, $\mathbf{V}^{th}$ and $\mathbf{V}^{PDG}$. From Eqs. (11) and (12), it follows that

$$\det \mathbf{V}^{th} = \det\left[\mathbf{X}_u^\dagger \mathbf{V}^{PDG} \mathbf{X}_d\right], \tag{14}$$

in this expression $\mathbf{X}_u$ and $\mathbf{X}_d$ are the diagonal unitary matrices of phases ocurring in Eq. (11). The quark field phases themselves are determined only up to a common additive constant. Since the quark field phases are unobservable, without loss of generality, we may fix one of them, and solve for the others. In this way, if we set $\chi_1^d = 0$, we get the diagonal matrices of phases required to compute the phase transformed $\tilde{\mathbf{V}}^{th}$

$$\mathbf{X}_u = diag[e^{iw_{11}^{th}}, e^{i(-w_{12}^{th}-w_{33}^{th}+2\Phi^*)}, e^{i(-w_{23}^{th}-w_{12}^{th}+2\Phi^*)}] \tag{15}$$

and

$$\mathbf{X}_d = diag[1, e^{i(w_{11}^{th}-w_{12}^{th})}, e^{i(w_{12}^{th}-w_{23}^{th}-w_{33}^{th}+2\Phi^*)}]. \tag{16}$$

Hence, with the help of Eqs. (15)-(16), we verify that

$$\mathbf{X}_u^\dagger \mathbf{V}^{th} \mathbf{X}_d = \mathbf{V}^{PDG}, \tag{17}$$

is satisfied as an identity, provided that $|V_{ij}^{th}| = |V_{ij}^{PDG}|$.
Then, the CP-violating phase ocurring in $\mathbf{V}^{PDG}$ is

$$\delta_{13} = w_{11}^{th} + w_{12}^{th} - w_{13}^{th} + w_{23}^{th} + w_{32}^{th} - 2\Phi^*. \tag{18}$$

From Eq. (17) and the equality of the moduli of corresponding entries in $\mathbf{V}^{PDG}$ and $\mathbf{V}^{th}$, the three mixing angles $\theta_{12}, \theta_{23}, \theta_{13}$ and the CP-violating phase δ_{13} which appear in the phenomenological parametrization $\mathbf{V}^{PDG}$ as free, linearly independent parameters are expressed as functions of four quark mass ratios and only two flavour symmetry breaking parameters $Z^{*1/2}$ and Φ^* [8]. The numerical values of the mixing angles computed from our expressions are

$$\sin\theta^*_{12} = 0.22, \quad \sin\theta^*_{23} = 0.040, \quad \sin\theta^*_{13} = 0.0034, \quad \sin\delta^*_{13} = 0.966. \quad (19)$$

These values coincide almost exactly with the central values of the same mixing parameters determined from a fit to the experimental data [5], The predicted value of the CP violating phase is

$$\delta^*_{13} = 75°. \quad (20)$$

For the three inner angles α, β and γ of the unitarity triangle, we get, $\alpha = 83°, \beta = 22°, \gamma = 75°$ in good agreement with current data on CP violation in the $K^o - \bar{K}^o$ mixing system [5], [6] and oscillations in the B^o_s-$\bar{B}^o_s$ system [5], [7].

The predictive power of the theoretical quark mixing matrix $\mathbf{V}^{th}$ implied by the good agreement of our results with the experimental data has its origin in the flavour permutational breaking pattern from wich the texture in the quark mass matrices and the quark mixing matrix were derived. A detailed account of this work may be found in [8].

ACKNOWLEDGMENT(S)

This work was partially supported by DGAPA-UNAM under contract No. PAPIIT-IN125298 and by CONACYT (México) under contracts 32238-E and 87156.

REFERENCES

1. A. Mondragón, E. Rodríguez-Jáuregui *Phys. Rev.* **D 59**, 093009, (1999). see also A. Mondragón and E. Rodríguez-Jáuregui, *Rev. Mex. Fis.* **44(S1)**, 33 (1998), hep-ph/9804267
2. H. Fritzsch, *Mass hierarchies, Hidden Symmetry and Maximal CP-violation"*, hep-ph/9807551 See also H. Fritzsch, *The symmetry and the Problem of Mass Generation. Proceedings of the XXI International Colloquium on Group Theoretical Methods in Physics (Group 21), Goslar, Germany,* (1996), edited by H.-D. Doebner, W. Scherer, and C. Schutte (World Scientific, Singapore, 1997), Vol. II, p. 543.
3. H. Fusaoka and Y. Koide *Phys. Rev.* **D 57**, 3986 (1998).
4. C. Jarlskog, *Phys. Rev. Lett.* **55**, 1039 (1985).
5. Particle Data Group, C. Caso *et al.*, *Eur. Phys. J.* **C3**, 1 (1998).
6. S. Mele *Phys. Rev.* **D 59**, 113011, (1999).
7. A. Ali and D. London, *Eur. Phys. J* **C9**, 687-703, (1999).
8. A. Mondragón, and E. Rodríguez-Jáuregui, *Phys. Rev.* **D 61**, 0730XX, (2000) see also hep-ph/9906429.

Atomic theory and tests of the Standard Model in atomic experiments

V.V. Flambaum

School of Physics, University of New South Wales, Sydney, 2052, Australia

Abstract.
Measurements of the weak charge characterizing the strength of the electron-nucleon weak interaction provide tests of the Standard Model and a way of searching for new physics beyond the Standard Model. Atomic experiments give limits on the extra Z-boson, leptoquarks, composite fermions, and radiative corrections produced by particles that are predicted by new theories. To extract the accurate value of the weak charge from atomic experiments one has to perform high precision atomic calculations of the PNC effects.

The experiments that were suggested in [1] for measuring parity nonconservation (PNC) in heavy atoms have provided an important confirmation [2–5] of the Standard Model of elementary particles. Now the problem is different. There is a general belief that the Standard Model is only an intermediate step in the process of the unification of all interactions and particles into a new theory. Also, the foundation of the Standard Model is incomplete. A very important element of this model, the Higgs boson, has not been found. It seems that "new physics" beyond the Standard Model is inevitable. New particles would contribute to PNC in atoms as intermediate particles in the electron-nucleus PNC interaction or as a contribution to the radiative corrections to the weak charge. The Standard Model radiative corrections to the weak charge are about 2.5%. This gives an estimate of the accuracy which is needed for atomic experiments to be sensitive to new physics. Of course, we need high accuracy in both the atomic experiments and the atomic calculations that allow one to extract the value of the weak charge from the atomic experiments. Currently, the most accurate results are for the Cs atom: the accuracy of the measurements is 0.35% [5], the accuracy of the atomic theory is 1% [6–8]. This gives one the chance to study new physics beyond the Standard Model.

The measured nuclear spin-independent part of the parity nonconserving (PNC) effect in Cs is of the form

$$-\frac{\operatorname{Im}(E_{PNC})}{\beta} = 1.5939(56) \frac{\mathrm{mV}}{\mathrm{cm}} \qquad (1)$$

The PNC E1-amplitude of the 6s - 7s transition E_{PNC} is due to the weak electron-nucleus interaction which admixes p-states to s-states. The vector polarizability of the transition β comes from the Stark amplitude of the transition, which appears due to an admixture of p-states to s-states by an external electric field. The measured PNC effect in Cs is due to the interference of the PNC E1-amplitude and the Stark amplitude. Therefore, the relative magnitude of the PNC effect is proportional to the ratio E_{PNC}/β.

The most accurate theoretical values of E_{PNC} are as follows:

$$E_{PNC} = -i|e|a_0 10^{-11} \left(-\frac{Q_W}{N}\right) \begin{cases} 0.908(9) & \text{Ref [6]} \\ 0.905(9) & \text{Ref [7, 8]} \end{cases} \qquad (2)$$

Here Q_W is the weak charge of the cesium nucleus and N is the number of neutrons.

The method for *ab initio* calculations of E_{PNC} that we used in [6] was based on an all-orders summation of the dominating diagrams of the many-body perturbation theory in the residual Coulomb interaction. This technique has been described in [6,9].

The residual Coulomb interaction is equal to $U = H - H_{HF}$. Here H is the exact Hamiltonian of the atom and H_{HF} is the Hartree-Fock Hamiltonian. We generate the complete zero-approximation set of the eigenvalues, wave functions and Green's functions using the Hartree-Fock Hamiltonian. The small parameter of this many-body perturbation theory is the ratio of the non-diagonal matrix element of the residual interaction U to the large energy denominator for excitation of the electron from the closed electron shell (electron core), e.g. 5p -electron: $U/E_{5p} \sim 10^{-2}$.

We took into account direct and exchange polarization of the atomic core by the external electric field and the weak nuclear potential using the Time-Dependent Hartree-Fock method (summation of the "RPA with exchange" chain of diagrams). Then we calculated all second-order correlation corrections and three series of dominating higher-order diagrams:

1. Screening of the electron-electron interaction. This is a collective phenomenon and so the corresponding chain of diagrams is enhanced by a factor approximately equal to the number of electrons in the external closed subshell (the 5p electrons in Cs). We stress that our approach takes into account screening diagrams with double, triple and higher core electron excitations [10], in contrast to popular pair equations (coupled cluster) method, where only double excitations were considered.

2. Hole-particle interaction. This effect is enhanced by the large zero-multipolarity diagonal matrix elements of the Coulomb interaction.

3. Iterations of the self-energy operator ("correlation potential"). This chain of diagrams describes the nonlinear effects of the correlation potential and is enhanced by the small denominator, which is the energy for the excitation of an external electron (in comparison with the excitation energy of a core electron).

The error in the theoretical value was tested in many different ways: by estimating the contribution of the unaccounted higher-order diagrams, by comparing the calculated and measured values of the energy levels, the fine and hyperfine structure intervals, the probabilities of electromagnetic transitions, etc. (see Ref. [6]). The result for the PNC amplitude hardly changed when we introduced factors into the correlation potential to fit the energy levels (in imitation of the unaccounted higher-order diagrams). Important tests of our method include predictions of the spectrum [11] and electromagnetic transition amplitudes for the Fr atom [12], which is an analogue of Cs. Recently the positions of many energy levels [13] and some transition rates [14] of Fr were measured and found to be in excellent agreement with our predictions.

Our calculations of PNC for atoms with electron structures more complex than those of the alkaline atoms were proved to be accurate as well. In a series of works done about ten years ago we claimed an accuracy of 3% for Tl [15], 8% for Pb and 11% for Bi [16]. All these PNC effects were recently measured to an accuracy of about 1% [3,4] and found to be in good agreement with our predictions. This means that our estimates for the theoretical accuracy were correct and probably even too pessimistic. For example, in our first calculation of the Fr energy levels [11] we claimed the accuracy of our predictions to be about 0.5% while the actual agreement with latter measurements was found to be 0.1%. The situation was similar for the electromagnetic transitions $6s$-$6p_{1/2}$ and $6s$-$6p_{3/2}$ in Cs (see below). These numerous tests give us firm ground to believe that the theoretical error in E_{PNC} (2) indeed does not exceed 1%.

Very careful calculations of the PNC effect have been done by a different method in Refs. [7,8] and the calculations were compared with numerous experimental results. The presentation in Ref. [8] is very detailed so I refer the reader to this excellent work. It is important to note that the difference between the theoretical results [6] and [7,8] is only 0.3%. There are also accurate semiempirical calculations in Refs. [17,18] which have an accuracy of a few per cent and are in agreement with the many-body calculations [6–8].

As can be seen from (1) an accurate value of the vector transition polarizability β is also required for the interpretation of the PNC measurements. The following values of β were obtaned:

$$\beta = a_0^3 \begin{cases} 27.30(40) & \text{Ref [19]}, \\ 27.17(35) & \text{Ref [20]}, \\ 27.20(40) & \text{Ref [17]}, \\ 27.00(20) & \text{Ref [8]}, \\ 27.15(13) & \text{Ref [21]}, \end{cases} \quad (3)$$

The last calculation of β used the measured ratio of the vector and scalar polarizabilities of the $6s$-$7s$ transition and the most accurate values of the E1 - electromagnetic amplitudes derived from both the calculations and measurements. Using the last value of β (which is also very close to the mean value $\beta = 27.16$),

TABLE 1. Limits on new physics beyond the standard model currently obtained from atomic PNC and directly from high-energy physics (HEP).

New Physics	Parameter	Constraint from atomic PNC	Direct constraints from HEP
Oblique radiative corrections	$S + 0.006T$	-1.0 ± 1.2	$S = -0.28 \pm 0.19$, $T = -0.20 \pm 0.26$
Z_x-boson in SO(10) model	$M(Z_x)$	> 550 GeV	> 425 GeV
Leptoquarks	M_S	> 0.7 TeV	> 0.28 TeV
Composite Fermions	L	> 14 TeV	> 6 TeV

the measurement (1), the mean value of the theoretical amplitudes (2) and $|e|/a_0^2 = 5.1422 \times 10^{12}$ mV/cm, we obtain

$$Q_W(\text{exper}) = -72.41(25)_{\text{exper}}(80)_{\text{theor}}. \qquad (4)$$

Comparing this result for Q_W with the theoretical value [22]

$$Q_W(\text{theor}) = -73.20(13) - 0.8S - 0.005T, \qquad (5)$$

we can find the Peskin-Takeuchi parameter S characterizing new physics beyond the Standard Model (the parameter S stands for weak isospin conserving radiative corrections produced by new particles).

$$S + 0.006T = -1.0(0.3)_{\text{exper}}(1.0)_{\text{theor}}. \qquad (6)$$

This result is already important for high energy physics. For example, the prediction of the Technicolor Model is $S \approx 2$ [22] therefore the atomic result above seems to rule out this model. We can also use the calculation of the extra Z_x-boson contribution in the $SO(10)$ model [22]

$$\Delta Q_W = 0.4(2N + Z)\left(\frac{M_W}{M_{Z_x}}\right)^2 = 84.4\left(\frac{M_W}{M_{Z_x}}\right)^2 \qquad (7)$$

to find the limit for the mass of this boson

$$M_{Z_x} > 550 \text{ GeV} \qquad (8)$$

I would like to present a table from the D. Budker talk at the WEIN-98 conference comparing the limits on new physics obtained from the single atomic Cs PNC experiment and all High-Energy Physics (HEP) data (see also [23]).

Note that the values of the parameters S are compared using the assumption that $S_Z = S_W$, where S_Z and S_W are the parameters for the Z and W bosons.

Recently there was an interesting new development in the study of PNC in atoms. Bennett and Wieman [24] used the more accurate value $E_{\text{PNC}} = 0.9065(36)$ which is the average of our result $E_{\text{PNC}} = 0.908(9)$ [6] and the result of the Notre-Dame group $E_{PNC} = 0.905(9)$ [7,8]. Note that Bennett and Wieman assumed 0.4% accuracy of the calculations contrary to the 1% accuracy claimed in both calculations. This assumption was based on the comparison of the calculated atomic quantities relevant to the PNC amplitude (electromagnetic transition amplitudes between lower s and p states and hyperfine structure intervals of these states) with the latest very accurate measurements which resolved major discrepancies between theory and experiment in favor of theory.

Bennet and Wieman [24] also obtained the most precise value of β, $\beta = 27.024(43)(67)a_0^3$, from the measurements of the ratio $M1_{hfs}/\beta$ where $M1_{hfs}$ is the $M1$ transition amplitude between the states $6S$ and $7S$ induced by the hyperfine structure (hfs) interaction. Semiempirical formula for the $M1_{hfs}$ amplitude derived in Ref. [20,25] was used in the analysis:

$$M1_{hfs} = - \left| \frac{\mu_B}{c} \right| \frac{\sqrt{A_{6s} A_{7s}}}{E_{7s} - E_{6s}} \frac{1}{2}(g_S - g_I)1.0024 \tag{9}$$

Here A_{6s} and A_{7s} are the hfs constants of the $6s$ and $7s$ states of Cs, $g_S = 2.0025$, $g_I = -0.0004$, the coefficient 1.0024 was introduced to account for the many-body effects. Values $\beta = 27.024(43)(67)a_0^3$ and $E_{PNC} = 0.9065(36)$ and measurements of the PNC effect [5] lead to the value of the weak charge of ^{133}Cs $Q_W = -72.06(28)(34)$ which differs from the prediction of the Standard Model $Q_W = -73.20(13)$ [22] by 2.5σ.

From the point of view of accurate atomic calculations, there are two major questions in the analysis above which should be considered. The first is whether the actual accuracy of the PNC calculations is really 0.4%. The second is whether the semi-empirical formula (9) is accurate. Recently we addressed the second question in Ref. [26], leaving the first one for later work. We performed an accurate many-body calculation of the ratio

$$R = \frac{\langle 6s|H_{hfs}|7s\rangle}{\sqrt{\langle 6s|H_{hfs}|6s\rangle\langle 7s|H_{hfs}|7s\rangle}}, \tag{10}$$

where all hfs matrix elements are calculated in the same approximation. The value of R can be calculated with very high accuracy because uncertainties in different matrix elements cancel each other almost exactly. We demonstrated that inclusion of different many body and relativistic effects leave the formula

$$\langle 6s|H_{hfs}|7s\rangle = \sqrt{\langle 6s|H_{hfs}|6s\rangle\langle 7s|H_{hfs}|7s\rangle} \tag{11}$$

valid to the accuracy 0.1%.

As a result the M_{hfs} amplitude, tensor polarizability β and weak charge of the ^{133}Cs nucleus become

$$M_{hfs} = |\frac{\mu_B}{c}|0.8074(8) \times 10^{-5},$$
$$\beta = 26.957(43)(27)a_0^3, \tag{12}$$
$$Q_W = -71.88(28)(29).$$

To stress the importance of the result here we used an estimate of the theoretical accuracy 0.4% [24] in the value of E_{PNC}. Our result for M_{hfs} is in very good agreement with the result of Derevianko et al [27]

$$M_{hfs} = |\frac{\mu_B}{c}|0.8070(73) \times 10^{-5}, \tag{13}$$

but has the better accuracy. The weak nuclear charge Q_W in (12) represents even larger deviation from the Standard Model value $Q_W = -73.20(13)$ [22] than the result presented by Bennett and Wieman [24]. The deviation is 2.9σ if 0.4% accuracy of calculations of the E_{PNC} is assumed. Note that even if 1% accuracy is assumed for the calculated value of E_{PNC} as it was claimed in both theoretical works [6-8] then there is still 1.5σ deviation from the Standard Model. However, we would like to stress once more that before making any conclusions about agreement or disagreement with the Standard Model the question about the accuracy of the atomic calculations of the PNC electronic matrix element E_{PNC} should be carefully re-analyzed.

This work was supported by the Australian Research Council.

REFERENCES

1. Bouchiat, M.A. and Bouchiat, C., *Phys. Lett. B* **48**, 111 (1974); Khriplovich, I.B., *Pis'ma v ZhETF* **20**, 686 (1974) [*Sov. Phys. JETP Lett.* **20**, 315 (1974)]; Sandars, P.G.H. in *Atomic Physics*, edited by Putlitz, G. zu, **4**, 71, New York: Plenum ,1975; Sorede, D.S. and Fortson, E.N., *Bull. Am. Phys. Soc.* **20**, 491 (1975).
2. Barkov, L.M. and Zolotorev, M.S., *Pis'ma v ZhETF* **27**, 379 and **28**, 544 (1978) [*Sov. Phys. JETP Lett.* **27**, 357 and **28**, 503 (1978)]; Drell, P.S. and Commins, E.D., *Phys. Rev. A* **32**, 2196 (1985); DeMille, D., Budker, D. and Commins, E.D., *Phys. Rev. A* **50**, 4657 (1994); Birich, G.N., Bogdanov, Yu.V., Kanorskii, S.I., Sobel'man, I.I., Sorokin, V.N., Struk, I.I. and Yukov, E.A., *Zhur. Eksp. Teor. Fiz.* **87**, 776 (1984) [*Sov. Phys. JETP* **60**, 442 (1984)]; Bouchiat, M.A., Guena, J., Pottier, L. and Hunter, L., *J. Phys. (Paris)* **47**, 1709 (1986).
3. Edwards, N.H., Phipp, S.J., Baird, P.E.G. and Nakayama, S., *Phys. Rev. Lett.* **74**, 2654 (1995); Vetter, P.A., Meekhof, D.M., Majumder, P.K., Lamoreaux, S.K. and Fortson, E.N., *Phys. Rev. Lett.* **74**, 2658 (1995).
4. Meekhof, D.M., Vetter, P.A., Majumder, P.K., Lamoreaux, S.K., Fortson, E.N., *Phys. Rev. A.* **52**, 1895 (1995); Phipp, S.J., Edwards, N.H., Baird, P.E.G., Nakayama, S., *J. Phys. B.* **29**, 1861 (1996); Macpherson, M.J.D., Zetie, K.P., Warrington, R.B., Stacey, D.N. and Hoare, J.P., *Phys. Rev. Lett.* **67**, 2784 (1991).
5. Wood, C.S., Bennett, S.C., Cho, D., Masterson, B.P., Roberts, J.L., Tanner, C.E. and Wieman, C.E., *Science* **275**, 1759 (1997).
6. Dzuba, V.A., Flambaum, V.V. and Sushkov, O.P., *Phys. Lett. A* **141**, 147 (1989).
7. Blundell, S.A., Johnson, W.R. and Sapirstein, J., *Phys. Rev. Lett.* **65**, 1411 (1990).
8. Blundell, S.A., Sapirstein, J. and Johnson, W.R., *Phys. Rev. D* **45**, 1602 (1992).

9. Dzuba, V.A., Flambaum, V.V., Silvestrov, P.G. and Sushkov, O.P., *J. Phys. B* **20**, 1399 (1987); *Phys. Lett. A* **131**, 461 (1988); Dzuba, V.A., Flambaum, V.V. and Sushkov, O.P., *Phys. Lett. A* **140**, 493 (1989); Dzuba, V.A., Flambaum, V.V., Kraftmakher, A.Ya., Sushkov, O.P., *Phys. Lett. A* **142**, 373 (1989).
10. The point is that we exploit the Feynman diagram technique which contains all possible "time ordering" of the loops and therefore screening diagrams contain any number of excited electrons.
11. Dzuba, V.A., Flambaum, V.V. and Sushkov, O.P., *Phys. Lett.* **95A**, 230 (1983).
12. Dzuba, V.A., Flambaum, V.V. and Sushkov, O.P., *Phys. Rev. A* **51**, 3454 (1995).
13. Liberman, S. *et al*, *C.R. Acad. Sci.* Paris **286B**, 253 (1978); Andreev, S.V., Letokhov, V.S. and Mishin, V.I., *JETP. Lett.* **43**, 736 (1986); *Phys. Rev. Lett.* **59**, 1274 (1987); *J. Opt. Soc. Am.* **B5**, 2190 (1988); Bauche, J. *et al*, *J. Phys. B* **19**, L593 (1986); Duong, H.T. *et al*, *Europhys. Lett.* **3**, 175 (1987); Arnold, E., Borchers, W., Carré, M. *et al*, *J. Phys. B* **22**, L391 (1989); Arnold, E. *et al*, *J. Phys. B* **25**, 3511 (1990); Simsarian, J.E., Shi, W., Orozco, L.A., Sprouse, G.D., Zhao, W.Z., *Optics Lett.* **21**, 1939 (1996).
14. Zhao, W.Z., Simsarian, J.E., Orozco, L.A., Shi, W. and Sprouse, G.D., *Phys. Rev. Lett.* **78**, 4169 (1997).
15. Dzuba, V.A., Flambaum, V.V., Silvestrov, P.G. and Sushkov, O.P., *J. Phys. B* **20**, 3297 (1987).
16. Dzuba, V.A., Flambaum, V.V., Silvestrov, P.G. and Sushkov O.P., *Europhys. Lett.* **7**, 413 (1988).
17. Bouchiat, C. and Piketty, C.A., *Europhys. Lett.* **2**, 511 (1986).
18. Hartley, A.C. and Sandars, P.G.H., *J. Phys. B: At. Mol. Opt. Phys.* **23**, 2649 (1990).
19. Gilbert, S.L. and Wieman, C.E., *Phys. Rev. A* **34**, 792 (1986).
20. Bouchiat, M.A. and Guena, J., *J. Phys. (Paris)* **49**, 2037 (1988).
21. Dzuba, V.A., Flambaum, V.V., Sushkov, O.P., *Phys. Rev. A* **141**, R4357 (1997).
22. Marciano, W.J. and Rosner, J.L., *Phys. Rev. Lett.* **65**, 2963 (1990); **68**, 898(E) (1992).
23. Ramsey-Musolf, M.J., hep-ph/9903264 (1999). Casabluoni, R., De Curtis S., Dominici, D., Gatto, R., and Riemann, S., hep-ph/0001215 (2000).
24. Bennett S. C. and Wieman C. E., *Phys. Rev. Lett.* **82**, 2484 (1999); **82**, 4153(E) (1999); **83**, 889(E) (1999).
25. Bouchiat C. and Piketty C.-A., *J. Phys. (Paris)* **49**, 1851 (1988).
26. Dzuba V.A., Flambaum V.V., submitted to Phys. Rev. A.
27. Derevianko A., Safronova M. S., and Johnson W. R., *Phys. Rev. A* **60**, R1741 (1999).

Precision Tests of the Standard Model at Electron Colliders

David Muller

Stanford Linear Accelerator Center, Stanford, California 94309, USA

Abstract. We review electroweak physics studies in high-energy e^+e^- collisions at CERN and SLAC. Studies of couplings of the Z^0 boson to many of the fundmental fermions are now quite detailed, and those of the $W^\pm$ bosons are well under way. Sensitivity to radiative corrections due to the massive top quark, the as yet undiscovered Higgs boson, and new physics at the TeV scale has been achieved. The Standard Model is consistent with all data, although further studies are indicated in several areas. In the absence of new physics, the Higgs mass is limited to <188 GeV/c² at 95% C.L.

INTRODUCTION

The Standard Model (SM) of elementary particle physics comprises 12 fundamental fermions f, three flavors each of charged leptons ($f = e^-, \mu^-, \tau^-$), neutrinos (ν_e, ν_μ, ν_τ), up-type quarks (u, c, t) and down-type quarks (d, s, b), that interact via three forces. Each force is described by a fundamental symmetry, and we hope to find a higher symmetry that encompasses all three. Here we consider the electroweak (EW) interaction, which is understood in terms of an SU(2)×U(1) gauge symmetry that is broken spontaneously by the 'Higgs mechanism'. Three of the guage bosons, the W^+, W^- and Z^0, become massive, the photon γ is massless and a scalar Higgs boson remains to be discovered.

The EW SM includes ~17 free parameters: 4 are generally considered fundamental constants of nature; 4 describe quark mixing; the rest are particle masses. It is convenient to consider three roughly independent parts of the EW interaction. The γ couples to the electric charge Q_f of a fermion f and accounts for ordinary electromagnetism, which is described by QED with a single fundamental constant $\alpha = 1/137.035989(6)$ [1]. The $W^\pm$ are 'purely weak', coupling to the third component of weak isospin T_3^f with equal and opposite vector (V) and axial-vector (A) parts proportional to $G_F = 11.6639(1)(\hbar c)^3 \text{TeV}^{-2}$ [1]. Lepton flavor changing is not allowed and the only experimental questions are whether $V-A$ suffices to describe the structure of $Wl\nu$ interactions, and if the $We\nu_e$, $W\mu\nu_\mu$ and $W\tau\nu_\tau$ couplings are equal (lepton universality). In the quark sector the Wq_uq_d coupling for any up-type quark $q_u = i$ and any down-type quark $q_d = j$ is proportional to the magnitude of the CKM matrix element V_{ij}. In the SM, V is unitary, so only four real parameters are needed to describe the nine complex observables.

The Zff couplings include both electromagnetic and weak components, with axial-vector and vector couplings $\quad a_f \propto -T_3 \quad$ and $\quad v_f \propto T_3 - 2Q\sin^2\theta_W \quad$ (1), respectively, where the weak mixing angle θ_W is predicted given α, G_F and a third fundamental constant, e.g. the Z^0 mass, m_Z. There are 24 experimental observables, conventionally $R_f = \Gamma_{f\bar{f}}/\Gamma_{had} \propto a_f^2 + v_f^2$ and $A_f = 2a_f v_f/(a_f^2 + v_f^2)$, calculable from eqns. (1) at tree level and expected universal ($R_e = R_\mu = R_\tau$, etc.). Measurements of R_f and A_f to a few percent precision test the structure of the theory; more interesting are measurements at the $<1\%$ level, at which sizeable radiative corrections are expected from known and potential new physics.

The large radiative corrections due to strong interactions are conventionally absorbed into a total hadronic cross section σ_{had}^0. Others are absorbed into an effective angle $\sin^2 \theta_W^{eff}$, which depends linearly on the top mass m_t and logarithmically on the Higgs mass m_H, such that eqns. (1) hold to a good approximation. The values of $\sin^2 \theta_W^{eff}$ and m_W are rather sensitive to m_H and a variety of new physics. New physics might also modify a set of R_f or A_f; R_{dsb} and A_{dsb} are particularly sensitive to new left- and right-coupled physics, respectively. New high-scale physics might violate universality, perhaps most strongly for the b-quark.

The R_f are observable as branching ratios, and the A_f via both asymmetric distributions of the angle θ_f between the f and the e^-, and the f-polarizations,

$$d\sigma_f/d\cos\theta_f \propto (1 - A_e P_e)(1 + \cos^2\theta_f) + 2A_f(A_e - P_e)\cos\theta_f, \quad (2)$$

$$P_f \propto (1 - A_e P_e)(1 + \cos^2\theta_f)A_f + 2(A_e - P_e)\cos\theta_f, \quad (3)$$

where the e^- beam has longitudinal polarization P_e and the e^+ beam is unpolarized. For $P_e = 0$, $d\sigma_f/d\cos\theta$ is only sensitive to the product of initial and final state couplings $A_e A_f$ and the asymmetries are small. Large $|P_e|$ induces large asymmetries, and using both left- and right-polarized beams, $P_e = -|P_e|$ and $+|P_e|$, allows direct measurements of both A_e and A_f through the difference in total cross sections (left-right asymmetry) and the weighted difference in asymmetries (left-right-forward-backward asymmetry), respectively. The polarization P_f of the outgoing f allows direct access to A_f, via the average $\langle P_f \rangle$, and A_e, via the dependence on $\cos\theta_f$. The expected P_f is large for the quarks, but the f must be spin analyzed, which has so far only been done for $f = \tau^-$.

Programs to study this physics in detail have been under way for a decade at the LEP ring at CERN and the SLAC Linear Collider (SLC). LEP is a conventional e^+e^- storage ring that delivered 4 million hadronic events at a center-of-mass energy $E_{CM} = m_Z$ to each of four experiments, and has since been increasing E_{CM}, delivering substantial data above W^+W^- threshold. A higher energy e^+e^- storage ring would require a dramatic increase in size from LEP's 27 km circumference. The SLC is a step in an alternative direction; it is a single-pass collider featuring high $|P_e| = 0.73$ and a very small transverse collision region, 0.8×1.8 μm, which both increases luminosity (enhanced by the recently observed pinch effect) and greatly enhances the flavor tagging capabilites of the experiment. SLC shut down last year after achieving impressive luminosity and delivering over 0.5 million hadronic Z^0s. The four LEP experiments, ALEPH, DELPHI, L3 and OPAL, and the SLD experiment at SLC, were designed with unprecedented hermeticity, luminosity monitoring, vertexing, and lepton (L3) and hadron (DELPHI, SLD) identification, which have been instrumental to the physics program.

PHYSICS OF THE Z^0 BOSON

The Z^0 resonance is a very clean experimental environment. Real Z^0 production is the dominant process and easy to separate from other physics and backgrounds. Negligible energy is lost to initial state radiation. The Z^0 decays into $\nu\bar{\nu}$ are not detected, but the charged lepton ($Z^0 \to l^+l^-$) decays are identified easily. At $E = 45.6$ GeV, $e^\pm$ and $\mu^\pm$ are stable; e^+e^- or $\mu^+\mu^-$ decays yield back-to-back 45.6 GeV/c tracks that deposit their energy in the EM calorimeter or penetrate the muon detectors, respectively. The $\tau^\pm$ decays include 1–5 fairly energetic collinear tracks. The lepton flavor and the l^- polar angle are thus measured readily.

The $q\bar{q}$ decays are problematic; since the strong interaction is confining, quarks are not observable directly, but appear as jets of $\sim$20 hadrons. At high energy, the multiplicity and topology of back-to-back, collimated jets identify $q\bar{q}$ decays and allow the measurement of $|\cos\theta_q|$. The experimental challenge is to identify the event flavor (for R_q), and to separate the q-jet from the $\bar{q}$ jet (for A_q).

The Total Hadronic Cross Section

The cross section for inclusive hadronic final states has been measured at many E_{CM} at LEP. Data from L3 are shown in fig. 1; the Z^0 resonance is prominent with a long radiative tail. Events with $E_{visible} > 0.85 E_{CM}$ (open circles) account for under half the data and show the expected $1/E_{CM}^2$ behaviour for $E_{CM} \gg m_Z$. A SM fit (solid line on fig. 1) describes the data well. Current LEP average parameter values from such fits are [2]:

$$m_Z = 91.1872 \pm 0.0021 \text{ GeV}/c^2$$
$$\Gamma_Z = 2.4944 \pm 0.0024 \text{ GeV}$$
$$\sigma_{had}^0 = 41.544 \pm 0.037 \text{ nb}.$$

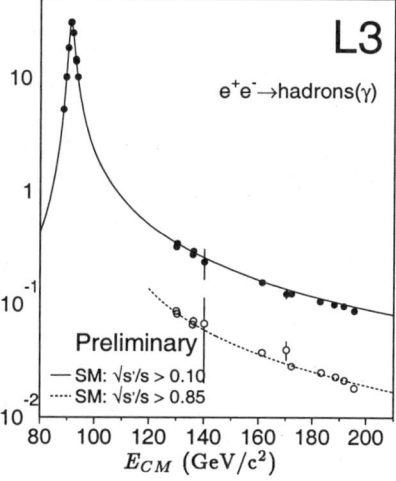

Figure 1: Total hadronic cross section in pb.

The relative precision on m_Z of 2.3×10^{-5} reflects important advances in luminosity (both measurement and theory), detector modelling, and most notably in beam energy monitoring to 1–2 MeV. This quantity is used as the third fundamental constant of the SM, so that any other observable can be predicted modulo corrections due to m_H and new physics. The precision on Γ_Z gives some sensitivity to m_H, with a 95% C.L. upper limit of $\sim$700 GeV/c^2; the exact limit depends on m_t and α_s, for which more precise measurements are desirable.

Couplings to the Charged Leptons

Cross sections for the $Z^0 \to l^+l^-$ have also been measured at LEP; the extracted m_Z and Γ_Z are included in the above averages. Cross section ratios yield [2]

$$1/R_e = 20.803 \pm 0.049, \quad 1/R_\mu = 20.786 \pm 0.033, \quad 1/R_\tau = 20.764 \pm 0.045,$$

consistent with the SM and providing a 0.3% test of lepton universality. Taking the SM R_ν, any non-SM decay width of the Z^0 is limited to $\Gamma_{inv} < 2.0$ MeV at 95% C.L. This implies no undiscovered particle with $m < m_Z/2$; in particular, if a fourth fermion generation exists, it must include a very massive neutrino.

SLD has made a precise measurement of A_e from the left-right asymmetry on the Z^0 peak, $A_{LR} = (\sigma_L - \sigma_R)/(\sigma_L + \sigma_R)$, where $\sigma_{L(R)}$ is the hadronic cross section for $P_e < 0 \, (> 0)$. A simple counting of hadronic events yields $A_e = 0.1514 \pm 0.0022$ [3], which is statistics dominated due to good understanding of the P_e measurement. The A_l can also be measured from the (left-right-)forward-backward asymmetries in $\cos\theta_l$ (eqn. 2). Distributions measured by SLD for e^-, μ^- and τ^- are shown in fig. 2. The asymmetries are small, but visible, and differences in both normalization and slope between left- and right-polarized e^- beams are evident. Measurements of A_e, A_μ and A_τ with 6–10% precision have been made by each experiment. The LEP experiments have also spin-analyzed the $\tau^\pm$ in several decay modes, measuring P_{τ^-} vs. $\cos\theta_{\tau^-}$ and extracting (eqn. 3) ~6% measurements of A_τ and A_e.

All measurements of lepton asymmetries are consistent, and the world average values [2], $A_e = 0.1513 \pm 0.0019$, $A_\mu = 0.1449 \pm 0.0090$ and $A_\tau = 0.1422 \pm 0.0042$, give a 7% test of lepton universality. Assuming universality, a grand average

$$A_{lepton} = 0.14979 \pm 0.00157, \quad \sin^2\theta_W^{eff} = 0.23117 \pm 0.00020,$$

is a robust measurement of the effective weak mixing angle with <0.1% precision. Assuming the SM and no new physics, a value of $\ln m_H$ can be extracted, with a 95% confidence interval of $\sim 10 < m_H < \sim 110$ GeV/c^2. There is theoretical debate at the few GeV/c^2 level; a better measurement of $\alpha(m_Z^2)$, via the $e^+e^- \to$ hadrons cross section at low energy, would be useful. The upper limit is tantalizingly near the lower limit of ~105 GeV/c^2 from direct searches.

Effective couplings have been measured vs. E_{CM} at LEP. The ALEPH asymmetry data in fig. 3 display the strong dependence on E_{CM}, especially near m_Z, expected in the SM (solid line) due to interference between γ and Z^0 exchange.

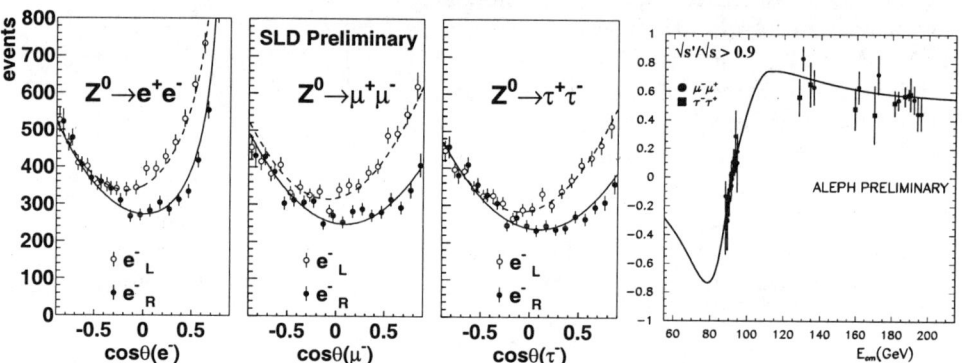

Figure 2: Lepton polar angle distributions with SM fits.

Figure 3: $A_{FB}^{\mu,\tau}$ vs. E_{CM}

Couplings to the Heavy Quarks

Great strides have been made at SLC and LEP in identifying heavy (b or c) jets. A b-jet contains a massive, energetic, leading bottom (B) hadron that, on average, flies 3 mm at the Z^0 and decays into 10 hadrons. Reconstruction of B hadrons is impractical, but modern vertex detectors can identify displaced vertices inclusively. Very few secondary B hadrons are produced in jet fragmentation, so all five experiments have obtained sample of 98% $b/\bar{b}$ purity with efficiencies of $\epsilon = 20$–60%. Separation of b from $\bar{b}$ jets has used the charge of identified $l^{\pm}$ from the decay (low ϵ, high analyzing power ap), identified $K^{\pm}$, or the net charge of tracks in the jet or vertex (high ϵ, low ap). Leading charmed (D) hadrons in c jets have lower mass, energy, decay multiplicity and lifetime, and are produced rarely in jets but copiously in B decays. Vertex detectors help separate these two sources, making both D reconstruction (unit ap) and inclusive tagging useful.

The key to making precise measurements of R_q (A_q) is to measure the relevant ϵ (and ap) from the data, rather than using a simulation. For R_b, such 'self-calibration' is routine. One can, e.g., define a $b/\bar{b}$ jet tag and take advantage of the presence of both a b and $\bar{b}$ jet by counting the number N_b of tagged jets and N_{bb} of events with a tag in each jet. One solves the two equations

$$N_b/2N_{event} = R_b\epsilon_b + \Sigma_{f=udsc}R_f\epsilon_f \quad \text{and} \quad N_{bb}/N_{event} = R_b\epsilon_b^2\lambda_b + \Sigma_{f=udsc}R_f\epsilon_f^2\lambda_f \quad (4)$$

for R_b and ϵ_b. For high purity, the backgrounds have small effects on the measured values; the correlation λ_b typically dominates the systematic error. The world average $R_b = 0.21642 \pm 0.00073$ [2] has 0.3% precision and is sensitive to m_t; it is consistent with the SM given the known m_t. A few R_c measurements are self-calibrated. Tags for $c/\bar{c}$ have substantial background only from $b/\bar{b}$, and one can, e.g., solve a set of five equations including (4) for R_c, R_b, ϵ_b and the c-tag rates η_c and η_b, since the remaining ϵ_q and η_q are small. The world average value [2] $R_c = 0.1674 \pm 0.0038$ is consistent with the SM within its 2.3% uncertainty.

There are many measurements of A_b and A_c (A_eA_b, A_eA_c at LEP) using a variety of methods. Some are self-calibrated, e.g. a new A_b from SLD that uses the net charge of tracks in a vertex, and derives both its purity and its ap from the data using doubly tagged events and comparing charges. The $\cos\theta_b$ distributions in fig. 4 show large asymmetries and this measurement is the world's most precise. An average A_b can be taken by dividing each LEP A_eA_b by the above A_{lepton}. The result, $A_b = 0.894 \pm 0.015$, is 2.7 standard deviations below the SM prediction $A_b = 0.935$. A similar average $A_c = 0.627 \pm 0.021$ is 2.1σ below the SM prediction $A_c = 0.673$. Further work is clearly needed in this area, especially any using new and/or self-calibrating methods.

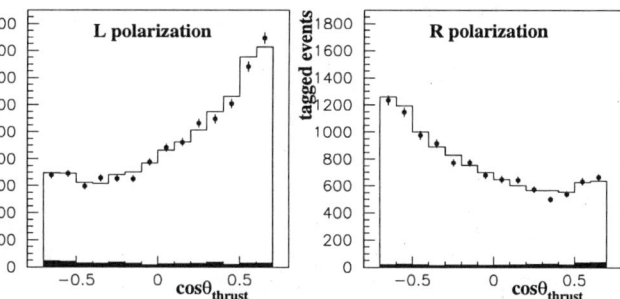

Figure 4: b-quark polar angle distributions.

Couplings to the Light Quarks

In contrast, the field of light-flavor Zuu, Zdd and Zss couplings is in its infancy. Light-flavor jets can also be identified using their leading particles, e.g. an s-jet may produce a leading K^- or Λ^0. However these particles have lower average energy than B or D hadrons and the same particle species are produced copiously in the fragmentation process, as well as in B and D decays, especially at low momentum. Most importantly, a leading K^- (Λ^0) may also be produced in a $\bar{u}$ (u or d) jet.

Since these aspects of jet fragmentation are not well measured, one must either rely on a model or perform a complicated self-calibration. The OPAL collaboration have done the latter [4], using five flavor tags, identified $\pi^\pm$, $K^\pm$, K_s^0, p/$\bar{p}$, and $\Lambda^0/\bar{\Lambda}^0$ with $p > 22.8$ GeV/c. A set of 20 equations similar to eqns. 4 can be solved for R_u, R_d and 15 ϵ. The equations are not independent, and statistics force a number of reasonable assumptions to be made, including $R_d = R_s$, however the resulting $R_u/(R_u + R_d + R_s) = 0.371 \pm 0.023$ is consistent with the SM prediction and a useful 6% test of up-type universality. To measure the A_q OPAL solves an additional 14 equations for 8 ap using 4 assumptions. The resulting $A_u = 0.36 \pm 0.65$ and $A_d = A_s = 0.61 \pm 0.33$ (where A_e has been divided out) are statistics limited, but this analysis is most encouraging and should be pursued by other experiments. Interesting related measurements include the first $\pi^\pm$ and $K^\pm$ spectra in u, d and s jets [5] shown in fig. 5, which already provide new and stringent model tests.

SLD and DELPHI have concentrated on precise measurements of A_s only, using high-momentum K^- (K^+) to tag s ($\bar{s}$) jets. DELPHI [6] relies on a fragmentation model to predict the purity (53%) and ap (74%), obtaining a statistics dominated $A_s = 0.909 \pm 0.108$. SLD [7] require an opposite-sign $K^\pm$ or a K_s^0 in the opposite jet to enhance signal, obtaining the $\cos\theta_s$ distributions shown in fig. 6. Jets and events with multiple identified kaons are used to calibrate the purity (66%) and ap (82%) with a modest model-dependence, yielding $A_s = 0.895 \pm 0.091$. These results provide an important 7% test of d-type universality, but full self-calibration will be needed for substantial improvement.

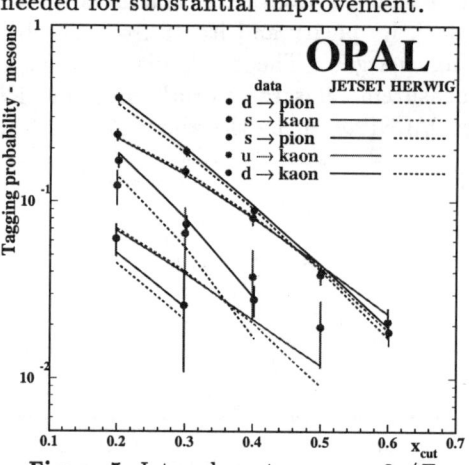

Figure 5: Integral spectra vs. $x = 2p/E_{CM}$.

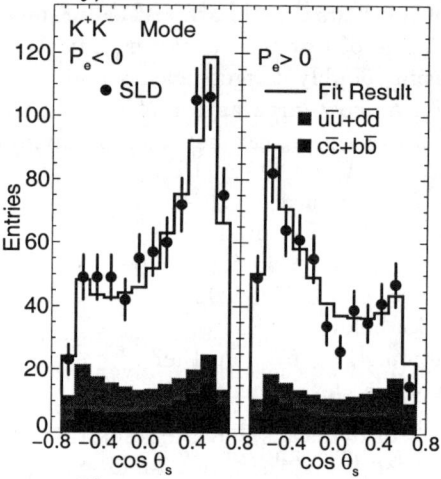

Figure 6: s-quark polar angles.

PHYSICS OF THE $W^\pm$ BOSONS

Much W physics is studied via τ decays and B^0-$\bar{B}^0$ mixing at the Z^0. LEP is now delivering the first monoenergetic $W^\pm$ samples in $e^+e^- \to W^+W^-$ at $E_{CM} > 2m_W$. Each W decays to 2 jets or an energetic $l\nu$, and progress has been rapid in distinguishing such events from the large backgrounds and measuring their properties.

Couplings to the Leptons

The constant G_F has been measured precisely at lower energies and is universal within 0.5%. Large samples of polarized $\tau^- \to W^- \nu_\tau$, $W^- \to e\nu_e, \mu\nu_\mu$ decays allow detailed studies of the structure of the $Wl\nu$ interaction in $e^+e^- \to \tau^+\tau^-$. A general Lorentz-invariant coupling gives a τ-rest-frame spectrum of $x = E_l/E_l^{max}$ (scaled $l^\pm$ energy) of the form $d\sigma/x^2 dx = 9 f_1 + \rho f_2 + 6\eta m_l f_1/xm_\tau - P_\tau \xi \cos\phi [3f_1 + \delta f_2]$, where $f_1 = 1-x$, $f_2 = 8x - 6$, ϕ is the angle between $\vec{p}_l$ and the τ spin, and in the SM the Michel parameters are $\rho = \delta = 0.75$, $\eta = 0$ and $\xi = 1$. Detailed analyses of lepton spectra at the Z^0 and $\Upsilon(4S)$ exploit the $\cos\theta$-dependence of P_τ and/or the anticorrelation between the τ^+ and τ^- helicities, and yield world averages [8] of

$$\rho = 0.752 \pm 0.008, \quad \eta = 0.035 \pm 0.031, \quad \xi = 0.978 \pm 0.031, \quad \xi\delta = 0.745 \pm 0.021,$$

limiting departures from $V - A$ structure of $Wl\nu$ interactions to the few % level.

Weak dipole moments probe CP-violation and τ substructure. In a general $Z\tau\tau$ coupling $\propto \gamma_\mu(v_\tau - a_\tau \gamma_5) + i\sigma^{\mu\nu}\{a_\tau^w/m_\tau - 2id_\tau^w \gamma_5/e\} \cos\theta_W \sin\theta_W$, the electric d_τ^w and anomalous magnetic moments a_τ^w modify transverse P_τ. All Z^0 experiments (L3 and SLD) have measured a_τ^w (d_τ^w, enhanced by $|P_e| \gg 0$). The averages [9]

$$Re(a^w\tau) = 0.22 \pm 1.13 \times 10^{-3}, \quad Re(d^w\tau) = -0.02 \pm 0.15 \times 10^{-17} \text{ e cm},$$
$$Im(a^w\tau) = -0.03 \pm 0.66 \times 10^{-3}, \quad Im(d^w\tau) = -0.13 \pm 0.29 \times 10^{-17} \text{ e cm},$$

are consistent with zero and becoming sensitive to a variety of new physics.

Production, Mass, Width and Branching Ratios

Selection of W^+W^- events is well advanced at LEP; the current average cross section, shown vs. E_{CM} in fig. 7, is consistent with the SM and shows that Z^0, γ, t-channel neutrino exchange, and their interference are all required to describe the data. First studies are under way of the structure of these events, i.e. W polar angles, helicities, etc., and have already set relative limits of 5–10% on some classses of non-SM contributions [2].

The W boson mass has received much attention due to its sensitivity (given m_Z) to m_H. Individual $W^\pm$ masses can be reconstructed from 2-jet or $l\nu$ decays, but the resolution is poor and in 4-jet events there is a 3-fold ambiguity in the jet pairing. Constraints such as 4-momentum

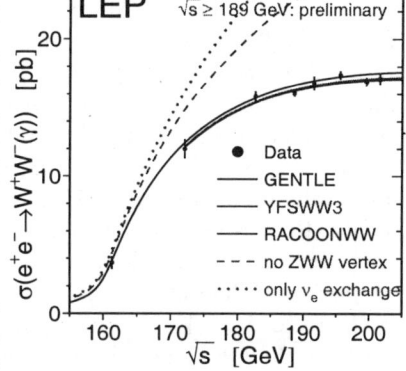

Figure 7: $e^+e^- \to W^+W^-$ cross section.

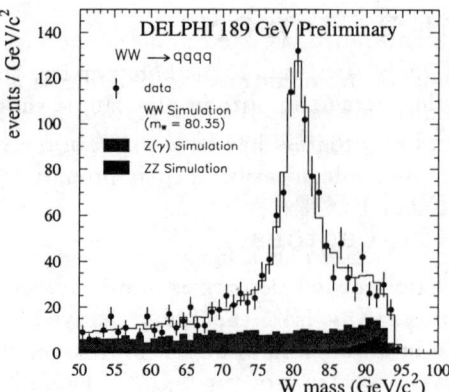

Figure 8: Sample $W^\pm$ mass distribution.

conservation, $m_{W^+} = m_{W^-}$, $\vec{p}_{W^+} = -\vec{p}_{W^-}$, etc., give improved resolution, as e.g. in fig. 8 for 4-jet events from DELPHI. The 4-jet events give the best resolution and statistics, but are subject to uncertainties in the mass scale due to final-state interactions. The current LEP average [2]

$$m_W = 80.401 \pm 0.049 \text{GeV}/c^2$$

corresponds to a low $m_H \approx 100$ GeV/c^2.

The SM predicts the W^- decay widths to $l^- \nu_l$ and, in terms of the CKM elements $|V_{ij}|$, to the six allowed $\bar{q}_u q_d$ modes $i=u, c$, $j=d, s, b$. As for Z^0s, the $l^- \nu_l$ decays are easy to identify; the $e\nu$, $\mu\nu$ and $\tau\nu$ branching ratios are measured to 3% [2]. They are consistent, averaging $BR(W \to l\nu) = 10.68 \pm 0.13\%$, consistent with the SM.

The ratios $R_{ij} = BR(W^- \to \bar{i}j)/BR(W^- \to$ hadrons) provide robust, theoretically clean probes of the relative $|V_{ij}|$. However one must identify the jet flavors, and the $\bar{i}b$ modes are rare. Three experiments have combined vertex and identified $K^\pm$ information to separate the two dominant modes, $\bar{u}d$ and $\bar{c}s$, and measure $R_{cs} = 0.50 \pm 0.03$ and $|V_{cs}| = 0.99 \pm 0.07$. The precision is not yet useful, however this result along with the heavy- and light-flavor results from the Z^0 give confidence that precise measurements could be made at a future e^+e^- collider.

B^0-$\bar{B}^0$ Mixing

The CKM element V_{td} (V_{ts}) can be probed via B_d^0-$\bar{B}_d^0$ (B_s^0-$\bar{B}_s^0$) mixing, but with an unfortunate theoretical uncertainty. To measure Δm_d, one must tag $B_d^0/\bar{B}_d^0$ decays and determine the flavor (B^0 or $\bar{B}^0$) at both production and decay time, preferably vs. proper decay time. Since 40% of b jets give a B_d^0, the above $b/\bar{b}$ and decay flavor tags are used. The flavor at production is tagged using information from the opposite jet, P_e, and same-jet tracks not assigned to the decay. Many measurements from LEP/SLC, and CDF and the $\Upsilon(4S)$ show clear, slow oscillations. The average $\Delta m_d = 0.476 \pm 0.016$ ps^{-1} [10] corresponds to $|V_{td}| = 0.0088 \pm 0.0002(\text{expt.}) \pm 0.0018(\text{theory})$.

There is no evidence for B_s^0-$\bar{B}_s^0$ mixing despite valiant effort. Only $\sim$10% of b jets contain a B_s meson and the oscillation is very fast, demanding excellent proper time resolution σ_τ. Theoretical errors partly cancel in the ratio $|V_{ts}|/|V_{td}|$, with implications (below) for the CKM matrix, so much work is going into B_s tagging and σ_τ in the race to

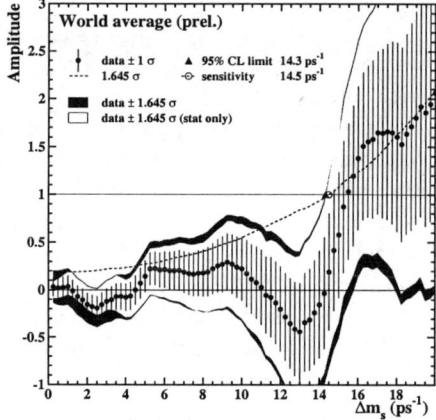

Figure 9: Combined world amplitude plot for B_s^0-$\bar{B}_s^0$ mixing.

observe B_s^0 mixing. In the 'amplitude method', Δm_s is fixed and the amplitude A of an oscillation fitted. If the $\langle A \rangle$ at a given Δm_s is $>1.65\sigma$ below unity, that Δm_s is excluded at the 95% C.L. The world average amplitude plot [10] is shown in fig. 9. The 95% C.L. limit of $\Delta m_s > 14.3$ ps^{-1} corresponds to $|V_{td}|/|V_{ts}| < 0.228$.

INTERPRETATION

Each result given above is consistent with the SM and a fairly low Higgs boson mass m_H. We now consider this body of data as a whole, and include some non-SLC/LEP measurements, notably m_t, m_W and Δm_s from the Tevatron, as well as world averages of the running strong coupling $\alpha_s(m_Z^2)$ and fine structure $\alpha(m_Z^2)$.

The Standard Model Higgs Boson

The fundamental question of consistency of the SM with the world's body of data is plagued by correlations among the measurements. The LEP ElectroWeak Working Group [2] performs the arduous task of considering all the data. Their global one-parameter fit gives $\ln m_H = 1.82 \pm 0.30$ and $\chi^2/\text{dof} = 22.9/15$. Thus the SM is consistent with the data at 8.5% C.L., and $m_H < 188$ GeV/c^2 at 95% C.L.

Smaller limits are given above, and the χ^2 is driven by the very precise A_{lepton} and A_b. If A_b is removed, $\chi^2/\text{dof} \approx 1$ and $m_H < 110$ GeV/c^2. Several general parametrizations of radiative corrections allow one to look beyond the SM, such as the S-T projection [11], fig. 10. A model gives a point on this plot; the SM with $m_H = 100$ GeV/c^2 and $m_t = 175$ GeV/c^2 defines the origin; δm_t gives width to an arc representing $0.1 < m_H < 1$ TeV/c^2. A measurement gives a band; three 'clean' measurements sensitive to m_H are shown, A_{lepton}, m_W and Γ_Z. All are consistent with the SM with a low value of m_H and their overlap restricts the parameter space considerably.

Figure 11 [12] shows the status of A_{lepton} and A_b. The A_{lepton} and SLD A_b bands are $\sim$orthogonal, with the LEP $A_e A_b$ band at $\sim 45°$. The measurements are consistent, and a fit is 2.0σ from the SM (thick line). The discrepancy is not significant, but is in the b sector, has persisted at 2-3.5σ for years, and the A_{lepton} and $A_e A_b$ bands cross the SM 2.7σ apart, so work continues to resolve this issue.

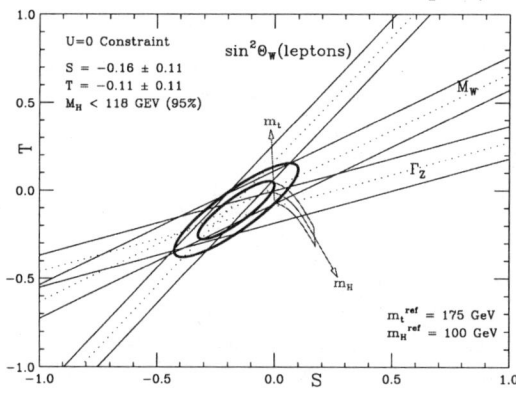

Figure 10: S-T plot; fit to non-hadronic data.

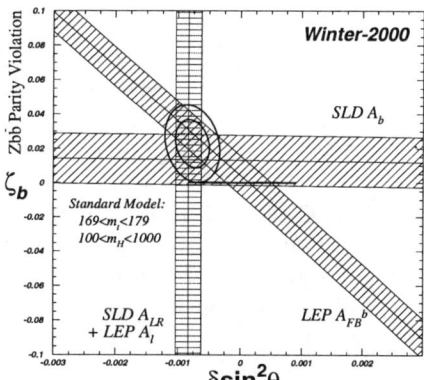

Figure 11: The 'TGR' plot.

The Unitarity Triangle

Figure 12 shows the 'ρ-η' plane, a conventional paramatrization of CP violation in the SM. Bands from K^0 decay (ϵ_K), B decay ($|V_{ub}|/|V_{cb}|$) and B_d^0 mixing overlap, but the errors do not restrict the angles of the unitarity triangle ($(0,0)$, (ρ,η), $(0,1)$) strongly. In the SM context, the B_s^0 mixing results give an outer limit indicated by the dashed line in fig. 12, that cuts the allowed region in half. As B_s^0 mixing is being pursued vigorously in Z^0 data, we should either observe it soon or restrict the allowed (ρ,η) region considerably.

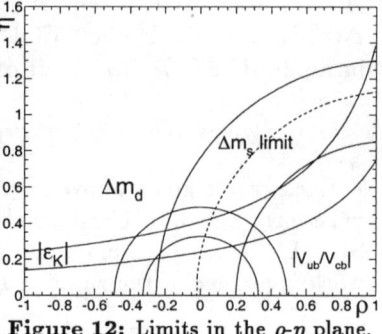

Figure 12: Limits in the ρ-η plane.

Summary and Conclusion

The high energy e^+e^- programs at CERN and SLAC have had tremendous success in testing the SM precisely, and LEP continues to do exciting $e^+e^- \to W^+W^-$, Z^0Z^0, etc. physics. The Z^0 mass is now known to 2.3×10^{-5} relative, but the Higgs boson remains unseen. Several measurements are sensitive to m_H and new physics via radiative corrections. No non-SM physics has been seen, although a 2σ issue remains in the b sector. Fitting the world's data yields $m_H < 188$ GeV/c^2 at 95% C.L.; the non-b data suggest a limit near the direct search limit, $m_H > 105$ GeV/c^2.

We are therefore on the verge of either discovering the Higgs boson or demonstrating the need for non-SM physics, and we will soon observe the very rapid oscillation of B_s mesons or demonstrate non-unitarity of the CKM matrix. The energy and luminosity of current accelerators is, alas, likely to give ambiguous answers to these questions. They can be addressed effectively, and new and existing tests performed precisely, at a future e^+e^- collider with E_{CM} of 90–300 GeV. The time has come for the community to converge on the choice of such a machine.

REFERENCES

1. Particle Data Group, *Eur. Phys. J.* **C3**, 1 (1998).
2. LEP EW Wkg. Gp., CERN-EP-2000-016, http://lepewwg.web.cern.ch/LEPEWWG
3. SLD Collab., K. Abe, et al., SLAC-PUB-8401, submitted to *Phys. Rev. Lett.*
4. OPAL Collab., K. Ackerstaff et al., *Z. Phys.* **C76**, 387 (1997).
5. OPAL Collab., G. Abbiendi et al., CERN-EP/99-164, submitted to *Eur. Phys. J.*
6. DELPHI Collab., P. Abreu et al. CERN-EP/99-134, submitted to *Eur. Phys. J.*
7. SLD Collab., K. Abe, et al., SLAC-PUB-8408, submitted to *Phys. Rev. Lett.*
8. A. Stahl, *Nucl. Phys. Proc. Suppl.* **76**, 173 (1999).
9. A. Zalite, *Nucl. Phys. Proc. Suppl.* **76**, 229 (1999);
    SLD Collab., K. Abe, et al., SLAC-PUB-8163, submitted to *Phys. Rev. Lett.*
10. LEP B-Oscillation Working Group, http://lepbosc.web.cern.ch/LEPBOSC.
11. M.E. Peskin and T. Takeuchi, *Phys. Rev.* **D46**, 381 (1992).
12. T. Takeuchi, A.K. Grant and J.L. Rosner, hep-ph/9409211.

Lorentz and CPT Tests in Atomic Systems

Robert Bluhm

Physics Department, Colby College, Waterville, ME 04901 USA

Abstract. A review of Lorentz and CPT tests performed in atomic systems is presented. A theoretical framework extending QED in the context of the standard model is used to analyze a variety of systems. Experimental signatures of possible Lorentz and CPT violation in these systems are investigated. Estimated bounds attainable in future experiments and actual bounds obtained in recent experiments are given.

INTRODUCTION

Many of the shapest tests of Lorentz and CPT symmetry have been performed in atomic systems [1, 2]. Although many of these experiments were performed in earlier decades, a new appreciation of them has emerged due to recent theoretical advances. The group of Kostelecký and collaborators have developed a consistent theoretical framework [3, 4, 5] that permits a detailed investigation of Lorentz and CPT tests in particle and atomic systems. In the context of this framework, it is possible to look for new signatures of Lorentz and CPT violation in atomic systems that were previously overlooked.

In the next section, I begin with a brief review of Lorentz and CPT symmetry and some of the atomic experiments that test these symmetries. This is followed by a discussion of some theoretical ideas that have been put forward over the years for ways in which Lorentz symmetry and CPT might be violated in nature. Different theoretical approaches to searching for Lorentz violation are also described. However, the main focus of this work will be on the standard-model extension of Kostelecký and collaborators, as it is this model which permits the most detailed investigation of Lorentz and CPT tests and is applicable across different experiments. The QED sector of this theory will be presented and used to examine five different atomic or macroscopic systems: experiments in Penning traps [6], clock-comparison tests [7], experiments with hydrogen and antihydrogen [8], Lorentz and CPT tests with macroscopic spin-polarized materials [9] and muon experiments [10].

As a result of our investigations, several atomic experimental groups have recently reanalyzed their data or taken new data to obtain improved bounds on

Lorentz and CPT violation. I will present summaries of the bounds that have been obtained as well as estimates of bounds that can be attained in future experiments.

Lorentz and CPT Symmetry

It appears that Lorentz symmetry and CPT are exact fundamental symmetries of nature [11]. All physical interactions appear to be invariant under continuous Lorentz transformations consisting of boosts and rotations and under the combined discrete symmetry CPT formed from the product of charge conjugation C, parity P, and time reversal T. These symmetries are linked by the CPT theorem [12], which states that under mild technical assumptions all local relativistic field theories of point particles are symmetric under CPT. The CPT theorem predicts that particles and antiparticles should have exactly equal lifetimes, masses, and magnetic moments.

Numerous experiments confirm CPT and Lorentz symmetry to extremely high precision. The best CPT test listed by the Particle Data Group [1] compares the masses of neutral K^0 mesons with their antiparticles and obtains the figure of merit,

$$r_K = \frac{|m_K - m_{\overline{K}}|}{m_K} \lesssim 10^{-18} \quad . \tag{1}$$

The Hughes-Drever type experiments are generally considered the best tests of Lorentz symmetry. These experiments place very tight bounds on spatially anisotropic interactions [13].

Experiments in Atomic Physics

Many of the sharpest tests of CPT and Lorentz symmetry are made in atomic systems. For example, the Hughes-Drever type experiments typically compare two clocks or high-precision magnetometers involving different atomic species. The best CPT tests for leptons and baryons cited by the Particle Data Group are made by atomic physicists using Penning traps. These experiments have obtained a bound on the g-factor difference for electrons and positrons given by [14]

$$r_g^e = \frac{|g_{e^-} - g_{e^+}|}{g_{\text{avg}}} \lesssim 2 \times 10^{-12} \quad , \tag{2}$$

while experiments with protons and antiprotons have obtained bounds on the difference in their charge-to-mass ratios given by [15]

$$r_{q/m}^p = \frac{|(q_p/m_p) - (q_{\overline{p}}/m_{\overline{p}})|}{(q/m)_{\text{avg}}} \lesssim 9 \times 10^{-11} \quad . \tag{3}$$

Similarly, two proposed experiments at CERN intend to make high-precision comparisons of the 1S-2S transitions in trapped hydrogen and antihydrogen [16]. The 1S-2S transition is a forbidden transition that can only occur as a two-photon transition. It has a long lifetime and a small relative linewidth of approximately 10^{-15}. Atomic experimentalists believe that ultimately the line center might be measured to 1 part in 10^3 yielding a CPT bound

$$r^H_{1S-2S} = \frac{|\Delta\nu_{1S-2S}|}{\nu_{1S-2S}} \lesssim 10^{-15} - 10^{-18} \quad . \tag{4}$$

It is interesting to note that of all the experiments testing CPT it is the atomic experiments which have the highest experimental precisions. For example, in the meson experiments quantities are measured with precisions of approximately 10^{-3}, while in the atomic experiments frequencies are typically measured with precisions of 10^{-9} or better. Nonetheless, the figure of merit in Eq. (1) is many orders of magnitude better than those in Eqs. (2) or (3). These differences are due in part to the fact that these experiments compare different physical quantities, such as masses, g factors, charge-to-mass ratios, and frequencies. What would obviously be desirable is a framework which puts these experiments on equal footing and allows better cross comparisons.

Ideas for Violation

Different ideas for violation of Lorentz or CPT symmetry have been put forward over the years since the proof of the CPT theorem. To evade the CPT theorem one or more of its assumptions must be disobeyed. A sampling of some of the theoretical ideas that have been put forward include the following: nonlocal interactions [17], infinite component fields [18], a breakdown of quantum mechanics in gravity [19], Lorentz noninvariance at a fundamental level [20], spontaneous Lorentz violation [3], and CPT violation in string theory [4, 21].

To explore the experimental consequences of possible Lorentz or CPT violation, a common approach is to introduce phenomenological parameters. Some examples include the anisotropic inertial mass parameters in the model of Cocconi and Salpeter [22], the δ parameter used in kaon physics [23], and the TH$\epsilon\mu$ model which couples gravity and electromagnetism [24]. Another approach is to introduce specific lagrangian terms that violate Lorentz or CPT symmetry [25, 26]. These approaches have the advantages that they are straightforward and are largely model independent. However, they also have the disadvantages that their relation to experiments can be unclear and they can have limited predictive ability. To make further progress, one would want a consistent fundamental theory with CPT and

Lorentz violation. This would permit the calculation of phenomenological parameters and the prediction of signals indicating symmetry violation. No such realistic fundamental theory is known at this time. However, a candidate extension of the standard model incorporating CPT and Lorentz violation does exist.

STANDARD-MODEL EXTENSION

The standard-model extension of Kostelecký and collaborators is an effective theory based on the idea of spontaneous Lorentz-symmetry breaking [3]. It is also motivated in part from string theory [4]. The idea is to assume the existence of a fundamental theory in which Lorentz and CPT symmetry hold exactly but are spontaneously broken at low energy. As in any theory with spontaneous symmetry breaking, the symmetries become hidden at low energy. The effective low-energy theory contains the standard model as well as additional terms that could arise through the symmetry breaking process. A viable realistic fundamental theory is not known at this time, though higher dimensional theories such as string or M theory are promising candidates. A mechanism for spontaneous symmetry breaking can be realized in string theory because suitable Lorentz-tensor interactions can arise which destabilize the vacuum and generate nonzero tensor vacuum expectation values.

Colladay and Kostelecký have derived the most general extension of the standard model that could arise from spontaneous Lorentz symmetry breaking of a more fundamental theory, maintains SU(3)×SU(2)×U(1) gauge invariance, and is renormalizable [5]. They have shown that the theory maintains many of the other usual properties of the standard model besides Lorentz and CPT symmetry, such as electroweak breaking, energy-momentum conservation, the spin-statistics connection, microcausality, and observer Lorentz covariance. In addition to the atomic experiments described here, the standard-model extension has been used to analyze Lorentz and CPT tests with neutral mesons [27, 28], photon experiments [5, 25, 29], and baryogenesis [30].

TESTS IN ATOMIC SYSTEMS

To consider experiments in atomic physics it suffices to restrict the standard-model extension to its QED sector. The modified Dirac equation for a four-component spinor field ψ of mass m_e and charge $q = -|e|$ in an electric potential A^μ is

$$\left(i\gamma^\mu D_\mu - m_e - a_\mu \gamma^\mu - b_\mu \gamma_5 \gamma^\mu - \tfrac{1}{2} H_{\mu\nu} \sigma^{\mu\nu} + i c_{\mu\nu} \gamma^\mu D^\nu + i d_{\mu\nu} \gamma_5 \gamma^\mu D^\nu \right)\psi = 0 \quad . \tag{5}$$

Here, natural units with $\hbar = c = 1$ are used, and $iD_\mu \equiv i\partial_\mu - qA_\mu$. The two terms involving the effective coupling constants a_μ and b_μ violate CPT, while the three terms involving $H_{\mu\nu}$, $c_{\mu\nu}$, and $d_{\mu\nu}$ preserve CPT. All five terms break Lorentz invariance. Each particle sector in the standard model has its own set of parameters which we distinguish using superscripts. Since no Lorentz or CPT violation has been observed, these parameters are assumed to be small. A perturbative treatment in the context of relativistic quantum mechanics can then be used. In this approach, all of the perturbations in conventional quantum electrodynamics are identical for particles and antiparticles. However, the interaction hamiltonians including the effects of possible Lorentz and CPT breaking are not the same.

Penning-Trap Experiments

Comparisons of the g factors and charge-to-mass ratios of particles and antiparticles confined within a Penning trap have yielded the CPT bounds in Eqs. (2) and (3). These quantities are obtained through measurements of the anomaly frequency ω_a and the cyclotron frequency ω_c. For example, $g - 2 = 2\omega_a/\omega_c$. These frequencies can be measured to $\sim 10^{-9}$ thereby determining g to $\sim 10^{-12}$.

We have analyzed Penning-trap experiments with electrons and positrons and with protons and antiprotons [6]. We find that to leading order in the Lorentz-violating parameters there are corrections to the anomaly frequencies which are different for particles and antiparticles, while the cyclotron frequencies receive corrections that are the same for particles and antiparticles. Both frequencies have corrections which cause them to exhibit sidereal time variations. We also find that to leading order the g factor has no corrections, and therefore the figure of merit $r_g^e \simeq 0$, even though there is explicit CPT breaking. Because of this, we have proposed using as an alternative figure of merit the relative relativistic energy shifts caused by Lorentz and CPT violation. This is a definition that can be used in any experiment and is consistent with neutral meson experiments, which use mass ratios.

Based on these observations, we proposed looking for two signals of Lorentz and CPT violation: one an instantaneous difference in anomaly frequencies for electrons and positrons, and the other sidereal-time variations in the anomaly frequency of electrons alone. Dehmelt's group at the University of Washington has recently published the results of these observations [31]. In the first case, they reanalyzed existing data and obtained a figure of merit $r_{\omega_a}^e \lesssim 1.2 \times 10^{-21}$ from a bound on the difference in the electron and positron anomaly frequencies. In the second case, they analyzed more recent data for the electron alone and obtained a bound on sidereal time variations given by $r_{\omega_a^-,\text{diurnal}}^e \lesssim 1.6 \times 10^{-21}$. This corresponds to a bound on the combination of components $\tilde{b}^e_J \equiv b^e_J - md^e_{J0} - \frac{1}{2}\varepsilon_{JKL}H^e_{KL}$ defined with respect to a nonrotating coordinate system [7] given by $|\tilde{b}^e_J| \lesssim 5 \times 10^{-25}\,\text{GeV}$.

Although no $g-2$ experiments have been made for protons or antiprotons, there have been recent bounds obtained on Lorentz violation in comparisons of cyclotron frequencies of antiprotons and H^- ions confined in a Penning trap [15]. In this case the sensitivity is to the parameters $c^p_{\mu\nu}$, and the figure of merit $r^{H^-}_{\omega_c} \lesssim 10^{-25}$ was obtained.

Clock-Comparison Experiments

The classic Hughes-Drever type experiments are atomic clock-comparison tests of Lorentz invariance [13]. These experiments look for relative changes between two "clock" frequencies as the Earth rotates. The "clock" frequencies are typically atomic hyperfine or Zeeman transitions. Using the standard-model extension, Kostelecký and Lane [7] have made an extensive analysis of these experiments. They have obtained approximate bounds on various combinations of the Lorentz-violating parameters from the published results of these experiments. For example, from the experiment of Berglund et al. the following bounds for the parameters $\tilde{b}_J$ have been found for the proton, neutron, and electron.

proton	$\tilde{b}^p_J$	$\sim 10^{-27}$ GeV
neutron	$\tilde{b}^n_J$	$\sim 10^{-30}$ GeV
electron	$\tilde{b}^e_J$	$\sim 10^{-27}$ GeV

Since certain assumptions about the nuclear configurations must be made to extract numerical bounds, these bounds should be viewed as good to within one or two orders of magnitude. To obtain cleaner bounds it is necessary to consider simpler atoms.

Hydrogen-Antihydrogen Experiments

We have analyzed the proposed experiments at CERN which will make high-precision spectroscopic measurements of the 1S-2S transitions in hydrogen and antihydrogen [8]. We find that the magnetic field plays an important role in the sensitivity of the 1S-2S transition to Lorentz and CPT breaking. For example, in free hydrogen in the absence of a magnetic field, the 1S and 2S levels shift by the same amount at leading order in hydrogen and antihydrogen. As a result of this, there are no leading-order corrections to the 1S-2S transition frequency in free H or $\bar{\text{H}}$.

In a magnetic trap, however, there are magnetic fields which mix the spin states in the four hyperfine levels. Since the Lorentz-violating couplings are spin-dependent, some of the 1S and 2S levels acquire energy corrections that are not equal. The transitions between these levels have leading-order sensitivity to Lorentz and CPT violation. However, these transitions are also field-dependent, making them prone to broadening in an inhomogeneous magnetic field. To be sensitive to leading-order Lorentz and CPT violation in 1S-2S transitions, experiments will have to overcome the difficulties associated with possible line broadening effects due to field inhomogeneities.

We have also considered measurements of the ground-state Zeeman hyperfine transitions in hydrogen and antihydrogen [8]. We find that certain transitions in a hydrogen maser are sensitive to leading-order Lorentz-violating effects. These measurements have now been made by Walsworth's group at the Harvard-Smithsonian Center using a double-resonance technique [32]. They have obtained bounds on the Lorentz-violation parameters for the electron and proton. The bound for the proton alone is $|\tilde{b}_J^p| \lesssim 10^{-27}$ GeV. This is an extremely clean bound and is now the most stringent test of Lorentz and CPT symmetry for the proton.

Spin-Polarized Matter

Experiments at the University of Washington using a spin-polarized torsion pendulum [33] are able to achieve very high sensitivity to Lorentz violation due to the combined effect of a large number of aligned electron spins [9]. The experiment uses stacked toroidal magnets with a net electron spin $S \simeq 8 \times 10^{22}$, but which have a negligible magnetic field. The apparatus is suspended on a turntable and a time-varying harmonic signal is sought. Our analysis shows that in addition to a signal with the period of the rotating turntable, the effects of Lorentz and CPT violation would induce additional time variations with a sidereal period caused by Earth's rotation. The University of Washington group has analyzed their data and have obtained a bound on the electron parameters equal to $|\tilde{b}_J^e| \lesssim 1.4 \times 10^{-28}$ GeV [34]. This is now the best Lorentz and CPT bound for the electron.

Muon Experiments

Despite the spectacular precision of recent Lorentz and CPT tests for protons and electrons, it is important to keep in mind that particle sectors in the standard-model extension might be independent of each other and should separately be tested. The situation is analogous to CP tests where violation is observed only in the neutral meson sector and not in the lepton or baryon sectors. A thorough investigation of Lorentz and CPT symmetry must therefore probe as many possible particle sectors as possible. For this reason, we also examine muon experiments, which involve second-generation leptons. We find that there are several different types of experiments that are sensitive to Lorentz and CPT. We have examined both muonium experiments [35] and $g-2$ experiments with muons that are being conducted at Brookhaven [36].

Our results are that experiments measuring the frequencies of ground-state Zeeman hyperfine transitions in muonium in a strong magnetic field are sensitive to Lorentz and CPT violation. If bounds on sidereal time variations are obtained at the 100 Hz level, then the Lorentz-violation parameter for the muon $\tilde{b}^\mu_J$ can be bounded at the level of $|\tilde{b}^\mu_J| \leq 5 \times 10^{-22}$ GeV. We also find that in relativistic $g-2$ experiments using positive muons with "magic" boost parameter $\delta = 29.3$, bounds on Lorentz-violation parameters are possible at a level of 10^{-25} GeV. Experiments looking for sidereal time variations in the muon anomaly frequency would yield stringent new Lorentz and CPT bounds.

SUMMARY AND CONCLUSIONS

In summary, by using a general framework we are able to analyze Lorentz and CPT tests in a variety of atomic experiments. We find that experiments that have traditionally been considered Lorentz tests are also sensitive to CPT and vice versa. We find that it is also possible to make very precise tests of CPT in matter alone. Many of the bounds that have been obtained are well within the range of suppression factors associated with the Planck scale. The atomic experiments complement those in particle physics and together they are able to test the robustness of the standard model to increasing levels of precision.

ACKNOWLEDGMENTS

I would like to acknowledge my collaborators Alan Kostelecký, Charles Lane, and Neil Russell. This work was supported in part by the National Science Foundation under grant number PHY-9801869.

REFERENCES

1. See, for example, R.M. Barnett et al., Review of Particle Properties, Phys. Rev. D **54** (1996) 1.
2. V.A. Kostelecký, ed., *CPT and Lorentz Symmetry* (World Scientific, Singapore, 1999).
3. V.A. Kostelecký and S. Samuel, Phys. Rev. Lett. **63** (1989) 224; Phys. Rev. Lett. **66** (1991) 1811; Phys. Rev. D **39** (1989) 683; Phys. Rev. D **40** (1989) 1886.

4. V.A. Kostelecký and R. Potting, Nucl. Phys. B **359** (1991) 545; Phys. Lett. B **381** (1996) 389; V.A. Kostelecký, M. Perry, and R. Potting, Phys. Rev. Lett., in press, hep-th/991243.

5. D. Colladay and V.A. Kostelecký, Phys. Rev. D **55** (1997) 6760; Phys. Rev. D **58**, 116002 (1998).

6. R. Bluhm, V.A. Kostelecký and N. Russell, Phys. Rev. Lett. **79** (1997) 1432; Phys. Rev. D **57** (1998) 3932.

7. V.A. Kostelecký and C.D. Lane, Phys. Rev. D **60**, 116010 (1999).

8. R. Bluhm, V.A. Kostelecký and N. Russell, Phys. Rev. Lett. **82** (1999) 2254.

9. R. Bluhm and V.A. Kostelecký, Phys. Rev. Lett. **84** (2000) 1381.

10. R. Bluhm, V.A. Kostelecký and C.D. Lane, Phys. Rev. Lett. **84** (2000) 1098.

11. For reviews of CPT, see R.F. Streater and A.S. Wightman, *PCT, Spin, and Statistics and All That*, Benjamin Cummings, Reading, 1964; R. G. Sachs, *The Physics of Time Reversal*, University of Chicago, Chicago, 1987.

12. J. Schwinger, Phys. Rev. **82** (1951) 914; J.S. Bell, Birmingham University thesis (1954); Proc. Roy. Soc. (London) **A 231** (1955) 479; G. Lüders, Det. Kong. Danske Videnskabernes Selskab Mat.fysiske Meddelelser **28**, No. 5 (1954); Ann. Phys. (N.Y.) **2** (1957) 1; W. Pauli, in W. Pauli, ed., *Neils Bohr and the Development of Physics*, McGraw-Hill, New York, 1955, p. 30.

13. V.W. Hughes, H.G. Robinson, and V. Beltran-Lopez, Phys. Rev. Lett. **4** (1960) 342; R.W.P. Drever, Philos. Mag. **6** (1961) 683; J.D. Prestage *et al.*, Phys. Rev. Lett. **54** (1985) 2387; S.K. Lamoreaux *et al.*, Phys. Rev. A **39** (1989) 1082; T.E. Chupp *et al.*, Phys. Rev. Lett. **63** (1989) 1541; C.J. Berglund *et al.*, Phys. Rev. Lett. **75** (1995) 1879.

14. P.B. Schwinberg, R.S. Van Dyck, Jr., and H.G. Dehmelt, Phys. Lett. A **81** (1981) 119; R.S. Van Dyck, Jr., P.B. Schwinberg, and H.G. Dehmelt, Phys. Rev. D **34** (1986) 722; L.S. Brown and G. Gabrielse, Rev. Mod. Phys. **58** (1986) 233; R.S. Van Dyck, Jr., P.B. Schwinberg, and H.G. Dehmelt, Phys. Rev. Lett. **59** (1987) 26.

15. G. Gabrielse et al., Phys. Rev. Lett. **82** (1999) 3198.

16. B. Brown et al., Nucl. Phys. B (Proc. Suppl.) **56A** (1997) 326; M.H. Holzscheiter et al., Nucl. Phys. B (Proc. Suppl.) **56A** (1997) 336.

17. P. Carruthers, Phys. Lett. B **26** (1968) 158.

18. A.I. Oksak and I.T. Todorov, Commun. Math. Phys. **11** (1968) 125.

19. S. Hawking, Commun. Math. Phys. **87** (1982) 395.

20. H.B. Nielsen and I. Picek, Nucl. Phys. **B211** (1983) 269.

21. J. Ellis, N.E. Mavromatos, and D.V. Nanopoulos, Int. J. Mod. Phys. A **11** (1996) 146.

22. G. Cocconi and E. Salpeter, Nuovo Cimento **10** (1958) 646.

23. See for example, T.D. Lee and C.S. Wu, Annu. Rev. Nucl. Sci **16** (1966) 511.

24. A.P. Lightman and D.L. Lee, Phys. Rev. D **8** (1973) 364.

25. S.M. Carroll, G.B. Field, and R. Jackiw, Phys. Rev. D **41** (1990) 1231.

26. S. Coleman and S.L. Glashow, Phys. Rev. D **59**, 116008 (1999).

27. B. Schwingenheuer et. al., Phys. Rev. Lett. **74** (1995) 4376; L.K. Gibbons et al., Phys. Rev. D **55** (1997) 6625; R. Carosi et al., Phys. Lett. B **237** (1990) 303; OPAL Collaboration, R. Ackerstaff et al., Z. Phys. C **76** (1997) 401; DELPHI Collaboration, M. Feindt et al., preprint DELPHI 97-98 CONF 80 (July 1997).

28. V.A. Kostelecký and R. Potting, in D.B. Cline, ed., *Gamma Ray–Neutrino Cosmology and Planck Scale Physics* (World Scientific, Singapore, 1993) (hep-th/9211116); Phys. Rev. D **51** (1995) 3923; D. Colladay and V. A. Kostelecký, Phys. Lett. B **344** (1995) 259; Phys. Rev. D **52** (1995) 6224; V.A. Kostelecký and R. Van Kooten, Phys. Rev. D **54** (1996) 5585; V.A. Kostelecký, Phys. Rev. Lett. **80** (1998) 1818; Phys. Rev. D **61**, 016002 (2000).

29. R. Jackiw and V.A. Kostelecký, Phys. Rev. Lett. **82** (1999) 3572; M. Pérez-Victoria, Phys. Rev. Lett. **83** (1999) 2518; J.M. Chung, Phys. Lett. B **461** (1999) 138.

30. O. Bertolami et al., Phys. Lett. B **395**, 178 (1997).

31. R.K. Mittleman, I.I. Ioannou, H.G. Dehmelt, and N. Russell, Phys. Rev. Lett. **83** (1999) 2116; H.G. Dehmelt, R.K. Mittleman, R.S. Van Dyck, Jr., and P. Schwinberg, Phys. Rev. Lett. **83** (1999) 4694.

32. R. Walsworth et al, in preparation; see also this volume.

33. E.G. Adelberger et al., in P. Herczeg et al., eds., *Physics Beyond the Standard Model*, p. 717, World Scientific, Singapore, 1999; M.G. Harris, Ph.D. thesis, Univ. of Washington, 1998.

34. B. Heckel et al., private communication.

35. W. Liu et al., Phys. Rev. Lett. **82** (1999) 711.

36. R.M. Carey et al., Phys. Rev. Lett. **82** (1999) 1632.

New Clock Comparison Searches for Lorentz and CPT Violation

Ronald L. Walsworth, David Bear, Marc Humphrey,
Edward M. Mattison, David F. Phillips, Richard E. Stoner,
and Robert F. C. Vessot

Harvard-Smithsonian Center for Astrophysics
Cambridge, MA 02138, U.S.A.

Abstract. We present two new measurements constraining Lorentz and CPT violation using the ^{129}Xe/^{3}He Zeeman maser and atomic hydrogen masers. Experimental investigations of Lorentz and CPT symmetry provide important tests of the framework of the standard model of particle physics and theories of gravity. The two-species ^{129}Xe/^{3}He Zeeman maser bounds violations of CPT and Lorentz symmetry of the neutron at the 10^{-31} GeV level. Measurements with atomic hydrogen masers provide a clean limit of CPT and Lorentz symmetry violation of the proton at the 10^{-27} GeV level.

INTRODUCTION

Lorentz symmetry is a fundamental feature of modern descriptions of nature. Lorentz transformations include both spatial rotations and boosts. Therefore, experimental investigations of rotation symmetry provide important tests of the framework of the standard model of particle physics and single-metric theories of gravity [1].

In particular, the minimal SU(3)×SU(2)×U(1) standard model successfully describes particle phenomenology, but is believed to be the low energy limit of a more fundamental theory that incorporates gravity. While the fundamental theory should remain invariant under Lorentz transformations, spontaneous symmetry-breaking could result at the level of the standard model in violations of local Lorentz invariance (LLI) and CPT (symmetry under simultaneous application of Charge conjugation, Parity inversion, and Time reversal) [2].

Clock comparisons provide sensitive tests of rotation invariance and hence Lorentz symmetry by bounding the frequency variation of a given clock as its orientation changes, e.g., with respect to the fixed stars [3]. In practice, the most precise limits are obtained by comparing the frequencies of two co-located clocks

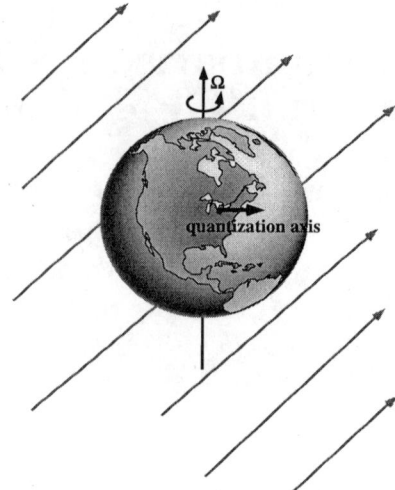

FIGURE 1. Bounds on LLI and CPT violation can be obtained by comparing the frequencies of clocks as they rotate with respect to the fixed stars. The standard model extension described in [3,9–17] admits Lorentz-violating couplings of noble gas nuclei and hydrogen atoms to expectation values of tensor fields. (Some of these couplings also violate CPT.) Each of the tensor fields may have an unknown magnitude and orientation in space, to be limited by experiment. The background arrows in this figure illustrate one such field.

as they rotate with the Earth (see Fig. 1). Atomic clocks are typically used, involving the electromagnetic signals emitted or absorbed on hyperfine or Zeeman transitions.

We report results from two new atomic clock tests of LLI and CPT:

(1) Using a two-species ^{129}Xe/^{3}He Zeeman maser [4–6] we placed a limit on CPT and LLI violation of the neutron of nearly 10^{-31} GeV, improving by more than a factor of six on the best previous measurement [7,8].

(2) We employed atomic hydrogen masers to set an improved clean limit on LLI/CPT violation of the proton, at the level of nearly 10^{-27} GeV.

MOTIVATION

Our atomic clock comparisons are motivated by a standard model extension developed by Kostelecký and others [3,9–17]. This theoretical framework accommodates possible spontaneous violation of local Lorentz invariance (LLI) and CPT symmetry, which may occur in a fundamental theory combining the standard model with gravity. For example, this might occur in string theory [18]. The standard model extension is quite general: it emerges as the low-energy limit of any underlying theory that generates the standard model and contains spontaneous Lorentz symmetry violation [19]. The extension retains the usual gauge structure and power-counting renormalizability of the standard model. It also has many other desirable properties, including energy-momentum conservation, observer Lorentz covariance, conventional quantization, and hermiticity. Microcausality and energy positivity are expected.

This well-motivated theoretical framework suggests that small, low-energy signals of LLI and CPT violation may be detectable in high-precision experiments. The dimensionless suppression factor for such effects would likely be the ratio of the low-energy scale to the Planck scale, perhaps combined with dimensionless coupling constants [3,9–19]. A key feature of the standard model extension of Kostelecký et al. is that it is at the level of the known elementary particles, and thus enables quantitative comparison of a wide array of tests of Lorentz symmetry. In recent work the standard model extension has been used to quantify bounds on LLI and CPT violation from measurements of neutral meson oscillations [9]; tests of QED in Penning traps [10]; photon birefringence in the vacuum [11,12]; baryogenesis [13]; hydrogen and antihydrogen spectroscopy [14]; experiments with muons [15]; a spin-polarized torsion pendulum [16]; observations with cosmic rays [17]; and atomic clock comparisons [3]. Recent experimental work motivated by this standard model extension includes Penning trap tests by Gabrielse et al. on the antiproton and H$^-$ [20], and by Dehmelt et al. on the electron and positron [21,22], which place improved limits on CPT and LLI violation in these systems. Also, a re-analysis by Adelberger, Gundlach, Heckel, and co-workers of existing data from the "Eöt-Wash II" spin-polarized torsion pendulum [23,24] sets the most stringent bound to date on CPT and LLI violation of the electron: approximately 10^{-28} GeV [25].

^{129}XE/^{3}HE MASER TEST OF CPT AND LORENTZ SYMMETRY

The design and operation of the two-species ^{129}Xe/^{3}He maser has been discussed in recent publications [4–6]. (See the schematic in Fig. 2.) Two dense, co-located ensembles of ^{3}He and ^{129}Xe atoms perform continuous and simultaneous maser oscillations on their respective nuclear spin 1/2 Zeeman transitions at approximately 4.9 kHz for ^{3}He and 1.7 kHz for ^{129}Xe in a static magnetic field of 1.5 gauss. This two-species maser operation can be maintained indefinitely. The population inversion for both maser ensembles is created by spin exchange collisions between the noble gas atoms and optically-pumped Rb vapor [26]. The ^{129}Xe/^{3}He maser has two chambers, one acting as the spin exchange "pump bulb" and the other serving as the "maser bulb". This two chamber configuration permits the combination of physical conditions necessary for a high flux of spin-polarized noble gas atoms into the maser bulb, while also maintaining ^{3}He and ^{129}Xe maser oscillations with good frequency stability: $\sim$ 100 nHz stability is typical for measurement intervals of $\sim$ 1

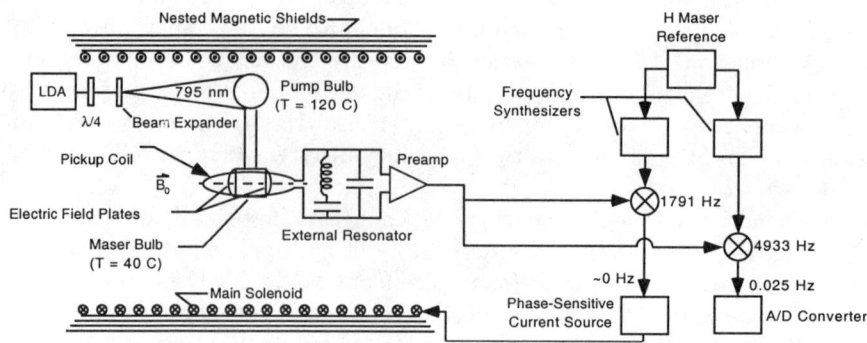

FIGURE 2. Schematic of the ^{129}Xe/^{3}He Zeêman maser

hour [6]. (A single-bulb ^{129}Xe/^{3}He maser does not provide good frequency stability because of the large Fermi contact shift of the ^{129}Xe Zeeman frequency caused by ^{129}Xe-Rb collisions [27].) Either of the noble gas species can serve as a precision magnetometer to stabilize the system's static magnetic field, while the other species is employed as a sensitive probe for LLI- and CPT-violating interactions or other subtle physical influences. (For example, we are also using the ^{129}Xe/^{3}He maser to search for a permanent electric dipole moment of ^{129}Xe as a test of time reversal symmetry; hence the electric field plates in Fig. 2.)

We search for a signature of Lorentz violation by monitoring the relative phases and Zeeman frequencies of the co-located ^{3}He and ^{129}Xe masers as the laboratory reference frame rotates with respect to the fixed stars. We operate the system with the quantization axis directed east-west on the Earth, the ^{3}He maser free-running, and the ^{129}Xe maser phase-locked to a signal derived from a hydrogen maser in order to stabilize the magnetic field. To leading order, the standard model extension of Kostelecký et al. predicts that the Lorentz-violating frequency shifts for the ^{3}He and ^{129}Xe maser are the same size and sign [3]. Hence the possible Lorentz-violating frequency shift in the free-running ^{3}He maser ($\delta \nu_{He}$) is given by:

$$\delta \nu_{He} = \delta \nu_{Lorentz} \left[\gamma_{He}/\gamma_{Xe} - 1 \right], \tag{1}$$

where $\delta \nu_{Lorentz}$ is the sidereal-day-period modulation induced in both noble gas Zeeman frequencies by the Lorentz-violating interaction, and $\gamma_{He}/\gamma_{Xe} \approx 2.75$ is the ratio of gyromagnetic ratios for ^{3}He and ^{129}Xe.

We acquired 90 days of data for this experiment over the period April, 1999 to April, 2000. We reversed the main magnetic field of the apparatus every ~ 4 days to help distinguish possible Lorentz-violating effects from diurnal systematic variations. In addition, we carefully assessed the effectiveness of the ^{129}Xe co-magnetometer, and found that it provides excellent isolation from possible diurnally-varying ambient magnetic fields, which would not average away with field reversals. Furthermore, the relative phase between the solar and sidereal

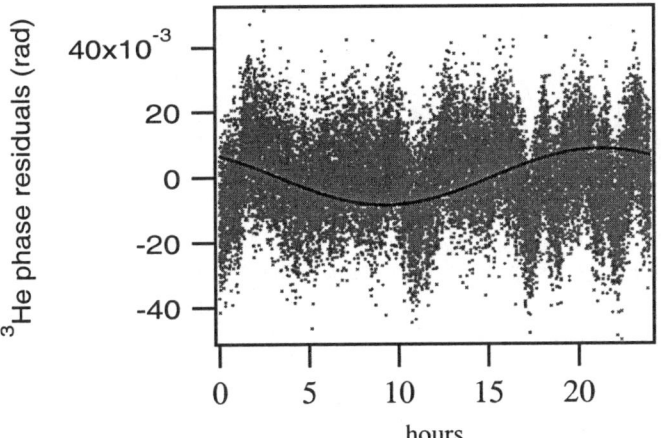

FIGURE 3. Typical data from the LLI/CPT test using the ^{129}Xe/^{3}He maser. ^{3}He maser phase data residuals are shown for one sidereal day. Larmor precession and drift terms have been removed, and the best-fit sinusoid curve (with sidereal-day-period) is displayed

day evolved about 2π radians over the course of the experiment; hence diurnal systematic effects from any source would be reduced by averaging the results from the measurement sets.

We analyzed each day's data and determined the amplitude and phase of a possible sidereal-day-period variation in the free-running ^{3}He maser frequency. (See Fig. 3 for an example of one day's data.) We employed a linear least squares method to fit the free-running maser phase vs. time using a minimal model including: a constant (phase offset); a linear term (Larmor precession); and cosine and sine terms with sidereal day period. For each day's data, we included terms corresponding to quadratic and maser amplitude-induced phase drift if they significantly improved the reduced χ^2 [28]. As a final check, we added a *faux* Lorentz-violating effect of known phase and amplitude to the raw data and performed the analysis as before. We considered our data reduction for a given sidereal day to be successful if the synthetic physics was recovered and there was no significant change in the covariance matrix generated by the fitting routine.

Using the 90 days of data, we found no statistically significant sidereal variation of the free-running ^{3}He maser frequency at the level of 90 nHz (two-sigma confidence). Kostelecký and Lane report that the nuclear Zeeman transitions of ^{129}Xe and ^{3}He are primarily sensitive to Lorentz-violating couplings of the neutron, assuming the correctness of the Schmidt model of the nuclei [3]. Thus our search for a sidereal-period frequency shift of the free-running ^{3}He maser ($\delta\nu_{He}$) provides a bound to the following parameters characterizing the magnitude of LLI/CPT violations in the standard model extension:

$$\left|-3.5\tilde{b}_J^n + 0.012\tilde{d}_J^n + 0.012\tilde{g}_{D,J}^n\right| \leq 2\pi\delta\nu_{He,J}(^{129}\text{Xe}/^3\text{He maser}) \qquad (2)$$

Here $J = X, Y$ denotes spatial indices in a non-rotating frame, with X and Y oriented in a plane perpendicular to the Earth's rotation axis and we have taken $\hbar = c = 1$. The parameters $\tilde{b}_J^n$, $\tilde{d}_J^n$, and $\tilde{g}_{D,J}^n$ describe the strength of Lorentz-violating couplings of the neutron to possible background tensor fields. $\tilde{b}_J^n$ and $\tilde{g}_{D,J}^n$ correspond to couplings that violate both CPT and LLI, while $\tilde{d}_J^n$ corresponds to a coupling that violates LLI but not CPT. All three of these parameters are different linear combinations of fundamental parameters in the underlying relativistic Lagrangian of the standard model extension [3,9–16].

It is clear from Eqn. (2) that the ^{129}Xe/^{3}He clock comparison is primarily sensitive to LLI/CPT violations associated with the neutron parameter $\tilde{b}_J^n$. Similarly, the most precise previous search for LLI/CPT violations of the neutron, the ^{199}Hg/^{133}Cs experiment of Lamoreaux, Hunter et al. [7,8], also had principal sensitivity to $\tilde{b}_J^n$ at the following level [3]:

$$\left|\frac{2}{3}\tilde{b}_J^n + \{\text{small terms}\}\right| \leq 2\pi\delta\nu_{Hg,J}(^{199}\text{Hg}/^{133}\text{Cs}). \qquad (3)$$

In this case, the experimental limit, $\delta\nu_{Hg,J}$, was a bound of 110 nHz (two-sigma confidence) on a sidereal-period variation of the ^{199}Hg nuclear Zeeman frequency, with the ^{133}Cs electronic Zeeman frequency serving as a co-magnetometer.

Therefore, in the context of the standard model extension of Kostelecký and co-workers [3], our ^{129}Xe/^{3}He maser measurement improves the constraint on $\tilde{b}_J^n$ to nearly 10^{-31} GeV, or more than six times better than the ^{199}Hg/^{133}Cs clock comparison [7,8]. Note that the ratio of this limit to the neutron mass ($10^{-31}\text{GeV}/m_n \sim 10^{-31}$) compares favorably to the dimensionless suppression factor $m_n/M_{Planck} \sim 10^{-19}$ that might be expected to govern spontaneous symmetry breaking of LLI and CPT originating at the Planck scale. We expect more than an order of magnitude improvement in sensitivity to LLI/CPT–violation of the neutron using a new device recently demonstrated in our laboratory: the ^{21}Ne/^{3}He Zeeman maser.

HYDROGEN MASER TEST OF CPT AND LORENTZ SYMMETRY

The hydrogen maser is an established tool in precision tests of fundamental physics [29]. Hydrogen masers operate on the $\Delta F = 1$, $\Delta m_F = 0$ hyperfine transition in the ground state of atomic hydrogen [30]. Hydrogen molecules are dissociated into atoms in an RF discharge, and the atoms are state selected via a hexapole magnet (Fig. 4). The high field seeking states, ($F = 1$, $m_F = +1, 0$) are focused into a Teflon coated cell which resides in a microwave cavity resonant with the $\Delta F = 1$ transition at 1420 MHz. The $F = 1$, $m_F = 0$ atoms are stimulated to

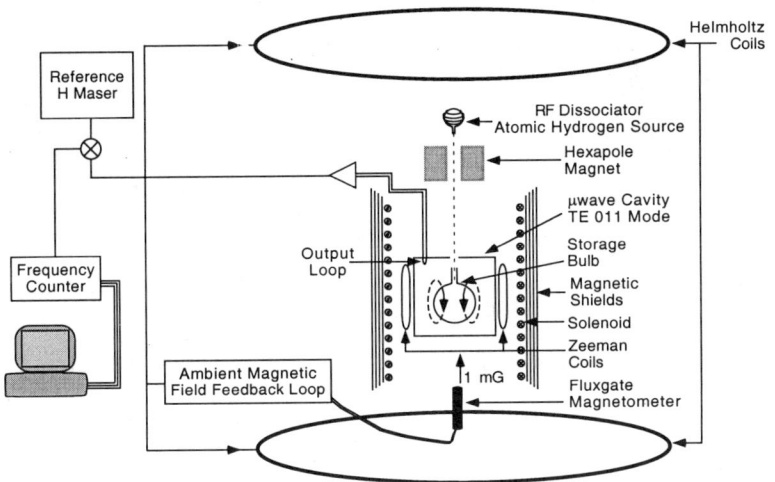

FIGURE 4. Schematic of the H maser in its ambient field stabilization loop.

make a transition to the $F = 0$ state by the field of the cavity. A static magnetic field of ~ 1 milligauss is applied to maintain the quantization axis of the H atoms.

The hydrogen transitions most sensitive to potential CPT and LLI violations are the $F = 1$, $\Delta m_F = \pm 1$ Zeeman transitions. In the 0.6 mG static field applied for these measurements, the Zeeman frequency is $\nu_Z \approx 850$ Hz. We utilize a double resonance technique to measure this frequency with a precision of ~ 1 mHz [31]. We apply a weak magnetic field perpendicular to the static field and oscillating at a frequency close to the Zeeman transition. This audio-frequency driving field couples the three sublevels of the $F = 1$ manifold of the H atoms. Provided a population difference exists between the $m_F = \pm 1$ states, the energy of the $m_F = 0$ state is altered by this coupling, thus shifting the measured maser frequency in a carefully analyzed manner [31] described by a dispersive shape (Fig. 5(a)). Importantly, the maser frequency is unchanged when the driving field is exactly equal to the Zeeman frequency. Therefore, we determine the Zeeman frequency by measuring the driving field frequency at which the maser frequency in the presence of the driving field is equal to the unperturbed maser frequency.

The $F = 1$, $\Delta m_F = \pm 1$ Zeeman frequency is directly proportional to the static magnetic field, in the small-field limit. Four layers of high permeability (μ-metal) magnetic shields surround the maser (Fig. 4), screening external field fluctuations by a factor of 32 000. Nevertheless, external magnetic field fluctuations cause remnant variations in the observed Zeeman frequency. As low frequency magnetic noise in the neighborhood of this experiment is much larger during the day than late at night, the measured Zeeman frequency could be preferentially shifted by this noise (at levels up to ~ 0.5 Hz) with a 24 hour periodicity which is difficult to distinguish

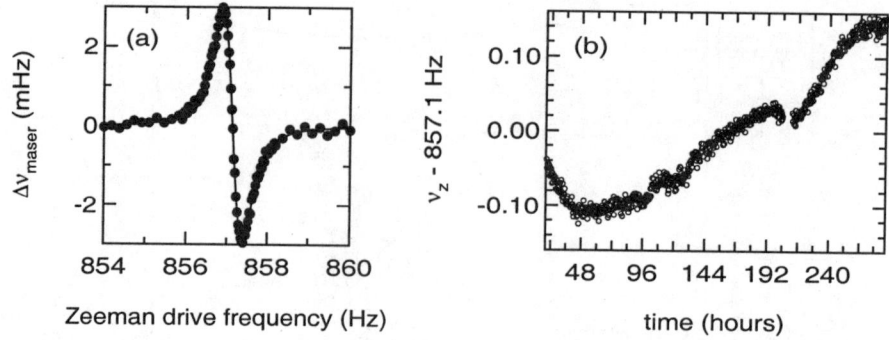

FIGURE 5. (a) An example of a double resonance measurement of the $F = 1$, $\Delta m_F = \pm 1$ Zeeman frequency in the hydrogen maser. The change from the unperturbed maser frequency is plotted versus the driving field frequency. (b) Zeeman frequency data from 11 days of the LLI/CPT test using the H maser.

from a true sidereal signal in our relatively short data sample. Therefore, we employ an active stabilization system to cancel such magnetic field fluctuations (Fig. 4). A fluxgate magnetometer placed as close to the maser cavity as possible controls large (2.4 m dia.) Helmholtz coils surrounding the maser via a feedback loop to maintain a constant ambient field. This feedback loop reduces the fluctuations at the sidereal frequency to below the equivalent of 1 μHz on the Zeeman frequency at the location of the magnetometer.

The Zeeman frequency of a hydrogen maser was measured for $\sim$ 31 days over the period Nov., 1999 to March, 2000. During data taking, the maser remained in a closed, temperature controlled room to reduce potential systematics from thermal drifts which might be expected to have 24 hour periodicities. The feedback system also maintained a constant ambient magnetic field. Each Zeeman measurement took approximately 20 minutes to acquire and was subsequently fit to extract a Zeeman frequency (Fig. 5(a)). Also monitored were maser amplitude, residual magnetic field fluctuation, ambient temperature, and current through the solenoidal coil which determines the Zeeman frequency (Fig. 4).

The data were then fit to extract the sidereal-period sinusoidal variation of the Zeeman frequency. (See Fig. 5(b) for an example of 11 days of data.) In addition to the sinusoid, piecewise linear terms (whose slopes were allowed to vary independently for each day) were used to model the slow remnant drift of the Zeeman frequency. No significant sidereal-day-period variation of the hydrogen $F = 1$, $\Delta m_F = \pm 1$ Zeeman frequency was observed. Our measurements set a bound on the magnitude of such a variation of $\delta \nu_Z^H \leq 0.3$ mHz. Expressed in terms of energy, this is a shift in the Zeeman splitting of less than $1 \cdot 10^{-27}$ GeV.

The hydrogen atom is directly sensitive to LLI and CPT violations of the proton and the electron. Following the notation of reference [14], one finds that a limit on a sidereal-day-period modulation of the Zeeman frequency ($\delta\nu_Z^H$) provides a bound to the following parameters characterizing the magnitude of LLI/CPT violations in the standard model extension of Kostelecký and co-workers:

$$|b_3^e + b_3^p - d_{30}^e m_e - d_{30}^p m_p - H_{12}^e - H_{12}^p| \leq 2\pi\delta\nu_Z^H \qquad (4)$$

for the low static magnetic fields at which we operate. (Again, we have taken $\hbar = c = 1$.) The terms b^e and b^p describe the strength of background tensor field couplings that violate CPT and LLI while the H and d terms describe couplings that violate LLI but not CPT [14]. The subscript 3 in Eqn. (4) indicates the direction along the quantization axis of the apparatus, which is vertical in the lab frame but rotates with respect to the fixed stars with the period of the sidereal day.

As in refs. [3,21], we can re-express the time varying change in the hydrogen Zeeman frequency in terms of parameters expressed in a non-rotating frame as

$$2\pi\delta\nu_{Z,J}^H = \left(\tilde{b}_J^p + \tilde{b}_J^e\right)\sin\chi. \qquad (5)$$

where $\tilde{b}_J^w = b_J^w - d_{j0}^w m_w - \frac{1}{2}\epsilon_{JKL}H_{KL}^w$, $J = X, Y$ refers to non-rotating spatial indices in the plane perpendicular to the rotation vector of the earth, w refers to either the proton or electron parameters, and $\chi = 42°$ is the latitude of the experiment.

As noted above, a re-analysis by Adelberger, Gundlach, Heckel, and co-workers of existing data from the "Eöt-Wash II" spin-polarized torsion pendulum [23,24] sets the most stringent bound to date on CPT and LLI violation of the electron: $\tilde{b}_J^e \leq 10^{-28}$ GeV [25]. Therefore, in the context of the standard model extension of Kostelecký and co-workers [14,3] the H maser measurement to date constrains LLI and CPT violations of the proton parameter $\tilde{b}_J^p \leq 2 \cdot 10^{-27}$ GeV at the one sigma level. This limit is comparable to that derived from the ^{199}Hg/^{133}Cs experiment of Lamoreaux, Hunter et al. [7,8] but in a much cleaner system (the hydrogen atom nucleus is a proton, compared to the complicated nuclei of ^{199}Hg and ^{133}Cs).

CONCLUSIONS

Precision comparisons of atomic clocks provide sensitive tests of Lorentz and CPT symmetries, thereby probing extensions to the standard model [3,9–17] in which these symmetries can be spontaneously broken. Measurements using the two-species ^{129}Xe/^{3}He Zeeman maser constrain violations of CPT and Lorentz symmetry of the neutron at the 10^{-31} GeV level. Measurements with atomic hydrogen masers provide clean tests of CPT and Lorentz symmetry violation of the proton at the 10^{-27} GeV level. Improvements in both experiments are being pursued.

ACKNOWLEDGMENTS

We gratefully acknowledge the encouragement and active support of these projects by Alan Kostelecký, and technical assistance by Marc Rosenberry and

Timothy Chupp. Development of the ^{129}Xe/^{3}He Zeeman maser was supported by a NIST Precision Measurement Grant. Support for the Lorentz violation tests was provided by NASA grant NAG8-1434, ONR grant N00014-99-1-0501, and the Smithsonian Institution Scholarly Studies Program.

REFERENCES

1. Will, C.M., *Theory and Experiment in Gravitational Physics*, Cambridge University Press, New York, 1981.
2. The discrete symmetries C, P, and T are discussed, for example, in Sachs, R.G., *The Physics of Time Reversal*, University of Chicago, Chicago, 1987.
3. Kostelecký, V.A., and Lane, C.D., *Phys. Rev. D* **60**, 116010/1-17 (1999).
4. Chupp, T.E., Hoare, R.J., Walsworth, R.L., and Wu, B., *Phys. Rev. Lett.* **72**, 2363-2366 (1994).
5. Stoner, R.E., Rosenberry, M.A., Wright, J.T., Chupp, T.E., Oteiza, E.R., and Walsworth, R.L., *Phys. Rev. Lett.* **77**, 3971-3974 (1996).
6. Bear, D., Chupp, T.E., Cooper, K., DeDeo, S., Rosenberry, M.A., Stoner, R.E., and Walsworth, R.L., *Phys. Rev. A* **57**, 5006-5008 (1998).
7. Berglund, C.J., Hunter, L.R., Krause, Jr., D., Prigge, E.O., Ronfeldt M.S., and Lamoreaux, S.K., *Phys. Rev. Lett.* **75**, 1879-1882 (1995).
8. Hunter, L.R., Berglund, C.J., Ronfeldt, M.S., Prigge, E.O., Krause, D., and Lamoreaux, S.K., "A Test of Local Lorentz Invariance Using Hg and Cs Magnetometers," in *CPT and Lorentz Symmetry*, edited by Kostelecký, V.A., World Scientific, Singapore, 1999, pp. 180-186.
9. Kostelecký, V.A., and Potting, R., "CPT, Strings, and the $K - \bar{K}$ System," in *Gamma Ray–Neutrino Cosmology and Planck Scale Physics*, edited by Cline, D.B., World Scientific, Singapore, 1993, hep-th/9211116; *Phys. Rev. D* **51**, 3923-3935 (1995); Colladay D., and Kostelecký, V.A., *Phys. Lett. B* **344**, 259-265 (1995); *Phys. Rev. D* **52**, 6224-6230 (1995); Kostelecký, V.A., and Van Kooten, R., *Phys. Rev. D* **54**, 5585 (1996); Kostelecký, V.A., *Phys. Rev. Lett.* **80**, 1818-1821 (1998); *Phys. Rev. D* **61**, 16002/1-9 (2000).
10. Bluhm, R., Kostelecký, V.A., and Russell, N., *Phys. Rev. Lett.* **79**, 1432-1435 (1997); *Phys. Rev. D* **57**, 3932-3943 (1998).
11. Carroll, S.M., Field, G.B., and Jackiw, R., *Phys. Rev. D* **41**, 1231-1240 (1990).
12. Colladay, D., and Kostelecký, V.A., *Phys. Rev. D* **55**, 6760-6774 (1997); *ibid.* **58**, 116002/1-23 (1998); Jackiw, R., and Kostelecký, V.A., *Phys. Rev. Lett.* **82**, 3572-3575 (1999);
13. Bertolami, O., Colladay, D., Kostelecký, V.A., and Potting, R., *Phys. Lett. B* **395**, 178-183 (1997).
14. Bluhm, R., Kostelecký, V.A., and Russell, N., *Phys. Rev. Lett.* **82**, 2254-2257 (1999).
15. Bluhm, R., Kostelecký, V.A., and Lane, C.D., *Phys. Rev. Lett.* **84**, 1098-1101 (2000).
16. Bluhm, R. and Kostelecký, V.A., *Phys. Rev. Lett.* **84**, 1381-1384 (2000).
17. Coleman, S., and Glashow, S.L., *Phys. Rev. D* **56**, 116008/1-14 (1999).
18. Kostelecký, V.A. and Samuel, S., *Phys. Rev. D* **39**, 683-685 (1989); *ibid.* **40**, 1886-

1903 (1989); Kostelecký, V.A. and Potting, R., *Nucl. Phys. B* **359**, 545-570 (1991); *Phys. Lett. B* **381**, 89-96 (1996).
19. Kostelecký, V.A., and Samuel, S., *Phys. Rev. Lett.* **63**, 224-227 (1989); *ibid.* **66**, 1811-1814 (1991).
20. Gabrielse, G., Khabbaz, A., Hall, D.S., Heimenn, C., Kalinowski, H., and Jhe, W., *Phys. Rev. Lett.* **82**, 3198-3201 (1999).
21. Mittleman, R.K., Ioannou, I.I., Dehmelt, H.G., and Russell, N., *Phys. Rev. Lett.* **83**, 2116-2119 (1999).
22. Dehmelt, H., Mittleman, R., Van Dyck, Jr., R.S., and Schwinberg, P., *Phys. Rev. Lett.* **83**, 4694-4696 (1999).
23. Adelberger, E.G., *et al.*, in *Physics Beyond the Standard Model,* edited by P. Herczeg *et al.,* World Scientific, Singapore, 1999, p. 717.
24. Harris, M.G., Ph.D. thesis, Univ. of Washington, 1998.
25. Heckel, B., presented at International Conference on Orbis Scientiae 1999: Quantum Gravity, Generalized Theory of Gravitation and Superstring Theory Based Unification (28th Conference on High Energy Physics and Cosmology Since 1964), Fort Launderdale, Florida, 16-19 Dec., 1999.
26. Walker, T.G., and Happer, W., *Rev. Mod. Phys.* **69**, 629-642 (1997).
27. Romalis, M.V., and Cates, G.D., *Phys. Rev. A* **58**, 3004-3011 (1998); Newbury, N.R., Barton, A.S., Bogorad, P., Cates, G.D., Mabuchi, H., and Saam, B., *Phys. Rev. A* **48**, 558-568 (1993); Schafer, S.R., Cates, G.D., Chien, T.R., Gonatas, D., Happer, W., and Walker, T.G., *Phys. Rev. A* **39**, 5613-5623 (1989).
28. We employed the F-test at the 99% confidence level to decide whether the addition of a new term to the fit model was justified. See Philip R. Bevington, *Data Reduction and Error Analysis for the Physical Sciences, Second Ed.,* McGraw-Hill, Boston, 1992, ch. 11.
29. Vessot, R.F.C., *et al.*, *Phys. Rev. Lett.* **45**, 2081-2084 (1980); Turneaure, R.J.P., Will, C.M., Farrell, B.F., Mattison, E.M., and Vessot, R.F.C., *Phys. Rev. D* **27**, 1705-1714 (1983); Walsworth, R.L., Silvera, I.F., Mattison, E.M., and Vessot, R.F.C., *Phys. Rev. Lett.* **64**, 2599-2602 (1990).
30. Kleppner, D., Goldenberg, H.M., and Ramsey, N.F., *Phys. Rev.* **126**, 603-615 (1962); Kleppner, D., Berg, H.C., Crampton, S.B., Ramsey, N.F., Vessot, R.F.C., Peters, H.E., and Vanier, J., *Phys. Rev.* **138**, A972-983 (1965).
31. Andresen, H.G., *Z. Physik,* **210**, 113-141 (1968).

Charges, Parity Doublets and Parity-Non-Conservation

Bertrand Desplanques*

*Institut des Sciences Nucléaires (UMR CNRS/IN2P3-UJF)
F-38026 Grenoble Cedex, France*

Abstract. Charges and Parity-Non-Conservation are two ingredients that, at present, are introduced by hand in the description of particles and their interactions. We elaborate on a suggestion which was made in the past by Landau and tends to relate them. Particles and their antiparticles are associated here to the parity partners appearing in a model with chiral and parity symmetry realized in the Wigner-Weyl mode, i.e. with parity doublets or, equivalently, states of opposite chiral charges. Parity-non-conservation in such a model is ascribed to the non-commutativity of the parity and the (chiral) charge operators, preventing one to have eigenstates of both at the same time.

INTRODUCTION

Discovered more than four decades ago [1], parity-non-conservation in physics [2] is still a mystery and its origin is unclear. At present, it is put by hand in the description of the standard model, which is based on the $SU(3)_c \otimes SU(2)_L \otimes U(1)_Y$ gauge group with spontaneous symmetry breaking. The explanation most currently evoked in the literature assumes it is a low energy property resulting from some spontaneous symmetry breaking. The complete gauge group would also involve $SU(2)_R$, the corresponding gauge bosons having a mass larger than the left-handed ones. At energies large enough in comparison of the gauge boson masses, parity conservation would be restored.

A different, much less known proposal is that one made by Landau soon after the discovery of parity-non-conservation [3]. He was considering that the difference between particles and antiparticles, as far as the space symmetry is concerned, is not much greater than that due to chemical stereo-isomerism. In other words, the charge conjugation, C, is nothing but a parity operation acting on an internal spatio-temporal structure that has a chiral character. In such a picture, parity-non-conservation occurs because charged particles, intrinsically, are not eigenstates of parity.

In the present work, we want to elaborate on this proposal, taking into account that parity-non-conservation is more or less systematic and affects most particles. After a few remarks relative to parity-non-conservation, we will present a general scheme that can account for it while preserving the parity symmetry at the level of the interaction. A tentative model is then considered in order to evidence the problems that a more specific description could encounter.

A FEW PRELIMINARY REMARKS

It is often considered that nature is preferentially left-handed, which has led to introduce the gauge group $SU(2)_L$ in its description. If one was describing the theory by using the charged conjugate fields, the gauge group would contain $SU(2)_R$ instead of $SU(2)_L$, in full correspondance with the fact that processes with antiparticles are rather right-handed. There is therefore some relativity in the definition of the handedness and, if one considers altogether particles and antiparticles, there is no special preference for one of them. To some extent, parity symmetry is present under our eyes but this is no more than the consequence of the PC symmetry that we assume to be conserved in the following.

The second remark concerns the mathematical definition of the parity operation. Strictly speaking, it only makes sense for interactions that remain invariant under it, such as the strong and electromagnetic ones. As noticed by Lee and Wick [4], if P is a parity operation, the product of P with any transformation which leaves these interactions invariant is also a parity operation. One cannot determine which of these parity operations is the physical one. This freedom can thus be taken to consider the product $\mathcal{P} = PC$ as a parity operation, since C leaves invariant the strong and electromagnetic interactions. This offers the further advantage that it is also a parity operation for weak interactions, as far as one can consider that the PC symmetry holds to a good accuracy.

While the above two remarks support the fact that the product $\mathcal{P} = PC$ could be a genuine parity operation, one has to understand why this parity operation factorizes into two parts that are separately conserved by strong or electromagnetic interactions. In the present approach, where we investigate the possibility that charges have a chiral character, this factorization most probably supposes that the dynamics relative to the internal chiral structure of a particle decouples from its external properties such as the spin and the momentum in particular. In this limit, it is quite conceivable that the charge conjugation be a parity operation restricted to this internal structure. At some point however, there should be some coupling and the parity asymmetry implied by the existence in the particle core of a chiral structure should show up in the processes where this particle enters.

GENERAL SCHEME

Particles and antiparticles have the same mass and most often carry quantum numbers such as "charges" that are conserved (electric charge, lepton and baryon numbers) or approximately conserved (strangeness,...). These two features, together with the possibility that these charges could have a chiral character, immediately suggest that particles and antiparticles may be the parity partners with opposite chiral charges that appear in a model with chiral symmetry and parity conservation. Such a model exhibits a conserved axial current, $J_\mu^A(x)$, while the associated axial charge, $Q^A = \int d\vec{x}\, J_0^A(x)$, anticommutes with the genuine parity operator, $\mathcal{P}$:

$$[\mathcal{P}, Q^A]_+ = \mathcal{P}\, Q^A + Q^A\, \mathcal{P} = 0. \tag{1}$$

This last equation can be used to show that the existence of a state with some parity and $Q^A |\text{state}> \neq 0$ leads to the existence of a degenerate state with the opposite parity. The partners of this parity doublet can also be combined so that to get degenerate solutions with opposite chiral charges which transform into each other in a parity operation, what can be shown directly from the above relation.

The above Wigner-Weyl realization of a model with chiral symmetry is generally discarded in favor of the Nambu-Goldstone realization. Indeed, by examining the spectra of particles, no candidate for parity doublets has been found. While looking for them, it was however assumed that they should have the same charges. The existence of parity doublets as combination of a particle and its antiparticle does not seem to have received particular attention. Notice that the present chiral symmetry has no relation to that one considered in QCD, which is well known to be realized in the Nambu-Goldstone mode. The existence of this chiral symmetry nevertheless makes less unlikely the possibility that another chiral symmetry may play a role in fundamental interactions.

On the other hand, Eq. (1) has an important consequence concerning the very origin of parity-non-conservation, while parity is conserved at the level of the interaction. Indeed, as $\mathcal{P}$ and Q^A do not commute, it is impossible to have at the same time eigenstates of both. By selecting in an experiment particles with a (chiral) charge automatically implies that they are not eigenstates of parity. The observed non-conservation of the parity symmetry could simply be a consequence of this important feature.

It is likely that the Landau's proposal to explain parity-non-conservation was only indicative. By relying on some underlying chiral symmetry, this proposal takes a more systematic character, as implied by observation and accounted for by the standard model.

While the above developments have a general character, they are far to provide a realistic description of the physical world. Many questions are open. Some are immediate. They obviously concern the physical nature of the chiral structure that underlies charges and their quantification. Another non-trivial question concerns the conservation of the axial current associated to the chiral symmetry under consideration here, knowing that particles have a non-zero mass. There are less immediate questions. The above general model for states with opposite chiral charges may just be realized by states with opposite helicity of a zero mass particle (like for a Majorana neutrino). To make this model of some interest, particles should acquire some mass, raising the problem of its origin. A further question concerns the equation that plays here the role of the Dirac equation. This last one is largely at the origin of the concept of antiparticles in physics, which have been historically interpretated as holes in a Fermi see filled up by negative energy states. This dissymmetry is absent in the present picture where particles and antiparticles are related by a unique parity operation, $\mathcal{P}$ (=PC).

To get insight about answers and evidence possible problems, we consider a particular model that is inspired from the Nambu-Jona-Lasinio model [5], but with an essential difference as to the role of spin and helicity.

A PARTICULAR MODEL

The simplest Lagrangian density one can think of for our purpose, with chiral symmetry and parity conservation, may be written as:

$$\mathcal{L}(x) = \frac{1}{2}\left(i\bar{\psi}(x)\gamma^\mu\partial_\mu\psi(x) - \frac{G}{2}\bar{\psi}(x)\gamma^\mu\gamma_5\psi(x)\ \bar{\psi}(x)\gamma_\mu\gamma_5\psi(x)\right), \qquad (2)$$

where the front factor $\frac{1}{2}$ here accounts for the fact we are dealing with a self-conjugate (Majorana) field.

Contrary to models with Dirac fields, there is in this model no conserved vector current of any interest and therefore no scalar charge (the fields in the current $\bar{\psi}(x)\gamma_\mu\psi(x)$ can be interchanged with a change in sign so that the operator part vanishes). The only conserved current and associated charge of interest stem from the chiral symmetry and have a behavior under parity opposite to the usual vector current and its associated charge. They are respectively given by:

$$J_\mu^A(x) = \frac{1}{2}\bar{\psi}(x)\gamma_\mu\gamma_5\psi(x), \qquad Q^A = \int d\vec{x}\, J_0^A(x). \qquad (3)$$

The last quantity, which has the character of a chiral charge, measures the helicity number carried by the particles that the field ψ describes.

The above Lagrangian, Eq. (2), has no mass term. To generate some, we assume that the vacuum expectation value of the current, $<|J_\mu^A|>$, takes a non-vanishing value, $<|J_\mu^A|> \propto m\,\epsilon_\mu$, with $\epsilon^2 = -1$, excluding in any case a uniform ϵ_μ. We will not extend on the vacuum structure which could be made of "ferromagnetic-type" domains, the main point here being that this assumption makes possible the existence of solutions for a "free particle" with a non-zero mass. After inserting the vacuum expectation of $<|J_\mu^A|>$ in (2), the following free Lagrangian is obtained:

$$\mathcal{L}(x) = \frac{1}{2}\left(i\bar{\psi}(x)\gamma^\mu\partial_\mu\psi(x) - m\,\bar{\psi}(x)\gamma.\epsilon\,\gamma_5\psi(x)\right). \qquad (4)$$

This Lagrangian has a structure different from that one leading to the Dirac equation. It is invariant under the chiral transformation, $\psi \to e^{i\alpha\gamma_5}\psi$, in accordance with the idea to associate charges to conserved chiral charges. Moreover, it does not contain the usual mass term $m\bar{\psi}(x)\psi(x)$. Nethertheless, it leads to solutions with a non-zero mass. This can be checked from the momentum-space equation derived from Eq. (4):

$$(\gamma.p - m\,\gamma.\epsilon\gamma_5)\,\psi_\epsilon(p) = 0. \tag{5}$$

Under the condition $\epsilon.p = 0$, two different solutions with $p^2 = m^2$ are obtained:

$$\psi_\epsilon(p) \propto (1+\gamma_5)(1-\gamma.\epsilon)\,u(p), \qquad \psi_\epsilon(p) \propto (1-\gamma_5)(1-\gamma.\epsilon)\,v(p), \tag{6}$$

where $u(p)$ and $v(p)$ represent the Dirac spinors describing particles and antiparticles to which the solutions are respectively associated. Their appearance is important as it makes it possible to recover the successful achievements obtained using the Dirac equation.

The front factors $(1+\gamma_5)$ and $(1-\gamma_5)$ in Eq. (6) are known to characterize weak interactions but here result from the underlying chiral nature assumed for the charges. The standard model of electroweak interactions would then represent the effective theory accounting for this intrinsic structure. The change in sign in front of γ_5 is not a problem since, as mentioned in preliminary remarks, it is expected to occur when the interaction is written in terms of the antiparticle rather than the particle fields.

Solutions given by Eq. (6) can be seen as wave functions describing a particle in an intrinsically deformed state. Dealing with this breaking of rotational invariance generally requires long and difficult theoretical developments. Solutions with correct spin quantum numbers can be built using rotation matrices. A particular application where the effect of their approximate character is probably minimized concerns the matrix element of the axial current between one-particle states with the same momentum. This is a rather limited result but it is of the utmost importance as it shows how the axial current is going to be conserved in the present approach. For a particle and an antiparticle, this matrix element respectively reads:

$$<S=\frac{1}{2},\,p\,|J^A_\mu(0)|\,S=\frac{1}{2},\,p> \;=\; \bar{u}(p)\gamma_\mu u(p) \quad \text{and} \quad -\bar{v}(p)\gamma_\mu v(p). \tag{7}$$

As seen from the right-hand side of the equation, the matrix elements of the axial current have turned into the matrix elements of a vector current. There is therefore no problem with the current conservation. As to the change in the parity, it is noticed that the effective current on the r.h.s. of Eq. (7) is odd under the charge conjugation. This operation being a parity operation in the present context, there is no contradiction in the apparent change of parity. Solutions used to get Eq. (7) assumed a decoupling between the intrinsic structure and the total angular momentum of the particle. Beyond this decoupling hypothesis, some contribution to an effective axial current, with a structure quite similar to that produced by an anapole moment, is expected. This is one mechanism that could lead to the occurence of the standard parity-non-conservation in the present approach.

CONCLUSION

Often, when trying to solve some problem, one has better to solve another one at the same time. Here, we considered that the origin of parity-non-conservation in physics and the very nature of charges may be deeply connected. In some sense, parity-non-conservation would teach us about the substructure underlying charges.

We presented a general scheme based on a model with chiral and parity symmetry. The realization of the model in the Wigner-Weyl mode provides a natural framework for accommodating particles and their antiparticles, which are associated to the parity partners then expected. In this scheme, where parity is conserved at the level of the interaction, parity-non-conservation occurs because particles carrying a chiral charge automatically involve a superposition of components with opposite parities. Thus, the problem here is not to explain the origin of parity-non-conservation, but rather to understand why parity symmetry turns out to be conserved by strong and electromagnetic interactions, allowing one to define the charge conjugation operation, C, as a parity operation acting on an internal chiral structure. Most probably, the answer in the present context should rely on the dynamics. An interesting consequence is that the charge conjugation is not a fundamental operation, which offers the advantage to divide by two the number of degrees of freedom required to construct a fundamental theory.

For some part, the above scheme may be simply realized by massless, Majorana-type particles. While looking for a non-trivial realization, with the help of the simplest model one can think of, we found suggestion for a mass generation mechanism different from the Higgs one. It supposes some spontaneous breaking of the rotational symmetry. We also found suggestion for how the axial current conservation could be fulfilled. The V-A structure of weak currents is recovered but this is a direct consequence of what the chiral structure underlying charges, built in in the model we considered, has been assumed to be. Many questions are however raised. They obviously concern the phenomenology of the standard model, which has to be reproduced. More fundamentally, they concern the mass generation mechanism and the associated spontaneous symmetry breaking. In principle, this can be dealt with by including RPA type correlations, though a consistent treatment is not an easy one. On the other hand, there are severe constraints on combining space-time and internal symmetries [6]. It is not sure that one can escape them.

REFERENCES

1. Wu C. S. et al., *Phys. Rev.* **105**, 1403 (1957).
2. Lee T.D. and Yang C.N., *Phys. Rev.* **104**, 254 (1956).
3. Landau L., *Nucl. Phys.* **3**, 127 (1957).
4. Lee T.D. and Wick G.C., *Phys. Rev.* **148**, 1385 (1966).
5. Nambu Y. and Jona-Lasinio G., *Phys. Rev.* **122**, 345 (1961).
6. Coleman S. and Mandula J., *Phys. Rev.* **159**, 1251 (1967).

Enhancement of Parity and Time Invariance Violation in Heavy Atoms

Vladimir A. Dzuba, Victor V. Flambaum, Jacinda S. M. Ginges

School of Physics, University of New South Wales, Sydney 2052, Australia

Abstract. Parity (P) and time (T) invariance violating effects are enhanced in atoms with close states of opposite parity, large nuclear charge Z, and collective P,T-odd nuclear moments. We have performed calculations of the atomic electric dipole moment (EDM) induced in radium by the electron EDM and the nuclear magnetic quadrupole and Schiff moments. We have also calculated the effects of parity nonconservation in radium produced by the nuclear anapole moment and the nuclear weak charge. Our results show that the values of parity and time invariance violating effects in radium are much larger than those considered so far in other atoms.

INTRODUCTION

Atoms are very good objects to study when searching for parity and time invariance violating effects, as these effects can be strongly enhanced in heavy atoms. The enhancement occurs due to several mechanisms. The three main contributing mechanisms to the enhancement in atoms are close electronic levels of opposite parity, large nuclear charge, and deformation of the nucleus.

If an atom has energy levels, E_1 and E_2, of opposite parity then the P- and T-odd effects associated with these levels are proportional to $1/(E_1 - E_2)$. So in the case when the energy interval $E_1 - E_2$ is very small the effect is strongly enhanced.

The matrix elements of P- and T-odd interactions increase with Z faster than $Z^2 R(Z)$, where Z is the nuclear charge and $R(Z)$ is a relativistic factor which also increases with Z. A large enhancement of P,T-odd effects can therefore occur in heavy atoms.

It has been demonstrated that deformed nuclei lead to enhanced nuclear moments compared to those associated with spherical nuclei [1–3]. A quadrupole deformed nucleus has a strongly enhanced magnetic quadrupole moment (MQM) [1]. The MQM is a P- and T-odd moment which can exist only if there are P- and T-odd interactions between nuclear particles. The parameter of enhancement is $A^{2/3}$, where A is the nuclear mass number. So effects in heavy atoms with nuclear quadrupole deformation can be enhanced by a factor ~ 10. An octupole deformed nucleus leads to strong enhancement of the nuclear Schiff moment (SM) [2,3], which is also P- and T-odd.

In the work of Flambaum [4] it was shown that radium is an excellent candidate for the study of P- and T-odd effects, and estimates for some of these effects were made. In the current work we make more accurate calculations of these and other effects. Below we present our results for the EDM induced in radium, ^{223}Ra and ^{225}Ra, by the electron EDM and the nuclear Schiff moment and MQM. We also present our results for the parity nonconserving electric dipole transition induced by the nuclear weak charge and that induced by the nuclear anapole moment.

P- AND T-ODD EFFECTS IN RADIUM

The radium atom satisfies all criteria listed above which lead to strong enhancement of P,T-odd effects: the states $7s7d\ ^3D_2$ and $7s7p\ ^3P_1$ with the respective energies $E = 13993.97$ cm^{-1} and $E = 13999.38$ cm^{-1} are separated by about 5 cm$^{-1} \sim 10^{-3}$ eV (it should also be noted that the states $7s7d\ ^3D_1$ and $7s7p\ ^3P_1$ are also quite close, $\Delta E \sim 283$ cm^{-1}) ; radium has a large nuclear charge ($Z = 88$); there is theoretical and experimental evidence that the odd isotopes of radium, e.g. ^{223}Ra with nuclear spin $I = 3/2$ and ^{225}Ra with $I = 1/2$, have nuclear octupole deformation, leading to enhanced Schiff moments [3]; and the isotope ^{223}Ra has nuclear quadrupole deformation which leads to an enhanced magnetic quadrupole moment (the MQM is zero for $I < 1$).

Method of Calculation

We used a relatively simple *ab initio* approximation which is a reasonable compromise between the simplicity of the calculations and the accuracy of the results. It is based on the relativistic Hartree-Fock (RHF) and configuration interaction (CI) methods. A minimum number of basis states were used at the CI stage of the calculations. However, important many-body effects, such as polarization of the atomic core by an external field and correlations between core and valence electrons, were included in the calculations of the single-electron matrix elements. These effects are important because core-valence correlations increase the electron density on the nucleus, strongly affecting the value of P,T-violating matrix elements, while core polarization makes $p-d$ single-electron matrix elements large due to the effective renormalization of the interaction of the external electron with the nucleus by the Coulomb field. To control the accuracy of the calculations we also calculated hyperfine structure intervals and lifetimes of the lower states of radium and its lighter analog barium. The accuracy for the energy is about 10% while the accuracy for the hfs and lifetimes is 10 - 30%.

Our calculations confirm the estimates done in the previous work [4] and show that the values of most P- and T-odd effects in radium are much higher than in other atoms considered before.

Atomic Electric Dipole Moment

The interaction of an electron EDM, d_e, with an atomic field mixes states with the same total electron momentum J and opposite parity. As a result, an atomic EDM, d_z, arises. The atomic EDM induced in radium by the electron EDM is strongly enhanced in the $7s7d\ ^3D_1$ state due to the admixture of the close opposite parity state $^3P_1$. In an approximation where only this admixture to the contribution of the EDM is considered, the result is

$$d_z = 5370 d_e. \tag{1}$$

This enhancement of the electron EDM is many times larger than corresponding values for the ground states of francium (910) and gold (260) [5].

The interaction of the atomic electrons with the nuclear Schiff moment produces an atomic EDM. In radium, the EDM arising due to the Schiff moment is strongly enhanced in the $^3D_2$ state. The results for the EDM of radium induced by the Schiff moment, in an approximation where only the admixture with the close opposite parity state $^3P_1$ is considered, are presented in Table 1. The results are presented in terms of the dimensionless constant of the P-,T-odd nucleon-nucleon interaction, η.

TABLE 1. EDM of the radium atom in the $^3D_2$ state induced by the nuclear Schiff moment

I	F[a]	d_z (a.u.)		d_z (e cm)
0.5	1.5	$-0.94 \times 10^8 S$	$-0.19 \times 10^{-11}\eta$[b]	$-0.36 \times 10^{-19}\eta$
1.5	0.5	$-0.16 \times 10^8 S$	$-0.42 \times 10^{-11}\eta$[c]	$-0.80 \times 10^{-19}\eta$
1.5	1.5	$-0.30 \times 10^9 S$	$-0.81 \times 10^{-11}\eta$[c]	$-0.15 \times 10^{-18}\eta$
1.5	2.5	$-0.28 \times 10^9 S$	$-0.76 \times 10^{-11}\eta$[c]	$-0.14 \times 10^{-18}\eta$

[a] F is the total atomic angular momentum, $\mathbf{F} = \mathbf{J} + \mathbf{I}$
[b] Nuclear Schiff moment S is assumed to be $S = 400 \times 10^8 \eta\ e\ \text{fm}^3$ [3]
[c] $S = 300 \times 10^8 \eta\ e\ \text{fm}^3$ [3]

An atomic EDM is also induced by the interaction of atomic electrons with the nuclear magnetic quadrupole moment. As with the Schiff moment, the EDM of radium induced by the MQM is strongly enhanced in the state $^3D_2$ by admixture with the state $^3P_1$. It should be noted that the MQM of isotopes where the nuclear spin $I < 1$ (like ^{225}Ra where $I = 1/2$) is zero. Results for the leading contribution to the radium EDM in the state $^3D_2$ induced by the MQM are presented in Table 2.

The results for the EDM induced in radium due to the Schiff and magnetic quadrupole moments are of the same order of magnitude and are about 10^5 times larger than the EDM of the mercury atom, which currently gives the best upper limit on η [6].

The EDMs of some other heavy atoms have also been estimated. The results are presented in Table 3.

TABLE 2. EDM of the ^{223}Ra isotope ($I = 3/2$) in the $^3D_2$ state induced by the nuclear magnetic quadrupole moment

F	d_z [a]	d_z [b]
0.5	$1344 M m_e$	$7.4 \times 10^{-20} \eta$ e cm
1.5	$1292 M m_e$	$7.0 \times 10^{-20} \eta$ e cm
2.5	$-806 M m_e$	$-4.4 \times 10^{-20} \eta$ e cm

[a] In terms of nuclear magnetic quadrupole moment M
[b] M is assumed to be $M = 10^{-19}(\eta/m_p)$ e cm, [1] where m_p is the proton mass

Parity Nonconservation

We found that parity nonconserving (PNC) effects produced by the nuclear weak charge (nuclear spin-independent) and those produced by the nuclear anapole moment (nuclear spin-dependent) are also strongly enhanced.

The interaction of an electron with the nuclear weak charge Q_W mixes states of the same total momentum J and opposite parity. So electric dipole transitions between states of initially equal parity become possible. In particular, the transition between the ground state $^1S_0$ and the excited $^3D_1$ state of radium is enhanced due to the closeness of the opposite parity state $^3P_1$. Taking into account only this dominating contribution to the PNC $E1$ transition, the results are:

$$^{225}\text{Ra}: E1_{PNC} = 0.77 \times 10^{-9}(Q_W/N)iea_0, \qquad (2)$$

$$^{223}\text{Ra}: E1_{PNC} = 0.76 \times 10^{-9}(Q_W/N)iea_0. \qquad (3)$$

The result for radium is about 100 times larger than the measured PNC amplitude in cesium [9], about 5 times larger than the corresponding amplitude in francium [10], and has the same order of magnitude as the PNC amplitude in ytterbium [11].

Similar to the spin-independent PNC interaction, the electron interaction with the nuclear anapole moment mixes states of opposite parity and leads to non-zero $E1$ transition amplitudes between states of initially equal parity. However, states with $\Delta J = 1$ can also be mixed and the interaction is dependent on the nuclear spin. The results for the dominating contributions to the transitions $^1S_0 - ^3D_1$ and $^1S_0 - ^3D_2$, due to admixture with $7s7p$, are presented in Table 4. The spin-dependent PNC transition amplitude $^1S_0 - ^3D_2$ is more than 1000 times larger than a similar amplitude in cesium [9].

TABLE 3. EDMs of some heavy atoms (in units $10^{-25}\eta$ e cm)

^{199}Hg	5.6[a]	^{129}Xe	0.47[b]	^{223}Rn	2000[c]
^{221}Fr	240[c]	^{223}Fr	2800[c]	^{225}Ac	~1000[c]

[a] Calculated in Ref. [7]
[b] Calculated in Ref. [8]
[c] Calculated in Ref. [3]

TABLE 4. PNC E1 transition amplitude between different hyperfine structure components F and F' induced by the nuclear anapole moment; κ_a is a dimensionless constant proportional to the strength of the PNC nucleon-nucleon interaction

I	F	F'	$E1_{PNC}$ in units $10^{-10}\kappa_a i e a_0$	
			$^1S_0 - {}^3D_1$	$^1S_0 - {}^3D_2$
0.5	0.5	1.5	2.05	-20.3
1.5	1.5	0.5	-0.58	5.7
	1.5	1.5	-1.4	13.8
	1.5	2.5	1.3	-12.9

CONCLUSION

The radium atom turns out to be a very promising candidate for the study of parity and time invariance violating effects, as these effects are hugely enhanced.

The contribution of different mechanisms to the P- and T-odd effects can be studied separately if measurements are performed for different states and different isotopes of the radium atom. For example, the atomic EDM induced by the electron EDM is strongly enhanced in the $^3D_1$ state, while contributions of the nuclear Schiff and magnetic quadrupole moments are strongly enhanced in the $^3D_2$ state. Also, the magnetic quadrupole moment is zero for isotopes with nuclear spin $I = 1/2$, like ^{225}Ra, while the Schiff moment for these isotopes is not zero.

The contribution of the anapole moment to the measured PNC amplitude can be determined by comparing the amplitudes between different hyperfine structure components similar to what was done for cesium [9]. It would be more efficient to measure the effect of the anapole moment in the $^1S_0 - {}^3D_2$ transition than the $^1S_0 - {}^3D_1$ transition as it is about ten times larger and because the nuclear spin-independent PNC interaction does not contribute to this amplitude at all due to the large change of the total electron angular momentum $\Delta J = 2$.

Calculations of parity and time invariance violating effects in radium reveal the importance of relativistic and many-body effects. The accuracy achieved in the present work is probably 20-30 %. However, a further improvement in accuracy is possible if such a need arises from the progress in measurements.

REFERENCES

1. Flambaum, V.V., *Phys. Lett.* **320B**, 211-215 (1994).
2. Auerbach, N., Flambaum, V.V., and Spevak, V., *Phys. Rev. Lett.* **76**, 4316-4319 (1996); Flambaum, V.V., Murray, D.W., and Orton, S.R., *Phys. Rev. C* **56**, 2820-2829 (1997).
3. Spevak, V., Auerbach, N., and Flambaum, V.V., *Phys. Rev. C* **56**, 1357-1369 (1997).
4. Flambaum, V.V., *Phys. Rev. A* **60**, R2611-R2613 (1999).
5. Byrnes, T.M.R., Dzuba, V.A., Flambaum, V.V., and Murray, D.W., *Phys. Rev. A* **59**, 3082-3083 (1999).

6. Jacobs, J.P., Klipstein, W.M., Lamoreaux, S.K., Heckel, B.R., and Fortson, E.N., *Phys. Rev. A* **52**, 3521-3540 (1995).
7. Flambaum, V.V., Khriplovich, I.B., and Sushkov, O.P., *Nucl. Phys. A* **449**, 750-760 (1986).
8. Dzuba, V.A., Flambaum, V.V., and Silvestrov, P.G., *Phys. Lett.* **154B**, 93-95 (1985).
9. Wood, C.S., Bennett, S.C., Cho, D., Masterson, B.P., Roberts, J.L., Tanner, C.E., and Wieman, C.E., *Science* **275**, 1759-1763 (1997).
10. Dzuba, V.A., Flambaum, V.V., and Sushkov, O.P., *Phys. Rev. A* **51**, 3454-3461 (1995).
11. DeMille, D., *Phys. Rev. Lett.* **74**, 4165-4168 (1995).

Symmetry Motivated Estimation of Nucleon Delta-excitation

S.I.Sukhoruchkin

Petersburg Nuclear Physics Institute, 188300, Gatchina, Russia

Abstract. Long-range correlations in nuclear excitations and binding energies are compared with empirical relations in particle mass spectrum.

I TUNING EFFECT IN NUCLEAR DATA

It was suggested by S.Devons [1] that fine nuclear effects originated from nucleon structure could be noticed and it would be difficult to observe them in high energy physics. One might expect that such effects as a tuning of nuclear excitations (E^*) and differences of nuclear binding energies (ΔE_B) with common parameters would be quantitatively connected with quark models. Parameters of tuning effect should be observed in sum distributions of E^*, spacings ($D=E_i^*-E_j^*$) or ΔE_B.

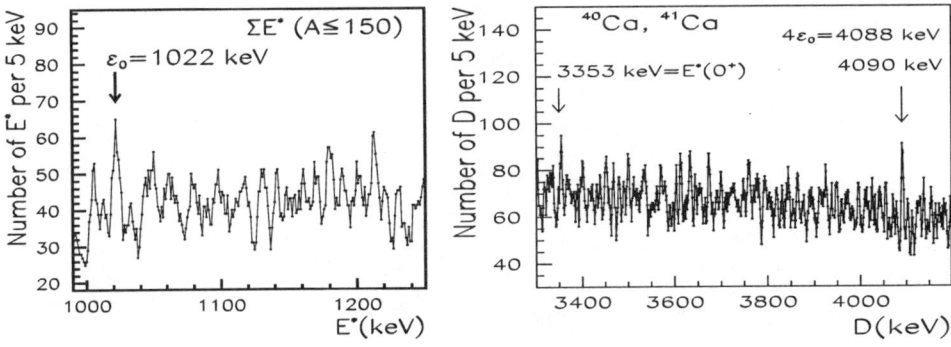

FIGURE 1. *Left*: Distribution of E^* in nuclei with $A \leq 150$; *Right*: Sum of spacing distributions in levels of ^{40}Ca and ^{41}Ca (data from ENSDF, neutron and transfer reactions.

Positions of maxima at $E^*=1022$ keV$=\varepsilon_o$ and at $E^*=1293$ keV in $\sum E^*$-distribution of all nuclei with $A \leq 150$ (Fig.1a [2]) and $A<70$ (Fig.2a [3]) coincide with electromagnetic mass differences of lepton 1022 keV$=2m_e$ and nucleon $m_n-m_p=1293.3$ keV [4]. Stable intervals $D_o=1293$ keV in level spacing were observed in ^{32}S,^{33}S [3] and in some other nuclei. The grouping at $E^*=\varepsilon_o$ is connected with the stable character of 0^+ excitations in some near-magic nuclei [5]. The maximum at 4090 keV in sum distribution of spacings in the magic nuclei ^{40}Ca and ^{41}Ca (seen at $D=4\varepsilon_o$ in Fig.1, right) could as well be connected with 0^+ excitations: one of such intervals is between the low-lying 0^+ states in ^{40}Ca and the maximum at

D=3355 keV includes 0_1^+ excitation of ^{40}Ca. Stable interval $D=3\frac{1}{2}\varepsilon_o$ in light nuclei [6] corresponds to 0_1^+ excitation in near-magic ^{18}Ne; its 0_2^+ excitation 4580(9) keV is close to $4\frac{1}{2}\varepsilon_o=9m_e=4599$ keV. The same period of $9m_e=\Delta$ was observed in distributions of ΔE_B of nuclei differing by $\Delta Z=2, \Delta N=4$ [7,8].

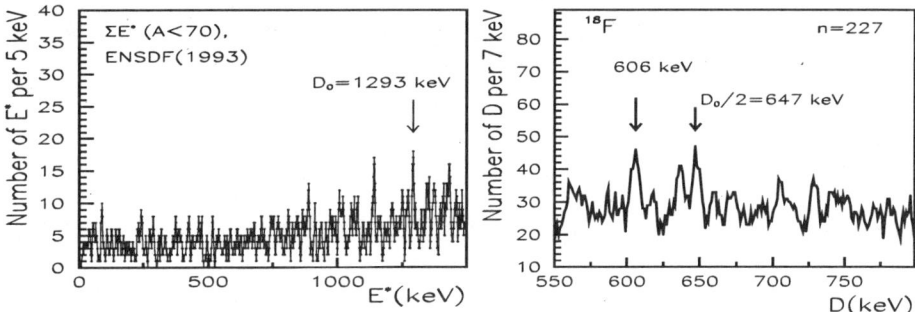

FIGURE 2. *Left*: Distribution of E^* in nuclei with A<70. *Right*: Spacing distribution in levels of ^{18}F, data from ENSDF and from resonances in reaction with α-particles.

The presence of two stable intervals (one with D=647 keV=1/2 D_o) in near-magic nucleus ^{18}F is shown in Fig.2, right. Stable character of the intervals multiple to 161 keV=$D_o/8$ (324 keV, 484 keV, 1940 keV etc.) was discussed elsewhere [5,9].

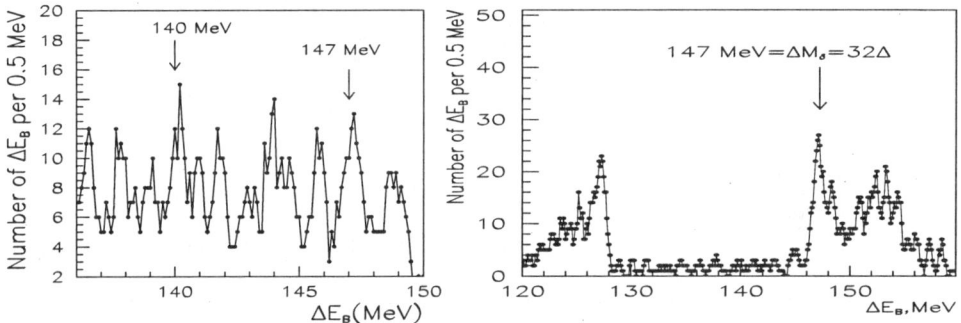

FIGURE 3. *Left*: Distribution of differences in binding energies (ΔE_B) of nuclei with $\Delta Z=\Delta N=8$ [7]. *Right*: Distribution of ΔE_B of nuclei with $\Delta Z=8, \Delta N=14$ [12].

The tuning effect in E_B [10] could be seen first of all as linear trends in the dependence of separation energies of valence nucleons on the nucleon numbers [11]. Derived from these trends parameters of valence nucleon residual interaction $\varepsilon_{N,2N}=340$ keV=1/3 ε_o were found to be nearly constant in a broad range of nuclei. Found in [2,8,12] discreteness in two- and four-proton separation energies has a period ε_o. The tuning effect in E_B consisted also in long-range correlations with parameters 140 MeV (close to m_π) and 147 MeV=$32\Delta=\Delta M_\Delta$ (close to the half of Δ-excitation of nucleon 294 MeV, both are marked by arrows in Fig.3). Integer ratio 1/27x32 between parameters 1/2 $\varepsilon_o=m_e$ and $3\Delta M_\Delta=3\times 147$ MeV=441 MeV is close to the radiative correction of QED $\alpha/2\pi=1.16\times 10^{-3}$. The stable nuclear interval 161 keV=$D_o/8$ forms the same ratio ($\alpha/2\pi$) with $m_{\pi\pm}=140$ MeV.

II TUNING EFFECT IN PARTICLE MASSES

The state of the Standard Model (which includes QCD as the base of hadronic physics) has been recently reviewed by Y.Nambu [13]. He named the mass problem as the first signal for the need of a further development of Standard Model (SM) while a search for empirical correlations in data is suggested as a first stage of such development. Appearance of nuclear parameters $\varepsilon_o = 2m_e$ and D_o allows to mention empirically introduced by several authors [14-16] relations in particle mass spectrum which contain multiple values of m_e. Mass splitting of pions $\delta m_\pi = m_{\pi^\pm} - m_\pi = 4593.6$ keV [4] is noticed to be close to $\Delta = 4599$ keV $= 9m_e$ [16] and deviates from $9m_e$ by a small value $1.17(11) \times 10^{-3}$ close to $\alpha/2\pi = 1.16 \times 10^{-3}$ [17-20].

TABLE 1. Comparison of particle masses with integers of $3m_e$ [14] and $16m_e$.

Part.	m_i, MeV	$m_i/3m_e$	$n \times 16m_e$	n	Difference	Comments
	1		2		1-2	(MeV)
μ	105.6584*	68.923*	106.2878	13	-0.6294	-0.511 -0.118
π^o	134.9764*	88.046*	138.9917	17	-4.0153	
$\pi^\pm$	139.5699*	91.043*		17	+0.5782	+0.511 +0.065
η^o	547.30(12)	357.11	547.7910	67	-0.49(12)	
ω	781.94(12)	510.08	784.8946	96	-2.96(12)	$6 \times 0.511 = 3.066$
η'	957.78(14)	624.76	956.5902	117	+1.19(14)	
φ	1019.413(8)	664.98	1021.998	125	-2.585(8)	$2D_o = 2.586$ MeV
p	938.2723*	612.051*	940.2383	115	-1.96596	-0.511 -1.455 = $\frac{9}{8}D_o$
n	939.5656*	612.894*		115	-0.67264	-0.511 -0.162 = $\frac{1}{8}D_o$
Δ^o	1233.3(6)	804.50	1234.573	151	-1.3(6)	
Σ^o	1192.64(2)	777.920	1193.694	146	-1.05(2)	(-0.511×2) [6,19]
Ξ^o	1314.9(6)	857.732	1316.333	161	-1.40(60)	(-0.511×3) [6,19]
Ξ^-	1321.3(1)	861.92	1324.510	162	-3.19(13)	$6 \times 0.511 = 3.066$

The empirical observation by R.Frosch [14] that majority of particle masses are correlated with period $3m_e$ was based on the analysis of recent mass data for 47 particles (nearly the same as in [4]). In Table 1 some well established particle masses are compared with periods of $3m_e$ [14] and $16m_e$ [15,16,19]. Masses of five particles noticed in [16] as being close to integers of m_e are marked in Table 1 by asterisks. Parameter of discreteness in particle masses $\delta = 16m_e = 2\Delta - 2m_e$ was introduced from the proximity of ΔM_Δ, muon and pion masses to the integers of $16m_e$ (n=18,17,13). Stable intervals in masses of isoscalar mesons: $m_{\eta'} - m_\eta = m_\eta - m_\pi = 409$ MeV (n=50) and $m_\phi - m_\omega = m_\omega - m_\eta = 234$ MeV (n=29) were discussed in [19,20]. Light quarks (forming nucleons and mesons) and the electron in the SM-framework are included in the same family of fermions. Relation 9:1 between δ_π and m_e as well as the ratio between masses of leptons ($L = m_\mu/m_e = 206.77$) become closer to the integers 9 and $9 \times 23 = (13 \times 16) - 1 = 207$ if radiative correction factor $(1 - \alpha/2\pi)$ would be applied to m_e (such part of m_e corresponds to the polarization effect in QED [21]).

TABLE 2. Baryon states in SU(6)$_{spin,flavor}$ representation. Comparison of baryon singlet masses [4,19] with integers of $m_{\pi^\pm}$=139.6 MeV.

J^P,D,L_N^P	Octet members(Mass,MeV [1])			Singlets	m×$m_{\pi^\pm}$	m
1/2$^+$,56, 0$_0^+$	N(939)	Λ(1116)*	Σ(1193)		1117	8
1/2$^+$,56, 0$_2^+$	N(1440)	Λ(1600)	Σ(1660)			
1/2$^-$,70, 1$_1^-$	N(1535)	Λ(1670)*	Σ(1620)	Λ(1405)*	1675, 1396	
3/2$^-$,70, 1$_1^-$	N(1520)	Λ(1690)*	Σ(1670)	Λ(1520)*	1675, 1535	
1/2$^-$,70, 1$_1^-$	N(1650)	Λ(1800)*	Σ(1750)		1814	
3/2$^-$,70, 1$_1^-$	N(1700)					
5/2$^-$,70, 1$_1^-$	N(1675)	Λ(1830)*	Σ(1775)		1814	
3/2$^+$,56, 0$_0^+$	Δ(1232)	Σ(1385)	Ξ(1530)	Ω(1672)*	1535, 1675	11, 12

It should be mentioned that current quark masses m_q are very uncertain and could not be used to check the periods $\Delta=9m_e$ and $\delta=16m_e$ (m_d=9.3 MeV, m_u=5.1 MeV and $m_d - m_u$=4.1 MeV [22,23] are close to 2Δ, $\Delta+m_e$ and $8m_e$). Deviation of neutron mass from n × $16m_e$-m_e is 162(2) keV (close to $D_o/8$).

Noticed in [24] proximity of the mass interval between nucleon and muon to $6m_{\pi^\pm}$ and of Λ-hyperon mass to $8m_{\pi^\pm}$ were later confirmed by the computer analysis [25] which showed that masses of some other baryon singlets are also close to m × $m_{\pi^\pm}$ (marked by asterisk in Table 2). The above mentioned interval 409 MeV (50×$16m_e$) between masses of pseudoscalar mesons is close to $m_{\pi^\pm}+2m_{\pi^o}$.

Shown in Table 2 integer relations in pion and strange baryon masses (m=1,8,11-12) are in accord with mentioned in review [26] proximity of the mean mass splitting in baryon decuplet (139 MeV) to m_π. However two mass splittings in baryon decuplet are close to ΔM_Δ= 147 MeV (n=18) [19]. In Table 3 the comparison of masses of several other nucleon excitations (of N-baryons [4,6]) with integer number of ΔM_Δ are given. The proximity to ΔM_Δ of some mass splittings was explained in Nonrelativistic Quark Models (NRQM) [19]. However the presence of rational relations in masses with periods m_π and ΔM_Δ (Tables 2,3) is the empirical fact.

TABLE 3. Symmetry motivated estimation of Δ-excitation (SYMMED) [4,6].

$J^\pi(N^*)$, m_{Ξ^-}	3/2$^+$	3/2$^-$	1/2$^-$	1/2$^-$	5/2$^-$	7/2$^-$	9/2$^-$	11/2$^-$	m_{Ξ^-}
M_i	1234	1524	1534	1659	1676	2127	2268	2577	1321
M_i-M_N	294	584	594	719	737	1187	1328	1637	–
M_N+m'x147	m_Δ	1529	1529	1676	1676	2117	2264	2559	1323
difference	-	-5	5	-17	0	10	4	18	9ΔM_Δ

We have used for estimations of these common parameters the masses of baryons which are usually described in the framework of SU(6) spin-flavor symmetry. Hence we call the relations in which two well-known parameters are involved as SYMMED (SYMmetry Motivated Estimation of nucleon Δ-excitation and pion mass).

We include into SYMMED also observations by R.Sternheimer [25] and P.Kropotkin [27] that interval in particle masses of 440 MeV appears several times (for example, as mass-difference between Σ-hyperon $(3/2^+)$, nucleon and kaon 1380-939=441 MeV, 939-498=441 MeV) and its value is close to $1/3\ m_{\Xi^-}=1321/3=440$ MeV (see in Table 3 a comparison of m_{Ξ^-} with $9\times\Delta M_\Delta$).

III STANDARD MODEL AND QUARK MODEL PARAMETERS

Recent quark model with Goldstone Boson Exchange GBE [28] (very successfully reproducing baryon masses) contains the estimation of the initial nucleon mass $M_N^{GBE} = 3M_q$ of about 1360 MeV (which leads to M_q=450 MeV close to $3\Delta M_\Delta$=441 MeV and to some other estimations [6,29]). The mass m_{Ξ^-} is close to M_N^{GBE} due to cancellation of the mass-decrease from quark interaction and the mass-increase from exchange of d-quarks by s-quarks. Parameter ΔM_Δ in NRQM is equal to $3R_{qqq}$, where R_{qqq} is a matrix element of one-gluon exchange [30].

In Table 1 masses of ω-meson and Ξ^--baryon are boxed: these masses with shifts (about $6m_e$) were used in estimations of the constituent quark mass (M_q) from assumptions $M_M=2M_q$ and $M_B=3M_q$. Noticed by P.Kropotkin and G.Wick stable intervals in particle masses 440 MeV=$m_{\Xi^-}/3$ and 391 MeV=$m_\omega/2$ could be considered as estimations of baryon's M_q and meson's M_q'' constituent quark masses (different due to the annihilation processes in M_M [29]).

Integer relations in SYMMED (numbers m=8,11-12 and m'=9 for hyperon masses boxed in Table 2-3) are the result of: 1) the above mentioned cancellation of mass intervals due to the spin-dependent quark interaction and the mass difference between strange and non-strange quarks and 2) the proximity of non-strange constituent quark masses to the three-fold values of common parameters ΔM_Δ and $m_{\pi^\pm}$ (n=18 and 17 in units of $16m_e$).

The masses of Ξ^--hyperon and ω-meson are well represented by the period $3m_e$ (Table 1). Estimations of $M_q=m_{\Xi^-}/3=3\times 18\delta-2m_e$ (and $M_q''=m_\omega/2=3\times 16\delta-3m_e$ [19]) could be combined with the above discussed ratio of $m_e/3\Delta M_\Delta=\alpha/2\pi$) and with the ratio between the accurately known masses of muon and vector boson m_μ/M_Z= 105.658 keV/91187(7) MeV= $1.1587(1)\times 10^{-3}$ which is close to QED radiative correction $\alpha/2\pi - 0.33\alpha^2/\pi^2 = 1.15965\times 10^{-3}$ [2,4,6,19,31]. The proximity of α/π to the parameter of CP-nonconservation in kaon decay was mentioned in [32]. Due to the above mentioned observation that lepton ratio L=m_μ/m_e becomes close to integer (after the small correction of m_e) one can obtain the lepton ratio as a ratio between the boson mass and parameter M_q derived from the Ξ-hyperon mass (still close to $3\Delta M_\Delta=3\times 18\times 16m_e$). The ratio 91187 MeV/440.4 MeV=207.05 is close to integer. The discussed relations might be considered as an indication of the specific properties of SM-condensate essential in the mass formation [33]). Tuning effects in masses and nuclear data show the relation between the SM-parameters (m_e, m_μ, M_Z) and the hadronic parameters $\Delta, m_\pi, \Delta M_\Delta$.

In connection with the integer ratios (n=13,17,18 between the muon mass, m_π and ΔM_Δ) we could mention the well-known cases [21] when other SM-parameters, namely $\alpha^{-1}=137.04$ (or even $\alpha_o^{-1}=128.90(9)$ for short distances [34]) are close to integers. The proximity of the parameter α^{-1} to the ratio $(m_\pi + m_e)/2m_e=137.06$ was noticed in [19,24]. The tuning effects in nuclear data and in particle masses with parameters close to the electromagnetic mass splittings of lepton and nucleon could indicate certain additional aspects of the mass problem mentioned by Y.Nambu.

REFERENCES

1. Devons S., *Proc. Rutherford Jubilee Int.Conf.,Ed.J.Birks, Heywood, 1961, p.611.*
2. Sukhoruchkin, S.I., *Proc. Int. Conf. Exp. Phys. Europe, Sevilla,1999*, AIP 495,p.482.
3. Sukhoruchkin, S.I., *Proc. Int. Symp. CGR-10, Santa Fe, 1999*, AIP, in press.
4. Particle Data Group, *Eur. Phys. J.* C **3**, 1 (1998).
5. Soroko Z.N., *Proc. Int. Conf. Quark Nucl. Phys.* Adelaide, 2000 (to be published).
6. Sukhoruchkin S.I., *Phys. At. Nucl.* **61**, 1825 (1998) and references therein.
7. Sukhoruchkin S.I., *J. Phys.G: Nucl. Part. Phys.* **25**, 921 (1999).
8. S.I.Sukhoruchkin, *Phys. At. Nucl.* **53**, 1266 (1991).
9. S.I.Sukhoruchkin, *ISINN-IV*, Dubna, 1996. JINR publ. E3-96-336, p.389 (1996).
10. G.Audi, A.H.Wapstra, *Nucl. Phys.* A, **595**, 409 (1995).
11. S.I.Sukhoruchkin, *Proc.9-th CGR*, Budapest, 1996. Springer, v.1, 358 (1997).
12. Sukhoruchkin, S.I., Sukhoruchkin, D.S., *ENAM98*, AIP Conf. Proc. 455, p.134.
13. Y.Nambu, *Nucl. Phys.* A **629**, 3c (1998).
14. Frosch, R., *Nuovo Cim.* **A104**, 913 (1991) and references therein.
15. A.Hautot, *Atti Fond. G.Ronchi*, **30**, No.4, p. 477 (1975).
16. S.I.Sukhoruchkin, *Stat. Prop. Nuclei*, Ed. J.B.Garg, Plenum Press, p.215, (1972).
17. S.I.Sukhoruchkin, *Progr. Thes. 40-th Meet. Nucl. Spectr.*, Leningrad, (1990), p.147.
18. G.Mac-Gregor, *Nuovo Cim.* A **58**, 159 (1980); **103**, 983 (1990).
19. S.I.Sukhoruchkin, *Symmetry Meth. Phys.*, Dubna,1993. JINR E2-94-347, pp.528,536.
20. S.I.Sukhoruchkin, Symmetry Methods Phys., Dubna, 1995. JINR E2-96-224, p.549.
21. R.Feynman, *QED-Strange Theory of Light and Matter*, Princ. Univ. Press, (1985).
22. H.Leutwyler, *The ratios of the light quark masses*, hep-ph/9602366, (1996).
23. R.F.Lebed, *Phys. Rev.* D **47**, 1134 (1993).
24. Y.Nambu, *Progr. Theor. Phys.*, **7**, 595 (1952).
25. R.Sternheimer, *Phys. Rev.*,**136**, 1364 (1964); **170**, 1267 (1968).
26. N.P.Samios et al., Rev. Mod. Phys., **46**, 49 (1974).
27. P.N.Kropotkin, *Field and Matter*, Moscow State Univ., p.106 (1971).
28. L.Ya.Glozman, *Nucl. Phys.* A **629**, 121c (1998).
29. C.Itoh et al., *Phys. Rev.* D, **40**, 3660 (1989).
30. M.Anselmino et al., *Z. fuer. Phys.* C **48**, 605 (1990).
31. S.I.Sukhoruchkin, *Nucl. Data Sci. Techn.*, Trieste, 1997. SIF, **59**, p.311 (1997).
32. J.Bernstein, *Elem. Part. Their Currents*, MIR-publ.(in rus.), p.383 (1970).
33. J.Gasser, H.Leutwyler, *Phys. Rep.* **87**, 77 (1982).
34. C.Quigg, *Acta Phys. Pol.* B. **30**, 2145 (1999).

Pion-Baryon Couplings and SU(3)

Alfons J. Buchmann* and Ernest M. Henley[†]

* Institute for Theoretical Physics, University of Tübingen,
D-72076 Tübingen, Germany

[†] Department of Physics and Institute for Nuclear Theory,
University of Washington, Box 351560, Seattle, WA 98195, USA

Abstract. We have extended and applied a general QCD parameterization method to the emission of pions from baryons. We use it to calculate the strength and sign of the coupling of pions to the octet and decuplet of baryons. We first use SU(3) and then include SU(3)-breaking effects.

INTRODUCTION

During the period of approximately the last decade Morpurgo and colleagues [1,2] have developed a simple but quite general and complete parameterization of QCD for the properties of hadrons. They have applied it to masses, magnetic moments, transition amplitudes, and other properties of the baryon octet and decuplet in terms of a few parameters. They have also extended the method to nucleon electromagnetic form factors and radii [3]. The method uses only general features of QCD and baryon descriptions in terms of quarks. Recently, Dillon and Morpurgo have shown that the method is independent of the choice of the quark mass renormalization point in the QCD Lagrangian [2]. Although they begin with a fully relativistic theory, they reduce it to a nonrelativistic, but nevertheless complete set of operator structures for the property being investigated. We have extended their methodology to pion-baryon couplings [4] and to quadrupole moments and transitions [5]. Here we will discuss pion-baryon couplings.

METHODOLOGY

For the lowest baryon octet and decuplet we consider quarks in an S-state. In this case we can write for the matrix element of an operator Ω

$$\langle B|\Omega|B\rangle = \langle \Phi_B|V^\dagger \Omega V|\Phi_B\rangle = \langle W_B|\mathcal{O}|W_B\rangle . \tag{1}$$

Here B is a complicated baryon state, which is reduced to a simple one, Φ_B by the complex unitary operator V [1]. The Φ_B are pure $L = 0$ three-quark states excluding any quark-antiquark or gluon components. W_B stands for the standard three-quark $SU(6)$ spin-flavor wave functions. The operator V contains a Foldy-Wouthuysen transformation and dresses the auxiliary states Φ_B with $q\bar{q}$ components and gluons and thereby generates the exact QCD eigenstate, B. The operator $\mathcal{O}$ is the most general operator with the symmetries of the property being investigated. It can depend on spin, isospin, flavor, and space. For magnetic moments, for instance, it must be linear in the charges of the quarks. With the assumption of SU(3) the operator $\mathcal{O}$ can broken into one-body, two-body, and three-body terms,

$$\vec{\mu}_{op} = A \sum_{i=1}^{3} Q_i \vec{\sigma}_i + B \sum_{i \neq j=1}^{3} Q_i \vec{\sigma}_j + C \sum_{i \neq j \neq k=1}^{3} Q_i \vec{\sigma}_i \vec{\sigma}_j \cdot \vec{\sigma}_k . \qquad (2)$$

Here A,B, and C are parameters which arise from space and color integrations. It has been shown by Morpurgo that C is not independent [1,2]; it can be expressed in terms of a constant dependent on the spin of the baryon and the one-body term. Morpurgo has also argued that that the two-body term, proportional to B is $\leq 1/3$ of the one-body one, due to the necessity of exchanging a gluon; actually, for the magnetic moments, $B/A \approx 0.05$. The three-body term is $\leq 1/9$ of the one-body term by the same argument as that for B/A. If we neglect this term, then with the assumption of SU(3) there are just 2 constants to predict all magnetic moments and magnetic transition moments for the lowest octet and decuplet. The fit gives the experimental magnetic moments to about 30%. The problem is that SU(3)-breaking corrections are also of the order of 30% of the SU(3) values and thus cannot really be neglected.

Without SU(3) symmetry, seven constants are needed,

$$\vec{\mu}_{op} = A \sum_{i=1}^{3} Q_i \vec{\sigma}_i + a \sum_{i=1}^{3} Q_i \vec{\sigma}_i P_i^\lambda + B \sum_{i \neq j=1}^{3} Q_i \vec{\sigma}_j + b_1 \sum_{i \neq j=1}^{3} Q_i \vec{\sigma}_j P_i^\lambda$$

$$+ b_2 \sum_{i \neq j=1}^{3} Q_i \vec{\sigma}_j P_j^\lambda + b_3 \sum_{i \neq j=1}^{3} Q_i \vec{\sigma}_i P_j^\lambda + C \sum_{i \neq j \neq k=1}^{3} Q_i \vec{\sigma}_j P_k^\lambda . \qquad (3)$$

Here P_i^λ is a projection operator which distinguishes between strange (s) and up, down (u,d)quarks. By fitting the known magnetic moments, Morpurgo obtains a solution with

$$A = 2.79, \ a = -0.94, \ (a/A \approx -1/3), \ B = -0.076, \ b_1 = 0.41, \ (b_1/A \approx 0.15),$$
$$b_2 = 0.097, b_3 = -0.134, \ C = 0.155 . \qquad (4)$$

He goes on to predict the transition magnetic moments such as $\Sigma^0 \to \Lambda^0 \gamma$ and $\Delta^+ \to p\gamma$ and obtains good fits.

We have found an approximate way to include SU(3) symmetry-breaking, which involves no parameters. One-gluon exchange between two quarks involves a magnetic term proportional to the product of the inverse masses of the two quarks.

The larger mass of the s quark thus gives a correction $r = (m_u + m_d)/2m_s \approx 330 MeV/550 MeV \approx 0.6$ or approximately a 40% correction. This would give $a/A \approx 0.4$ vs. the fit of 0.33 and $b_1/\frac{A}{3} \approx 0.45$ versus a fit of 0.39. In the case of the magnetic moments, the same factor of r should be applied to the one-body term since the magnetic moments are inversely proportional to the masses of the quarks. When this reduction is applied to the one-body terms, a fit to within approximately 10% of all measured magnetic moments and transition moments within the octet is obtained. However, we do not satisfy the so-called Lipkin sum rule [6], but obtain $R_{\Sigma/\Lambda} \equiv \frac{\mu_{\Sigma^+} + 2\mu_{\Sigma^-}}{\mu_{\Lambda^0}} = -1$, as for the non-relativistic quark model. With the inclusion of the two-body term and its SU(3) symmetry-breaking factor r we also obtain a fit to the magnetic moments and transition moment of about 10% accuracy, and lower $R_{\Sigma/\Lambda}$ to about -0.6, but not to the experimental value of -0.23. Because μ_{Σ^+} is almost canceled by $2\mu_{\Sigma^-}$, $R_{\Sigma/\Lambda}$ is very sensitive to these magnetic moments. If we were to introduce the effect of the larger strange quark mass more appropriately, we would be able to fit all magnetic moments and $R_{\Sigma/\Lambda}$, but we would be back to the use of seven constants [1].

PION-BARYON COUPLINGS

The investigation of πBB couplings parallels that of magnetic moments, which is why we used it as an example. For the octet, the coupling can be written as

$$H_{\pi BB} = -\frac{f}{m_\pi} \bar{\psi} \gamma_\mu \gamma_5 \psi \partial^\mu \vec{\phi} \cdot \vec{\tau}, \tag{5}$$

$$\approx \frac{f}{m_\pi} \phi_B \vec{\sigma} \cdot \vec{\nabla} \vec{\tau} \phi_B \cdot \vec{\phi}. \tag{6}$$

This equation defines the coupling constant f of the (point) pion field $\vec{\phi}$ to the spin 1/2 baryon, ϕ_B; $\vec{\sigma}$ is the baryon spin operator and $\vec{\tau}$ the isospin matrix. We then obtain for the parameterization

$$\langle B|H_{\pi BB}|B\rangle = \vec{k} \cdot \left\langle B \left| A_1 \sum_{i=1}^{3} \vec{\sigma}_i \vec{\tau}_i + A_2 \sum_{i\neq j}^{3} \vec{\sigma}_i \vec{\tau}_j \right| B \right\rangle. \tag{7}$$

We neglect three-body terms such as $A_3 \langle B|\sum_{i\neq j\neq k} \vec{k} \cdot \vec{\sigma}_i \vec{\tau}_j \vec{\sigma}_j \cdot \vec{\sigma}_k |B\rangle$. The operator $\vec{\tau}_i \times \vec{\tau}_j \vec{\sigma}_i \times \vec{\sigma}_j$ is allowed, but it is not independent of the other terms if its expectation value is taken between baryon states [1,2].

The radial and color matrices are absorbed in the constants A_1 and A_2. We use the third component of the spin and isospin operators to compute the coupling constants in the states of maximal spin and isospin projection. If we neglect A_2, the $\pi^0 p$ coupling, $f^2/(4\pi) = 0.08$, fixes $A_1 = \frac{3}{5}f$. We then obtain for the $\Delta^+ \to p\pi^0$ decay constant $f^2_{\Delta N\pi}$, $\frac{f^2_{\Delta N\pi}}{f^2} = \frac{72}{25}$, in accord with the additive quark model. When we include A_2, we fix its value from the Δ decay described by the Hamiltonian,

$$H_{\Delta N \pi} = -\frac{f_{N\Delta\pi}}{m_\pi} \mathbf{S}^\dagger \cdot \nabla \vec{\mathbf{T}}^\dagger \cdot \vec{\phi} + \text{h.c.}, \tag{8}$$

where $\mathbf{S}^\dagger$ and $\vec{\mathbf{T}}^\dagger$ are transition spin and isospin operators; they are defined such that their matrix elements are simple Clebsch-Gordan coefficients. The coupling $f_{N\Delta\pi}$ is taken to be 2 f, which gives the $\Delta(1232)$ its experimental width of about 130 MeV.

We include SU(3)-breaking effects due to the more massive s quark by means of the factor r introduced earlier. We obtain various relations between couplings, such as

$$f - \frac{1}{4} f_{\pi^0 \Delta^+ \Delta^+} = \frac{\sqrt{2}}{3} f_{\pi^0 p \Delta^+}. \tag{9}$$

TABLE 1. Table of quark model matrix elements of the operator in Eq.(7) to first order ($\mathcal{O}_1$) and second order corrections ($\mathcal{O}_2$).

Baryon	First order	Second order
p	$\frac{5}{3} A_1$	$-\frac{2}{3} A_2$
Σ^+	$\frac{4}{3} A_1$	$\frac{2(2-r)}{3} A_2$
$\Sigma^0 \to \Lambda^0$	$-\frac{2\sqrt{3}}{3} A_1$	$\frac{2\sqrt{3}}{3} A_2$
Ξ^0	$-\frac{1}{3} A_1$	$\frac{4r}{3} A_2$
$\Delta^+ \to p$	$\frac{4\sqrt{2}}{3} A_1$	$-\frac{4\sqrt{2}}{3} A_2$
$\Sigma^{*+} \to \Sigma^+$	$\frac{2\sqrt{2}}{3} A_1$	$\frac{2\sqrt{2}(1-2r)}{3} A_2$
$\Sigma^{*0} \to \Lambda^0$	$\frac{2\sqrt{6}}{3} A_1$	$-\frac{2\sqrt{6}}{3} A_2$
$\Xi^{*0} \to \Xi^0$	$\frac{2\sqrt{2}}{3} A_1$	$-\frac{2\sqrt{2}r}{3} A_2$
Δ^+	A_1	$2 A_2$
Σ^{*+}	$2 A_1$	$2(1+r) A_2$
Ξ^{*0}	A_1	$2r A_2$

We summarize our results in Table I and II for (i) one-body terms alone, (ii) two-body together with one-body terms with r = 1 and (iii) with r=0.6. In Table III we compare our results with SU(3) breaking taken into account with the values obtained by Stoks and Rijken (SR) [7] from fits to baryon-nucleon scattering data and one-boson exchange potentials. The columns labeled KDOL [8] are obtained from QCD sum rules with the use of SU(3) and "beyond" SU(3) by correcting for mass effects. Our values tend to be closer to those of Stoks and Rijken [7].

CONCLUSIONS

In summary, we have used the Morpurgo formalism to predict pion-baryon coupling constants. The inclusion of two-body operators and SU(3)-breaking effects

TABLE 2. Coupling constants of the pion to various members of the octet and decuplet, and the decuplet-octet transitions in terms of $f = f_{\pi^0 p}$. The * indicates an input. The ratio $r = m_u/m_s$ of non-strange and strange quark masses indicates the degree of flavor symmetry breaking.

Baryon	First order ($A_2 = 0$)	Total r=1	Total r=0.6
p	1	1	1
Σ^+	0.80	0.59	0.54
$\Sigma^0 \to \Lambda^0$	-0.69	-0.82	-0.82
Ξ^0	-0.20	-0.42	-0.32
$\Delta^+ \to p$	1.70	2*	2*
$\Sigma^{*+} \to \Sigma^+$	0.98	1.16	0.92
$\Sigma^{*0} \to \Lambda^0$	-1.20	-1.42	-1.42
$\Xi^{*0} \to \Xi^0$	-1.20	-1.42	-1.28
Δ^+	0.80	0.23	0.23
Σ^{*+}	0.80	0.23	0.32
Ξ^{*0}	0.80	0.23	0.42

TABLE 3. Coupling constants of the pion to baryons in terms of $f_{\pi^0 p}$ as found by various authors.

	SR [7]	KDOL [8]		this work
		SU(3)	broken SU(3)	
Σ^+	0.71	0.28	0.83	0.54
$\Sigma^0 \to \Lambda^0$	-0.75	—	—	-0.82
Ξ^0	-0.29	-0.46	-2.05	-0.32

lead to significant corrections of the additive quark model values. Finally, we hope that this work will stimulate experimental efforts to determine some of the unmeasured decay rates or couplings.

REFERENCES

1. G. Morpurgo, Phys. Rev. **D40**, 2997 (1989); Phys. Rev. **D40**, 3111 (1989).
2. G. Morpurgo, Phys. Rev. Lett. **68**, 139 (1992); Phys. Rev. **D46**, 4068 (1992); G. Dillon and G. Morpurgo, Phys. Rev. **D53**, 3754 (1996).
3. G. Dillon and G. Morpurgo, Phys. Lett. **B448**, 107 (1999); Phys. Lett. **B459**, 321 (1999); Z. Phys. **C73**, 547 (1997).
4. A.J. Buchmann and E.M. Henley, LANL Archives nucl-th 9912044, (1999).
5. A.J. Buchmann and E.M. Henley, (in preparation)
6. H.J. Lipkin, LANL arch. hep-ph/9911261 (1999); A.W. Thomas and G. Krein, LANL arch. nucl-th/0004008 (2000).
7. V. G. J. Stoks and Th. A. Rijken, Phys. Rev. **C59**, 3009 (1999).
8. H. Kim, T. Doi, M. Oka, and S.H. Lee, Nucl. Phys. **A662**, 371 (2000), LANL Arch. nucl-th/9909007

Direct CP Violation in Charmed Hadron Decays via $\rho - \omega$ Mixing

X.-H. Guo and A.W. Thomas

*Department of Physics and Mathematical Physics,
and Special Research Center for the Subatomic Structure of Matter,
University of Adelaide, SA 5005, Australia* [1]

Abstract. We study direct CP violation in the charmed hadron decays $D^+ \to \rho^+ \rho^0(\omega) \to \rho^+ \pi^+ \pi^-$, $D^+ \to \pi^+ \rho^0(\omega) \to \pi^+ \pi^+ \pi^-$, $D^0 \to \phi \rho^0(\omega) \to \phi \pi^+ \pi^-$, $D^0 \to \eta \rho^0(\omega) \to \eta \pi^+ \pi^-$, $D^0 \to \eta' \rho^0(\omega) \to \eta' \pi^+ \pi^-$, $D^0 \to \pi^0 \rho^0(\omega) \to \pi^0 \pi^+ \pi^-$, and $\Lambda_c \to p \rho^0(\omega) \to p \pi^+ \pi^-$ via $\rho-\omega$ mixing. The CP violation parameter depends on the effective parameter, N_c, which is relevant to hadronization dynamics of each decay channel. It is found that for fixed N_c the CP violation parameter reaches its maximum value when the invariant mass of the $\pi^+\pi^-$ pair is in the vicinity of the ω resonance. For most of the parameter space explored the CP violating asymmetry is of order 10^{-4}. However, over a small range, $1.98 \leq N_c \leq 1.99$ and $1.95 \leq N_c \leq 2.02$, the asymmetries for $D^0 \to \pi^0 \rho^0(\omega) \to \pi^0 \pi^+ \pi^-$ and $\Lambda_c \to p \rho^0(\omega) \to p \pi^+ \pi^-$ (respectively) can exceed 1% - at the cost of a small branching ratio, $\sim 2 \times 10^{-8}$ and 7×10^{-9}, respectively.

INTRODUCTION

Recent studies of direct CP violation in the B meson system [1] have suggested that large CP-violating asymmetries may be observed in forthcoming experiments. However, in the charm sector, due to the suppression of the Cabbibo-Kobayashi-Maskawa (CKM) matrix elements, the CP violation is usually estimated to be smaller than 10^{-3} [2]. Experimental measurements in some decay channels are consistent with zero asymmetry within large errors [3].

Direct CP violation occurs through the interference of at least two amplitudes with different weak and strong phases. The weak phase difference is determined by the CKM matrix elements and the strong phase is usually very uncertain. In Refs. [4,5], the authors used $\rho - \omega$ mixing to obtain large strong phase while studying direct CP violation in hadronic B decays. It was found that the CP violating asymmetry reaches its maximum value in the vicinity of the ω resonance. This mechanism was also applied to Λ_b hadronic decays, $\Lambda_b \to \Lambda(n)\rho^0(\omega) \to \Lambda(n)\pi^+\pi^-$ and we found even larger possible CP violation [6]. In this talk I will present our

---

[1] Supported by the Australian Research Council.

recent study on direct CP violation in the hadronic decays of charmed hadrons, $H_c \to f\pi^+\pi^-$ (H_c could be $D^\pm$, D^0, or Λ_c, and f could be $\rho^\pm, \pi^{\pm,0}, \eta^{(\prime)}$, and p), with the same mechanism [7].

FORMALISM

The amplitude, A, for the decay $H_c \to f\pi^+\pi^-$ can be divided into two parts:

$$A = \langle \pi^+\pi^- f|\mathcal{H}^T|H_c\rangle + \langle \pi^+\pi^- f|\mathcal{H}^P|H_c\rangle, \tag{1}$$

where $\mathcal{H}^T$ and $\mathcal{H}^P$ are the Hamiltonians for the tree and penguin operators, respectively. The relative magnitude and phases of these two parts are defined as: $A = \langle \pi^+\pi^- f|\mathcal{H}^T|H_c\rangle[1 + re^{i\delta}e^{i\phi}]$, $\bar{A} = \langle \pi^+\pi^- \bar{f}|\mathcal{H}^T|\bar{H}_c\rangle[1 + re^{i\delta}e^{-i\phi}]$, where δ and ϕ are strong and weak phases, respectively. ϕ is $\arg[V_{ub}V_{cb}^*/(V_{uq}V_{cq}^*)]$ for the $c \to q$ transition ($q = d$ or s). r is defined as: $r \equiv \left|\frac{\langle \pi^+\pi^- f|\mathcal{H}^P|H_c\rangle}{\langle \pi^+\pi^- f|\mathcal{H}^T|H_c\rangle}\right|$. Then the CP-violating asymmetry, a, can be written as:

$$a \equiv \frac{|A|^2 - |\bar{A}|^2}{|A|^2 + |\bar{A}|^2} = \frac{-2r\sin\delta\sin\phi}{1 + 2r\cos\delta\cos\phi + r^2}. \tag{2}$$

It can be seen from Eq.(2) that large strong phase is needed to produce large CP violation. $\rho - \omega$ mixing has the dual advantages that the strong phase difference is large at the ω resonance and well known. In this scenario, to the first order of isospin violation, one has [5–7]

$$\langle \pi^+\pi^- f|\mathcal{H}^T|H_c\rangle = \frac{g_\rho}{s_\rho s_\omega}\tilde{\Pi}_{\rho\omega}t_\omega + \frac{g_\rho}{s_\rho}t_\rho, \tag{3}$$

$$\langle \pi^+\pi^- f|\mathcal{H}^P|H_c\rangle = \frac{g_\rho}{s_\rho s_\omega}\tilde{\Pi}_{\rho\omega}p_\omega + \frac{g_\rho}{s_\rho}p_\rho, \tag{4}$$

where t_V ($V=\rho$ or ω) is the tree and p_V is the penguin amplitude for producing a vector meson, V, by $H_c \to fV$; g_ρ is the coupling for $\rho^0 \to \pi^+\pi^-$; $\tilde{\Pi}_{\rho\omega}$ is the effective $\rho-\omega$ mixing amplitude and s_V^{-1} is from the propagator of V, $s_V = s - m_V^2 + im_V\Gamma_V$, with $\sqrt{s}$ being the invariant mass of the $\pi^+\pi^-$ pair. The numerical values for the $\rho-\omega$ mixing parameter are [5,8,9]: $\text{Re}\tilde{\Pi}_{\rho\omega}(m_\omega^2) = -3500 \pm 300 \text{MeV}^2$, $\text{Im}\tilde{\Pi}_{\rho\omega}(m_\omega^2) = -300 \pm 300 \text{MeV}^2$. The direct coupling $\omega \to \pi^+\pi^-$ is effectively absorbed into $\tilde{\Pi}_{\rho\omega}$ [9]. This leads to the explicit s dependence of $\tilde{\Pi}_{\rho\omega}$.

Defining

$$\frac{p_\omega}{t_\rho} \equiv r'e^{i(\delta_q+\phi)}, \quad \frac{t_\omega}{t_\rho} \equiv \alpha e^{i\delta_\alpha}, \quad \frac{p_\rho}{p_\omega} \equiv \beta e^{i\delta_\beta}, \tag{5}$$

where δ_α, δ_β and δ_q are strong phases, one has the following expression for r and δ,

$$re^{i\delta} = r'e^{i\delta_q}\frac{\tilde{\Pi}_{\rho\omega} + \beta e^{i\delta_\beta}s_\omega}{s_\omega + \tilde{\Pi}_{\rho\omega}\alpha e^{i\delta_\alpha}}. \tag{6}$$

In order to obtain the CP violating asymmetry in Eq.(2), we need to know $\alpha e^{i\delta_\alpha}$, $\beta e^{i\delta_\beta}$, and $r'e^{i\delta_q}$. These quantities can be extracted from the matrix elements $\langle f\rho^0(\omega)|\mathcal{H}^{T(P)}|H_c\rangle$. The effective weak Hamiltonian, which is Cabibbo first-forbidden, has the following form:

$$\mathcal{H}_{\Delta C=1} = \frac{G_F}{\sqrt{2}}[\sum_{q=d,s} V_{uq}V_{cq}^*(c_1 O_1^q + c_2 O_2^q) - V_{ub}V_{cb}^* \sum_{i=3}^{6} c_i O_i] + H.C.. \quad (7)$$

Here c_i ($i = 1,...,6$) are the Wilson coefficients and the operators O_i have the following expressions:

$$O_1^q = \bar{u}_\alpha \gamma_\mu(1-\gamma_5)q_\beta \bar{q}_\beta \gamma^\mu(1-\gamma_5)c_\alpha, \quad O_2^q = \bar{u}\gamma_\mu(1-\gamma_5)q\bar{q}\gamma^\mu(1-\gamma_5)c,$$
$$O_3 = \bar{u}\gamma_\mu(1-\gamma_5)c\sum_{q'}\bar{q}'\gamma^\mu(1-\gamma_5)q', \quad O_4 = \bar{u}_\alpha\gamma_\mu(1-\gamma_5)c_\beta\sum_{q'}\bar{q}'_\beta\gamma^\mu(1-\gamma_5)q'_\alpha,$$
$$O_5 = \bar{u}\gamma_\mu(1-\gamma_5)c\sum_{q'}\bar{q}'\gamma^\mu(1+\gamma_5)q', \quad O_6 = \bar{u}_\alpha\gamma_\mu(1-\gamma_5)c_\beta\sum_{q'}\bar{q}'_\beta\gamma^\mu(1+\gamma_5)q'_\alpha,$$

$$(8)$$

where α and β are color indices, and $q' = u$, d, s. O_1 and O_2 are the tree operators, while $O_3 - O_6$ are QCD penguin operators.

The Wilson coefficients, c_i ($i = 1,...,6$), are calculable in renormalization group modified perturbation theory. The solution has the following form,

$$\mathbf{C}(\mu) = U(\mu, m_W)\mathbf{C}(m_W), \quad (9)$$

where $U(\mu, m_W)$, which describes the QCD evolution, was evaluated to next-to-leading order in Refs. [10,11], and its explicit form was given in b- or s-decays. Since the strong interaction is independent of quark flavors, we can use the formulas in Refs. [10,11] to obtain $\mathbf{C}(m_c)$. In general, the Wilson coefficients depend on the renormalization scheme. In this work we have chosen to use the scheme-independent Wilson coefficients, which is defined as

$$\bar{\mathbf{C}}(\mu) = \left(1 + \frac{\alpha_s}{4\pi}R^T\right)\mathbf{C}(\mu), \quad (10)$$

where R is the renormalization matrix associated with the four-quark operators $O_i (i = 1,...,6)$ at the scale m_W. While calculating the Wilson coefficients, we have taken $\alpha_s(m_Z) = 0.118$. To be consistent, the matrix elements of the operators O_i should also be renormalized to the one-loop order since we are working to the next-to-leading order for the Wilson coefficients. This results in effective Wilson coefficients, c'_i, which satisfy the constraint

$$c_i(m_c)\langle O_i(m_c)\rangle = c'_i \langle O_i\rangle^{\text{tree}}, \quad (11)$$

where $\langle O_i(m_c)\rangle$ are the matrix elements, renormalized to the one-loop order. c'_i are μ and renormalization scheme independent. However, they do depend on gauge and infrared regulator. More details on the scale, scheme, gauge, and infrared regulator dependence of the Wilson coefficients can be found in Ref. [12]. The numerical results for c'_i can be found in Ref. [7]

$\alpha e^{i\delta_\alpha}$, $\beta e^{i\delta_\beta}$, and $r'e^{i\delta_q}$ can be obtained from $\langle f\rho^0(\omega)|\mathcal{H}^{T(P)}|H_c\rangle$. With the aid of Eq.(11), we use the factorization approximation to evaluate the tree-level matrix elements of the operators $O_i(i=1,...,6)$, $\langle O_i\rangle^{\text{tree}}$ [7,13,14]. Consequently, one current in O_i generates a meson which has the same quantum numbers as the current. Thus the decay amplitude of the two body nonleptonic decay of H_c becomes the product of two matrix elements, one related to the decay constant of the factorized meson and the other to the weak transition matrix element between H_c and the other hadron. Take $D^+ \to \rho^+\rho^0(\omega) \to \rho^+\pi^+\pi^-$ as an example. We found $\alpha e^{i\delta_\alpha} = -1$, $\beta e^{i\delta_\beta} = 0$, and $r'e^{i\delta_q} = 2\frac{(c'_3+c'_4)(1+\frac{1}{N_c})+c'_5+\frac{1}{N_c}c'_6}{(c'_1+c'_2)(1+\frac{1}{N_c})}\left|\frac{V_{ub}V^*_{cb}}{V_{ud}V^*_{cd}}\right|$, where N_c arises from Fierz transformation[2]. Similarly, for $D^0 \to \phi\rho^0(\omega) \to \phi\pi^+\pi^-$, $D^0 \to \eta\rho^0(\omega) \to \eta\pi^+\pi^-$, $D^0 \to \eta'\rho^0(\omega) \to \eta'\pi^+\pi^-$, and $\Lambda_c \to p\rho^0(\omega) \to p\pi^+\pi^-$, $\alpha e^{i\delta_\alpha}$, $\beta e^{i\delta_\beta}$, and $r'e^{i\delta_q}$ only depend on c'_i and the CKM Matrix elements. For $D^0 \to \eta\rho^0(\omega) \to \eta\pi^+\pi^-$ and $D^0 \to \eta'\rho^0(\omega) \to \eta'\pi^+\pi^-$, the strong phase, δ, happens to be zero in this approach. Hence we do not have CP violation in these two processes to the approximation we are working with. For $D^+ \to \pi^+\rho^0(\omega) \to \pi^+\pi^+\pi^-$, two kinds of matrix element products are involved after factorization, i.e., $\langle \rho^0(\omega)|(\bar{d}d)|0\rangle\langle\pi^+|(\bar{u}c)|D^+\rangle$ and $\langle \pi^+|(\bar{u}d)|0\rangle\langle\rho^0(\omega)|(\bar{d}c)|D^+\rangle$. These two quantities cannot be related to each other by symmetry. We evaluate them in the Bauer-Stech-Wirbel [15] phenomenological quark model. The process $D^0 \to \pi^0\rho^0(\omega) \to \pi^0\pi^+\pi^-$ is similar.

NUMERICAL RESULTS

In the numerical calculations, we have several parameters: q^2 (q is the momentum transfer of the gluon in the penguin diagram), N_c, and the CKM matrix elements. The value of q^2 is conventionally chosen to be in the range $0.3 < q^2/m_c^2 < 0.5$. For the CKM matrix elements, we use $\lambda = 0.221$, $\eta = 0.34$ and $\rho = -0.12$ in the Wolfenstein parametrization [16]. Since the hadronization information is included in N_c, the value of N_c may be different for different decay channels. Furthermore, since the color-octet contribution associated with each operator in the Hamiltonian (7) can vary, the effective N_c in the Fierz transformation for each operator may be different. However, since we do not have enough information about the operator dependence of N_c, we assume N_c is universal for each operator [7,14]. The value of N_c should be determined by the experimental data.

In the numerical calculations, it is found that for a fixed N_c there is a maximum point, $a_{\max}$, for the CP violating parameter a, when the invariant mass of the $\pi^+\pi^-$ pair is in the vicinity of the ω resonance. We have calculated $a_{\max}$ in the range $N_c > 0$ for different decay channels. In the calculations we use two sets of

---

[2] We have ignored the difference between $\langle \rho^0|(\bar{d}d)|0\rangle\langle\pi^+|(\bar{u}c)|D^+\rangle$ and $\langle \pi^+|(\bar{u}d)|0\rangle\langle\rho^0|(\bar{d}c)|D^+\rangle$. This isospin violating effect has been checked to be negligible.

form factors which correspond to taking the average transverse momentum of the constituents in the meson to be 400MeV or 500MeV, respectively [15].

The decay widths for nonleptonic decays of D-meson can be calculated straightforwardly in the quark model of Ref. [15]. Comparison of these theoretical results with the experimental data lead to constraints on N_c. For $D^0 \to \phi\rho^0$, the data for the branching ratio $(6 \pm 3) \times 10^{-4}$ [17] corresponds to $1.31 \leq N_c \leq 1.53$ ($1.41 \leq N_c \leq 1.60$) for the first (second) set of form factors. The branching ratio for $D^+ \to \pi^+\rho^0$ is $(1.05 \pm 0.31) \times 10^{-3}$ [17], corresponding to $2.1 \leq N_c \leq 2.9$ ($2.5 \leq N_c \leq 3.4$) for the first (second) set of form factors.

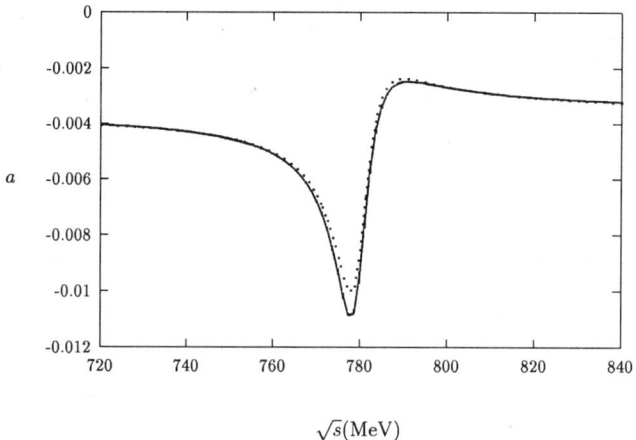

FIGURE 1. The CP-violating asymmetry for $\Lambda_c \to p\rho^0(\omega) \to p\pi^+\pi^-$, with $N_c = 2.02$. The solid (dotted) line is for $q^2/m_c^2 = 0.3\ (0.5)$.

The numerical results show that for $D^+ \to \rho^+\rho^0(\omega) \to \rho^+\pi^+\pi^-$, in the whole range $N_c > 0$, we have $a_{\max} \leq 3 \times 10^{-4}$. In the region of N_c allowed by the experimental data, $a_{\max}$ is $(1.2 \sim 1.8) \times 10^{-4}$ for $D^0 \to \phi\rho^0(\omega) \to \phi\pi^+\pi^-$ and $(3.0 \sim 6.1) \times 10^{-4}$ for $D^+ \to \pi^+\rho^0(\omega) \to \pi^+\pi^+\pi^-$. For the decay processes $D^0 \to \pi^0\rho^0$ and $\Lambda_c \to p\rho^0$ there are no experimental data at present [17] to constrain N_c. In Tables 1 and 2 we list numerical results for $a_{\max}$ and branching ratios for $D^0 \to \pi^0\rho^0$ and $\Lambda_c \to p\rho^0$ (respectively), with different values of N_c and q^2/m_c^2. It should be noted that $Br(H_c \to f\rho^0)$ is almost same for $q^2/m_c^2 = 0.3$ and $q^2/m_c^2 = 0.5$. We can see from Table 1 that when $1.98 \leq N_c \leq 1.99$, $a_{\max} \geq 1\%$ for $D^0 \to \pi^0\rho^0(\omega) \to \pi^0\pi^+\pi^-$. Table 2 shows that for $\Lambda_c \to p\rho^0(\omega) \to p\pi^+\pi^-$, $a_{\max} \geq 1\%$ when $1.95 \leq N_c \leq 2.02$, and $a_{\max} \geq 10^{-3}$ when $1.9 \leq N_c \leq 2.1$. In fact, the reason why we can find large CP violation in some range of N_c is that in this range t_ρ becomes small enough so that r', and hence r, becomes large. On the other hand, small t_ρ leads to small branching ratios for $H_c \to f\rho^0$. This makes it difficult to observe large CP violation in experiments. The branching ratio for $\Lambda_c \to p\rho^0$ shown in Table 2 is calculated in the heavy quark limit $m_c \to \infty$ and with the diquark model hadronic wavefunctions for both the heavy baryon, Λ_c, and the proton, p ($\langle k_\perp^2 \rangle^{\frac{1}{2}}$ is the average transverse momentum of the c quark in the Λ_c) [6,7]. It can be seen from Tables 1 and 2 that in order to have large CP violation ($a_{max} \geq 1\%$) the branching ratios are small, i.e., $\sim 2 \times 10^{-8}$ and 7×10^{-9} for $D^0 \to \pi^0\rho^0$ and $\Lambda_c \to p\rho^0$, respectively. The behaviour of a as a function of the invariant mass of the $\pi^+\pi^-$ is plotted in Fig.1, for $\Lambda_c \to p\rho^0(\omega) \to p\pi^+\pi^-$, with $N_c = 2.02$ and $q^2/m_c^2 = 0.3, 0.5$ (for $N_c = 1.95$ we have similar results). We can see from this plot that we can have $a_{max} \geq 1\%$. The plot for $D^0 \to \pi^0\rho^0(\omega) \to \pi^0\pi^+\pi^-$ is similar.

TABLE 1. Values of $Br(D^0 \to \pi^0\rho^0)$ with the first (second) set of form factors and a_{max} for $D^0 \to \pi^0\rho^0(\omega) \to \pi^0\pi^+\pi^-$, with $q^2/m_c^2 = 0.3(0.5)$

N_c	$Br(D^0 \to \pi^0\rho^0)$	a_{max} (Set 1)	a_{max} (Set 2)
0.5	$2.1(2.8)\times 10^{-2}$	$1.0(0.94)\times 10^{-4}$	$1.1(0.93)\times 10^{-4}$
1.0	$2.3(3.0)\times 10^{-3}$	$9.5(8.2)\times 10^{-5}$	$9.3(8.1)\times 10^{-5}$
1.5	$2.5(3.3)\times 10^{-4}$	$9.7(8.7)\times 10^{-5}$	$9.3(8.4)\times 10^{-5}$
1.9	$4.8(6.2)\times 10^{-5}$	$-7.1(-8.3)\times 10^{-4}$	$-7.4(-8.6)\times 10^{-4}$
1.98	$1.4(1.8)\times 10^{-8}$	$-1.4(-1.6)\times 10^{-2}$	$-1.5(-1.7)\times 10^{-2}$
1.99	$1.7(2.1)\times 10^{-8}$	$1.4(1.6)\times 10^{-2}$	$1.4(1.7)\times 10^{-2}$
2.1	$7.3(9.4)\times 10^{-6}$	$7.2(8.1)\times 10^{-4}$	$7.5(8.3)\times 10^{-4}$
2.5	$1.0(1.3)\times 10^{-4}$	$2.5(2.8)\times 10^{-4}$	$2.7(2.9)\times 10^{-4}$
3.0	$2.8(3.6)\times 10^{-4}$	$2.0(2.1)\times 10^{-4}$	$2.1(2.1)\times 10^{-4}$
10.0	$1.6(2.0)\times 10^{-3}$	$1.5(1.4)\times 10^{-4}$	$1.5(1.4)\times 10^{-4}$

TABLE 2. Values of $Br(\Lambda_c \to p\rho^0)$ with $\langle k_\perp^2 \rangle^{\frac{1}{2}} = 400\text{MeV}$ (600MeV) and a_{max} for $\Lambda_c \to p\rho^0(\omega) \to p\pi^+\pi^-$, with $q^2/m_c^2 = 0.3(0.5)$

N_c	$Br(\Lambda_c \to p\rho^0)$	a_{max}
0.5	$2.2(1.9)\times 10^{-4}$	$2.0(1.8)\times 10^{-4}$
1.0	$2.4(2.1)\times 10^{-5}$	$3.3(3.0)\times 10^{-4}$
1.5	$2.6(2.3)\times 10^{-6}$	$7.3(6.7)\times 10^{-4}$
1.9	$4.9(4.3)\times 10^{-8}$	$3.9(4.1)\times 10^{-3}$
1.95	$7.9(6.9)\times 10^{-9}$	$1.0(1.1)\times 10^{-2}$
2.02	$7.5(7.0)\times 10^{-9}$	$-1.1(-1.0)\times 10^{-2}$
2.1	$7.5(6.5)\times 10^{-8}$	$-3.4(-3.1)\times 10^{-3}$
2.5	$1.1(0.92)\times 10^{-6}$	$-8.0(-7.4)\times 10^{-4}$
3.0	$2.8(2.5)\times 10^{-6}$	$-4.3(-4.0)\times 10^{-4}$
10.0	$1.6(1.4)\times 10^{-5}$	$-1.1(-1.1)\times 10^{-4}$

In summary, we have studied direct CP violation in several charmed hadron decays via $\rho - \omega$ mixing. For $D^+ \to \rho^+\rho^0(\omega) \to \rho^+\pi^+\pi^-$, $D^+ \to \pi^+\rho^0(\omega) \to \pi^+\pi^+\pi^-$, $D^0 \to \phi\rho^0(\omega) \to \phi\pi^+\pi^-$, $D^0 \to \eta\rho^0(\omega) \to \eta\pi^+\pi^-$, $D^0 \to \eta'\rho^0(\omega) \to \eta'\pi^+\pi^-$, the CP violating asymmetry is smaller than a few$\times 10^{-4}$. However, over a small range of N_c, the asymmetries for $D^0 \to \pi^0\rho^0(\omega) \to \pi^0\pi^+\pi^-$ and $\Lambda_c \to$

$p\rho^0(\omega) \to p\pi^+\pi^-$ can exceed 1%. It will be intersting to study direct CP violation in these channels in experiments.

REFERENCES

1. Carter, A.B., and Sanda, A.I., *Phys. Rev. Lett.* **45**, 952 (1980); *Phys. Rev.* **D23**, 1567 (1981); Bigi, I.I., and Sanda, A.I., *Nucl. Phys.* **B193**, 85 (1981).
2. Burdman, G., " Potential for Discoveries in Charm Meson Physics", Workshop on the Tau/Charm Factory, Argonne (6/95), hep-ph/9508349.
3. Frabetti, P.L., *et al.*, *Phys. Rev.* **D50**, 2953 (1994); E791 collaboration, *Phys. Lett.* **B403**, 377 (1997), **B421**, 405 (1998); CLEO collaboration, *Phys. Rev.* **D52**, 4860 (1995).
4. Enomoto, R., and Tanabashi, M., *Phys. Lett.* **B386**, 413 (1996).
5. Gardner, S., O'Connell, H.B., and Thomas, A.W., *Phys. Rev. Lett.* **80**, 1834 (1998).
6. Guo, X.-H., and Thomas, A.W., *Phys. Rev.* **D58**, 096013 (1998).
7. Guo, X.-H., and Thomas, A.W., *Phys. Rev.* **D**, to be published.
8. Gardner, S., and O'Connell, H.B., *Phys. Rev.* **D57**, 2716 (1998).
9. O'Connell, H.B., Thomas, A.W., and Williams, A.G., *Nucl. Phys.* **A623**, 559 (1997); Maltman, K., O'Connell, H.B., and Williams, A.G., *Phys. Lett.* **B376**, 19 (1996).
10. Buras, A.J., Jamin, M., Lautenbacher, M., and Weisz, P., *Nucl. Phys.* **B400**, 37,75 (1993), **B370**, 69 (1992), **B375**, 501(A) (1992); Buras, A.J., Jamin, M., and Lautenbacher, M., *Nucl. Phys.* **B408**, 209 (1993).
11. Ciuchini, M., Franco, E., Martinelli, G., and Reina, L., *Nucl. Phys.* **B415**, 403 (1994).
12. Buras, A.J., and Silvestrini, L., *Nucl. Phys.* **B548**, 293 (1999); Cheng, H.-Y., Li, H.-n., and Yang, K.-C., *Phys. Rev.* **D60**, 094005 (1999).
13. Chen, Y.-H., Cheng, H.-Y., Tseng, B., and Yang, K.-C., *Phys. Rev.* **D60**, 094014 (1999).
14. Ali, A., and Greub, C., *Phys. Rev.* **D57**, 2996 (1998); Ali, A., Chay, J., Greub, C., and Ko, P., *Phys. Lett.* **B424**, 161 (1998).
15. Bauer, M., Stech, B., and Wirbel, M., *Z. Phys.* **C34**, 103 (1987); Wirbel, M., Stech, B., and Bauer, M., *Z. Phys.* **C29**, 637 (1985).
16. Wolfenstein, L., *Phys. Rev. Lett.* **51**, 1945 (1983).
17. The Particle Data Group, Caso, C., *et al.*, *Eur. Phys. J.* **C3**, 1 (1998).

Large N_c QCD Sum Rules

W-Y. Pauchy Hwang

Department of Physics, National Taiwan University
Taipei, Taiwan 106, R.O.C.

Abstract. I wish to stress that, in large N_c QCD (with N_c the number of color), a closed set of coupled differential equations may be derived for nonlocal condensates which, in the presence of the nontrivial QCD ground state or vacuum, are used to characterize the quark or gluon propagator, or other Green functions of higher order. It is pointed out that, by solving the coupled equations so obtained, the method of QCD sum rules, in its generalized sense, is free of additional parameters whenever we modify or generalize the method such as by considering the QCD vacuum in the presence of external fields. In addition, I wish to demonstrate a soft-pion theorem for the parton distributions of Goldstone pions.

INTRODUCTION

As of today, quantum chromodynamics (QCD) has been taken universally as the underlying theory of strong interaction physics. Although the asymptotically free nature of QCD allows us to test the candidate theory at high energies, the nonperturbative feature dominates for hadrons or nuclei at low energies, So far, it remains almost impossible to solve problems related to hadrons or nuclei.

The ground state, or the vacuum, of QCD is known to be nontrivial, in the sense that there are non-zero condensates, including gluon condensates, quark condensates, and perhaps infinitely many higher-order condensates. In such a theory, propagators, i.e. causal Green's functions, such as the quark propagator

$$iS_{ij}^{ab}(x) \equiv <0 \mid T(q_i^a(x)\bar{q}_j^b(0)) \mid 0>, \qquad (1)$$

carry all the difficulties inherent in the theory. Specifically, we may write the quark propagator as a sum of a perturbative part and a nonperturbative part:

$$iS_{ij}^{ab}(x) = iS_{ij}^{(0)ab}(x) + i\tilde{S}_{ij}^{ab}(x), \qquad (2)$$

where the perturbative part assumes the form of a free propagator:

$$iS_{ij}^{(0)ab}(x) \equiv \int \frac{d^4p}{(2\pi)^4} e^{-ip\cdot x} iS_{ij}^{(0)ab}(p), \qquad (3a)$$

$$iS^{(0)ab}_{ij}(p) = \delta^{ab}\frac{i(\hat{p}+m)_{ij}}{p^2 - m^2 + i\epsilon}, \tag{3b}$$

with $\hat{a} \equiv \gamma^\mu a_\mu$ for a four-vector a_μ, and the nonperturbative part may be parametrized as follows:

$$i\tilde{S}^{ab}_{ij}(x) \equiv \delta^{ab}\{\delta_{ij}f(x^2) + i\hat{x}_{ij}g(x^2)\}, \tag{4}$$

with $x^2 \equiv x_0^2 - \vec{x}^2$. The functions $f(x^2)$ and $g(x^2)$ are referred to as "nonlocal condensates". The solution to these functions would characterize the quark propagation completely and would thus represent the solution to the problem of nonperturbative QCD.

LARGE N_C QCD

In the well-known many-body theory developed for atomic or nuclear physics, the equations for Green functions of a given order usually involve Green functions of higher order, giving rise to an open-ended hierarchy of equations (which is often useless in practice). In what follows, however, we wish to show that, in large N_c QCD, there is in fact a natural way of setting up closed sets of differential equations which govern the inter-related Green functions to a given order.

In light of the nontrivial QCD vacuum, we begin by considering the feasibility of working directly with the various matrix elements such as the quark propagator of Eq. (1). Useful relations may be derived if we regard the equations for interacting fields [1],

$$\{i\gamma^\mu(\partial_\mu + ig\frac{\lambda^a}{2}A^a_\mu) - m\}\psi = 0; \tag{5}$$

$$\partial^\nu G^a_{\mu\nu} - 2gf^{abc}G^b_{\mu\nu}A^c_\nu + g\bar{\psi}\frac{\lambda^a}{2}\gamma_\mu\psi = 0, \tag{6}$$

as the equations of motion for *quantized* interacting fields, subject to the standard rule for quantization that the equal-time (anti-)commutators among these quantized interating fields are identical to those among non-interacting quantized fields.

Allowing the operator $\{i\gamma^\mu\partial_\mu - m\}$ to act on the matrix element defined by Eq. (1), we obtain

$$\{i\gamma^\mu\partial_\mu - m\}_{ik}iS^{ab}_{kj}(x) = i\delta^4(x)\delta^{ab}\delta_{ij} + <0|T(\{g\frac{\lambda^n}{2}A^n_\mu\gamma^\mu q(x)\}^a_i \bar{q}^b_j(0))|0>. \tag{7}$$

Making use of Eqs. (2)-(4), we find

$$\{i\gamma^\mu\partial_\mu - m\}_{ik}i\tilde{S}^{ab}_{kj}(x) = <0|T(\{g\frac{\lambda^n}{2}A^n_\mu\gamma^\mu q(x)\}^a_i \bar{q}^b_j(0))|0>, \tag{8}$$

an equation which may be used to investigate the nonlocal condensates $f(x^2)$ and $g(x^2)$.

As a useful benchmark, we shall work with the fixed-point gauge,

$$A_\mu^n(x) = -\frac{1}{2}G_{\mu\nu}^n x^\nu + \cdots. \tag{9}$$

As a result, the nonperturbative part $i\tilde{S}_{ij}^{ab}(x)$ may be solved immediately as a power series in x^μ,

$$i\tilde{S}_{ij}^{ab}(x) = -\frac{1}{12}\delta^{ab}\delta_{ij} <\bar{q}q> +\frac{i}{48}m\hat{x}_{ij}\delta^{ab} <\bar{q}q>$$

$$+\frac{1}{192}<\bar{q}g\sigma \cdot Gq> \delta^{ab} x^2 \delta_{ij} +\cdots, \tag{10}$$

The first term is the integration constant which defines the so-called "quark condensate", while the mixed quark-gluon condensate appearing in the third term arises because of Eqs. (8) and (9). It is obvious that the series (10) is a short-distance expansion, which converges for sufficiently small x_μ. We note that Eq. (10) is just the standard quark propagator cited in most papers in QCD sum rules [2]. However, the aim of this paper is to show that we may do a much better job regarding how to obtain the solution to $f(x^2)$ and $g(x^2)$.

The approach which we suggest here [3] is based upon two key elements, namely, the set of interacting field equations *plus* the rule of canonical quantization (for interacting fields). The equations which we obtain, such as Eq. (7) or (8), are much the same as the set of Schwinger-Dyson equations (for the matrix elements). An important aspect in our derivation is that the nontriviality of the vacuum $|0>$ is observed at every step — a central issue in relation to QCD.

To proceed further, we shall work only with the leading term in the fixed-point gauge and introduce

$$<: \{gG_{\mu\nu}q(x)\}_i^a \bar{q}_j^b(0) :>$$
$$= \delta^{ab}\{(\gamma_\mu x_\nu - \gamma_\nu x_\mu)A(x^2) + i\sigma_{\mu\nu}B(x^2) + (\gamma_\mu x_\nu - \gamma_\nu x_\mu)\hat{x}C(x^2) + i\sigma_{\mu\nu}\hat{x}D(x^2)\}, \tag{11}$$

with $G_{\mu\nu} \equiv (\lambda^n/2)G_{\mu\nu}^n$, an antisymmetric operator. The invariant functions $A(x^2)$, $B(x^2)$, $C(x^2)$, and $D(x^2)$ are additional nonlocal condensates which we must deal with explicitly in this paper.

Under the assumption that we keep only the leading term in the fixed-point gauge (a simplifying assumption which can be removed if necessary), we have

$$\{i\gamma^\alpha\partial_\alpha - m\}_{ik} <: \{gG_{\mu\nu}q(x)\}_k^a \bar{q}_j^b(0) :>$$

$$= -\frac{1}{2}x^\beta <: \{g^2 G_{\mu\nu}G_{\alpha\beta}\gamma^\alpha q(x)\}_i^a \bar{q}_j^b(0) :> \tag{12a}$$

$$= -\frac{1}{2} \cdot \frac{1}{96} \cdot \frac{4}{3} <g^2 G^2> <: \{(\gamma_\mu x_\nu - \gamma_\nu x_\mu)q(x)\}_i^a \bar{q}_j^b(0) :> \tag{12b}$$

$$= -\frac{1}{144} <g^2 G^2> \delta^{ab}(\gamma_\mu x_\nu - \gamma_\nu x_\mu)\{f(x^2) + i\hat{x}g(x^2)\}. \tag{12c}$$

Here the second line, (12a), follows from the field equation and the third line, (12b), is based on the factorization property of large N_c QCD that the contribution in which $G_{\mu\nu}$ and $G_{\alpha\beta}$ do not couple to color-singlet is suppressed by at least a factor of $1/N_c^2$.

Now, we may use Eqs. (8) and (12c) and obtain a closed set of equations, six ordinary differential equations for six functions:

$$2f'(x^2) - mg(x^2) = -\frac{3}{2}i(B - x^2 C), \tag{13a}$$

$$2x^2 g'(x^2) + 4g(x^2) + mf(x^2) = \frac{3}{2}x^2(A - D), \tag{13b}$$

$$4iB' - 2iC - 2ix^2 C' - mA = -\frac{1}{144} < g^2 G^2 > f(x^2), \tag{13c}$$

$$2iA + 2ix^2 D' - mB = 0, \tag{13d}$$

$$-2iA' + 4iD' - mC = -\frac{i}{144} < g^2 G^2 > g(x^2), \tag{13e}$$

$$2iB' + 2iC - mD = 0, \tag{13f}$$

where the derivatives are with respect to the variable x^2.

Treating m as an expansion parameter,

$$F(x^2) = \sum_{k=0}^{\infty} m^k F_k(x^2), \tag{14}$$

we may solve the coupled equations, (13a)-(13f), order by order in m. To leading order in m, we obtain

$$x^2 f_0''' + 3 f_0'' - \xi_0^2 x^2 f_0' - 2\xi_0^2 f_0 = 0, \tag{15}$$

$$(x^2)^3 g_0''' + 5(x^2)^2 g_0'' + \{2x^2 - \xi_0^2 (x^2)^3\} g_0' - \{2 + 2\xi_0^2 (x^2)^2\} g_0 = 0, \tag{16}$$

with $\xi_0^2 \equiv < g^2 G^2 > /384$. The equations for A_0, B_0, C_0, and D_0 can easily be solved once we obtain f_0 and g_0.

Eq. (16) can be simplified considerably by introducing

$$g_0(x^2) \equiv (x^2)^{-2} \tilde{g}_0(x^2), \tag{17a}$$

which leads to the equation:

$$x^2 \tilde{g}_0''' - \tilde{g}_0'' - \xi_0^2 x^2 \tilde{g}_0' = 0. \tag{17b}$$

Of course, Eqs. (15) and (17b) can be solved by iteration, again reproducing Eq. (10). However, it is often of critical importance to have analytic expressions for $f_0(x^2)$ and $g_0(x^2)$. This turns out to be possible by way of Laplace transforms. To indicate such possibility, we consider Eq. (15) in the timelike region specified by $t \equiv x^2 > 0$:

$$\bar{f}_0(s) \equiv \int_0^\infty ds\, e^{-st} f_0(t). \tag{18}$$

We obtain

$$\bar{f}_0(s) = -\frac{2f_0'(0)}{\xi_0^2} - \frac{f_0(0)}{\xi_0} \frac{s}{\sqrt{s^2 - \xi_0^2}} \sec^{-1}\frac{s}{\xi_0} + \frac{\gamma_0 s}{\sqrt{s^2 - \xi_0^2}}. \tag{19}$$

The spacelike region with $t \equiv x^2 < 0$ may be treated similarly. Analogously, Eq. (17b) may also be solved via Laplace transforms.

QCD SUM RULES

Thus far, we have described how to obtain a closed set of coupled equations for the nonlocal condensates which are relevant in the description of the quark propagator. We have also shown how these equations can be solved explicitly. Furthermore, some of the assumptions underlying our equations can be relaxed and more elaborate equations may then be obtained. Of course, some of our results are gauge dependent as the quark propagator (1) has been analyzed in a specific gauge (9). Nevertheless, our primary motivation for studying the quark propagator stems from our interest in the method of QCD sum rules [1], which may be regarded as the various approaches in which one tries to understand the roles played by the quark and gluon condensates for problems involving hadrons.

There are many variations in applying the method of QCD sum rules to specific problems in hadron physics, but the primary objective is quite clear and is to unravel the role of the nontrivial QCD vacuum in the problem. As the first approach, we may consider the Belyaev-Ioffe nucleon mass sum rules [5,4], where the short-distance expansion for the quark propagator is needed up to a certain (high) dimension. In this context, our analytical results on nonlocal condensates may be used to justify if the resultant series converges rapidly. The second approach is to consider the response of the QCD vacuum to some external fields, such as the method of QCD sum rules in the presence of an external axial field $Z_\mu(x)$ [6]. In this context, certain induced condensates are introduced (previously as new parameters) but the method offers a simple extension of the first approach in calculating magnetic moments, axial coupling constants, and other quantities by avoiding a need to treat explicitly the three-point Green's functions - a need which would involve a good deal of uncertainties. What is of great interest is that our analytical expressions for nonlocal condensates help to determine the induced condensates previously treated as new parameters, thereby making the external-field QCD sum rule method more powerful than what it used to be. To illustrate the point, we consider the external axial field Z_μ with the interaction,

$$\delta\mathcal{L}(x) = g Z^\mu(x) \bar{q}(x) \gamma_\mu \gamma_5 q(x). \tag{20}$$

For a constant Z^μ field, there are two major induced condensates [6]:

$$< 0 \mid \bar{q}(0)\gamma_\mu \gamma_5 q(0) \mid 0 >_{Z^\alpha} \quad \text{and} \quad < 0 \mid \bar{q}(0) g_c \tilde{G}_{\mu\nu} \gamma^\nu q(0) \mid 0 >_{Z^\alpha}.$$

We now have

$$< 0 \mid \bar{q}(0)\gamma_\mu \gamma_5 q(0) \mid 0 >_{Z^\alpha}$$

$$= i \int d^4 x g Z^\alpha(x) < 0 \mid T(\bar{q}(x)\gamma_\alpha \gamma_5 q(x) \bar{q}(0)\gamma_\mu \gamma_5 q(0)) \mid 0 >$$

$$= i \int d^4 x g Z^\alpha(x) \{ Tr[iS^{(0)}(-x)\gamma_\alpha\gamma_5 iS^{(0)}(x)\gamma_\mu\gamma_5]$$

$$+ Tr[i\tilde{S}(-x)\gamma_\alpha\gamma_5 iS^{(0)}(x)\gamma_\mu\gamma_5]$$

$$+ Tr[iS^{(0)}(-x)\gamma_\alpha\gamma_5 i\tilde{S}(x)\gamma_\mu\gamma_5]$$

$$+ Tr[i\tilde{S}(-x)\gamma_\alpha\gamma_5 i\tilde{S}(x)\gamma_\mu\gamma_5] \}. \quad (21)$$

The first term is the one-loop result which can be regularized (e.g., in d dimensions) and, as expected for a perturbative contribution, its finite part is small compared to the second and third terms. The last term, which can be treated numerically, involves products of two condensates and it has a dimension higher than the second or third term by at least three (and is likely of less numerical significance).

SOFT-PION THEOREM FOR THE PARTON DISTRIBUTIONS OF GOLDSTONE PIONS

Consider the amplitude given by

$$T_{\mu\nu}(q^2, p \cdot q) = i \int d^4 x e^{-iq \cdot x} < \pi^+(p) \mid T(J_\mu(x) J_\nu(0)) \mid \pi^+(p) > \quad (22a)$$

$$\equiv (-g_{\mu\nu} + \frac{q_\mu q_\nu}{q^2}) T_1(q^2, p \cdot q)$$

$$+ \frac{1}{p^2}(p_\mu - \frac{p \cdot q}{q^2} q_\mu)(p_\nu - \frac{p \cdot q}{q^2} q_\nu) T_2(q^2, p \cdot q), \quad (22b)$$

which characterizes the forward Compton scattering off π^+ (a Goldstone boson) and also the parton distributions of π^+. Applying the soft-pion theorem (together with current algebra), we find, in the limit of $p_\mu \to 0$,

$$T_{\mu\nu}(q^2, 0)$$

$$= \frac{i}{f_\pi^2} \int d^4 x e^{-iq \cdot x} < 0 \mid T(\{A_\mu^1(x) - iA_\mu^2(x)\}\{A_\nu^1(0) + iA_\nu(0)\} - 2V_\mu^3(x) V_\nu^3(0)) \mid 0 >$$

$$= \frac{i}{f_\pi^2} \int d^4 x e^{-iq \cdot x} Tr \{ iS_d^{ba}(-x)\gamma_\mu\gamma_5 iS_u^{ab}(x)\gamma_\nu\gamma_5$$

$$-\frac{1}{2}iS_u^{ba}(-x)\gamma_\mu iS_u^{ab}(x)\gamma_\nu - \frac{1}{2}iS_d^{ba}(-x)\gamma_\mu iS_d^{ab}(x)\gamma_\nu\}. \qquad (23)$$

The structure functions $W_i(q^2, p.q)$ (in the description of deep inelastic scattering off the π^+ target) is the imaginary part of $T_i(q^2, p \cdot q)$ divided by the factor π. Eq. (23) suggest that our analytical expressions for the nonlocal condensates may be useful for analyzing properties of Goldstone pions. We find, as a soft-pion limit,

$$W_1(q^2, p \cdot q) = \frac{1}{\pi}\mathrm{Im}\, T_1(q^2, p \cdot q)$$

$$\longrightarrow \frac{m <\bar{q}q>}{2f_\pi^2 \xi_0}, \quad \text{as} \quad p_\mu \to 0 \text{ and } q_\mu \to 0. \qquad (24)$$

This limit is derived making use of our analytical expressions on the nonlocal condensates, again with Wick's rotation on the time integration.

ACKNOWLEDGEMENT

The author wishes to acknowledge the National Science Council of R.O.C. for its partial support (NSC89-2112-M002-001Y) towards the present research.

REFERENCES

1. For notations, see, e.g., T.-P. Cheng and L.-F. Li, *Gauge Theory of Elementary Particle Physics* (Clarendon Press, Oxford, 1984).
2. M. A. Shifman, A.J. Vainshtein, and V.I. Zakharov, Nucl. Phys. **147**, 385, 448 (1979).
3. W-Y. P. Hwang, Preprint hep-ph/9601219 & MIT-CTP-2498, Z. Physc. **C**, accepted for publication.
4. K.-C. Yang, W-Y. P. Hwang, E.M. Henley, and L.S. Kisslinger, Phys. Rev. **D47**, 3001 (1993).
5. B. L. Ioffe, Nucl. Phys. **B188**, 317 (1981); (E) **B191**, 591 (1981); V. M. Belyaev and B. L. Ioffe, Zh. Eksp. Teor. Fiz. **83**, 876 (1982) [Sov. Phys. JETP **56**, 493 (1982)].
6. V. M. Belyaev and Ya. I. Kogan, Pis'ma Zh. Eksp. Teor. Fiz. **37**, 611 (1983) [JETP Lett. **37**, 730 (1983]; C. B. Chiu, J. Pasupathy, and S.J. Wilson, Phys. Rev. **D32**, 1786 (1985); E. M. Henley, W-Y. P. Hwang, and L.S. Kisslinger, Phys. Rev. **D46**, 431 (1992); Chinese J. Phys. (Taipei) **30**, 529 (1992).
7. S. Narison, Phys. Lett. **B 387**, 162 (1996).

Is Time Reversal Invariance Violated in Muon Decay? A Measurement of the Transverse Positron Polarization

K. Bodek[1,2], A. Budzanowski[3], N. Danneberg[1], W. Fetscher[1],
C. Hilbes[1], L. Jarczyk[2], K. Kirch[1], S. Kistryn[2], J. Klement[1],
K. Köhler[1], A. Kozela[1,3], J. Lang[1], G. Llosá Llácer[1],
M. Markiewicz[1], X. Morelle[4], T. Schweizer[1], J. Smyrski[2],
J. Sromicki[1], E. Stephan[5], A. Strzałkowski[2], J. Zejma[1,2]

[1] Institut für Teilchenphysik, ETH Zürich, CH 8093 Zürich, Switzerland[1]
[2] Institute of Physics, Jagellonian University, Kraków, Poland
[3] H. Niewodniczanski Institute of Nuclear Physics, Kraków, Poland
[4] Paul Scherrer Institut, CH-5232 Villigen-PSI, Switzerland
[5] Institute of Physics, University of Silesia, Katowice, Poland

Abstract.

In the standard model (SM) of electroweak interactions the positron from the decay of polarized positive muons is mainly longitudinally polarized. However the model also predicts a small transverse polarization component P_{T_1}, which lies in the plane spanned by muon-spin and positron momentum. Interference with additional, scalar couplings would result in substantial values for P_{T_1} as well as in a non-zero value of the transverse component P_{T_2} which is perpendicular to the above mentioned plane. A nonzero component P_{T_2}, proportional to the imaginary part of a possible scalar coupling, would be the first observation of time reversal violation in a purely leptonic decay. Measuring P_{T_1}, which is proportional to the real part, amounts to a model independent determination of the Fermi-coupling constant. The μP_T experiment [1] at the Paul Scherrer Institute will improve the current experimental limits $P_{T_1} = (16 \pm 22) \times 10^{-3}$, $P_{T_2} = (7 \pm 23) \times 10^{-3}$ by almost one order of magnitude. First preliminary results of the experiment are given.

---

[1] This project is supported in part by the Swiss National Science Foundation and by the Polish Committee for Scientific Research under Grant No. 2P03B05111.

INTRODUCTION

Measurements of muon decay are low energy tests of the standard model. In fact, only a few years ago it has been shown that $V - A$, as one of the basic assumptions of the standard model, *follows* from the results of a selected set of muon decay experiments (including inverse muon decay) [2]. The experimental limits obtained up to now, however, still allow for substantial contributions from non-standard couplings which differ in their spin structure from the $V - A$ interaction. The limits on these couplings can be efficiently improved by performing experiments with polarized muons and positrons. The measurement of the transverse positron polarization P_{T_1} as a function of the positron energy, in particular, offers the possibility to obtain the low energy parameter η without the suppression factor m_e/m_μ, which makes the determination of η from the electron energy spectrum extremely difficult. The simultaneous measurement of the polarization component P_{T_2} allows one to test time reversal invariance.

I OBSERVABLES AND INTERACTIONS

Fig. 1 shows the kinematic variables for muon decay. While the e^+ from μ^+ decay is mainly longitudinally polarized (polarization P_L), there also is a transverse polarization component P_{T_1} lying in the plane of muon polarization $\mathbf{P}_\mu$ and positron momentum $\mathbf{k_e}$. Within the standard model P_{T_1} is negligibly small at large positron energies, but substantial at lower energies and reaches the value $-1/3$ in

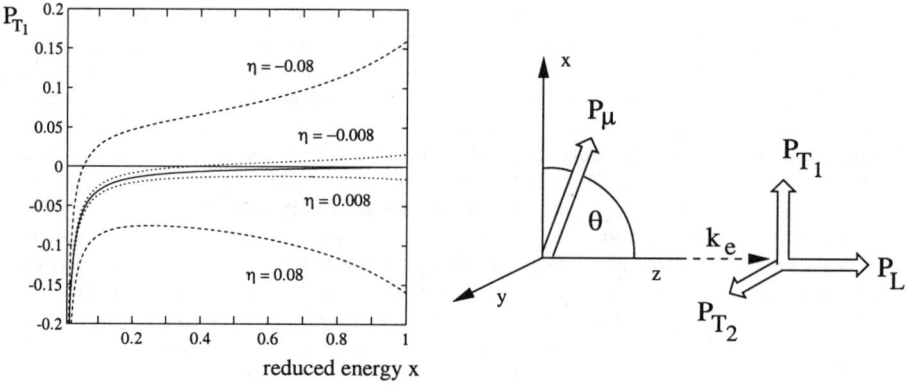

FIGURE 1. Transverse positron polarization P_{T_1} as a function of the reduced positron energy. The standard model predicts $\eta = 0$ (solid curve); the experimental limit from the direct measurement is $\eta = (11 \pm 85) \times 10^{-3}$, the fit to all available data is $\eta = (-7 \pm 13) \times 10^{-3}$. The definition of the kinematic variables is given in the right-hand picture.

the limiting case of a positron at rest (see Fig. 1, $\eta = 0$). Due to the low rate at small positron energies the energy averaged transverse polarization predicted by the standard model is $<P_{T_1}> = 0.003$ and therefore at present cannot be detected.

One can, however, obtain large transverse polarizations by including an additional interaction which renders the positrons with opposite chirality so that for a given final state the two interactions will interfere. In the representation of ref. [2] the matrix element for muon decay is given by

$$\mathcal{M} = \frac{4G_F}{\sqrt{2}} \sum_{\substack{\gamma=S,V,T \\ \varepsilon,\mu=R,L}} g^\gamma_{\varepsilon\mu} \langle \bar{e}_\varepsilon | \Gamma^\gamma | (\nu_e)_n \rangle \langle \bar{\nu}_m | \Gamma_\gamma | (\mu)_\mu \rangle \quad (1)$$

The index γ labels the type of interaction:

$$\Gamma^S = \text{4-scalar}$$
$$\Gamma^V = \text{4-vector}$$
$$\Gamma^T = \text{4-tensor}$$

The indices ε, μ indicate the chiralities of the spinors of the observed (charged) leptons. The chiralities n, m of the neutrinos are uniquely determined for given γ, ε and μ.

The transverse polarization component P_{T_1} yields the low energy parameter η *without* the suppression factor m_e/m_μ of η in the energy spectrum of the decay positron. In terms of the coupling constants defined above one obtains:

$$\eta = \frac{1}{2} \text{Re} \left\{ g^V_{LL} g^{S*}_{RR} + g^V_{RR} g^{S*}_{LL} + g^V_{LR}(g^{S*}_{RL} + g^{T*}_{RL}) + g^V_{RL}(g^{S*}_{LR} + g^{T*}_{LR}) \right\} \quad (2)$$

The standard model predicts

$$g^V_{LL} = 1, \quad g^\gamma_{\varepsilon\mu} = 0 \text{ (all other interactions)}. \quad (3)$$

In the general case there will be a phase between $V - A$ and an additional interaction which leads to a transverse component P_{T_2} *perpendicular* to the plane of muon polarization and positron momentum, and which violates time reversal invariance. With the experimental knowledge that $V - A$ is dominant [2], and neglecting exotic contributions in second order, one obtains

$$\eta \approx \frac{1}{2} \text{Re}\{g^S_{RR}\} \quad (4)$$

Correspondingly one derives a value for $Im\{g^S_{RR}\}$ from the energy dependence of P_{T_2}. Here g^S_{RR} represents a scalar, charge-changing interaction with right-handed charged leptons [4].

A more precise value of η is urgently needed for a model-independent determination of the Fermi coupling constant G_F: The influence of the uncertainty in the experimental value of η on the value of G_F is at present 20 times larger than the one of the more precisely known muon life time [4].

II EXPERIMENTAL SETUP

The experimental setup is shown in Fig. 2. A beam of highly polarized muons ($P_\mu \approx 91\%$) enters the beryllium stop target with bunches every 20 ns. The polarization of the stopped muons precesses in a homogeneous magnetic field with the same frequency as the accelerator RF. Thus every new muon bunch is added coherently with the same direction of the polarization vector. Decay e^+ emitted parallel to the B-field are tracked by drift chambers and can annihilate with polarized e^- in a magnetized foil. The two annihilation quanta are then detected by a hexagonal

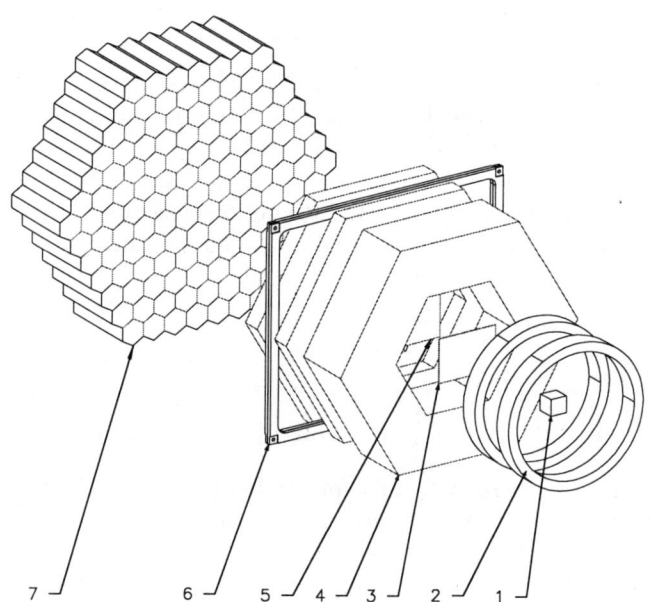

FIGURE 2. Experimental setup: 1 - Be target, 2 - spin precession magnet, 3 and 5 - plastic trigger counters, 4 - drift chamber (10 planes), 6- iron yoke of the magnetized Vacoflux foil, 7 - BGO calorimeter. Two additional drift chambers (2 planes each) sandwich the magnetized foil. An array of plastic veto counters (ANTI) in front of and cosmic trigger scintillators on top and below the BGO wall are not shown.

array of 127 BGO crystals. A valid annihilation event requires a coincidence of two plastic scintillator counters before the magnetized foil with two separated clusters of BGO detectors and an anticoincidence with a plastic counter array in front of the BGO wall. A possible transverse polarization would be detected as a harmonic time dependence of the annihilation rate for a given detector pair.

III EXPERIMENTAL RESULTS

In fall of 1999 we had the first data taking run. Valid annihilation events are identified with 13 % trigger efficiency; with appropriate geometry and energy cuts, however, the background is efficiently reduced. Part of the data (about 10%) has been analysed and preliminary results are given. The resulting energy spectrum of annihilating positrons is shown in Fig. 3.

The time distribution of the annihilation events contains two effects:

1. Since the accepted decay positrons are emitted into a cone whose axis coincides with the symmetry axis of the apparatus and is perpendicular to the precession plane of the muon polarization, there is a small remnant μSR effect (i.e., a time-dependent rate variation due to the decay asymmetry with respect to the precessing muon spin) of a few % in amplitude. This effect depends on the azimuthal angle of emission φ of the positron so that at a larger φ the maximum positron emission rate will be reached later than at a smaller angle.

2. The effect due to a possible transverse polarization P_T, in contrast, does not depend on φ, but only on the relative orientation of $\boldsymbol{P}_T$ and the electron

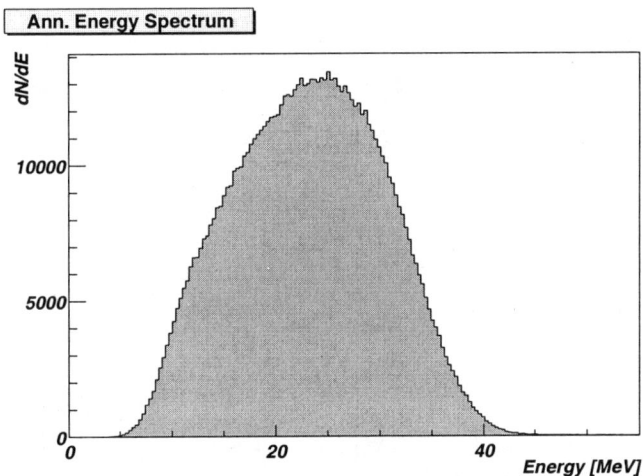

FIGURE 3. Energy distribution of positrons from accepted annihilation events after all hard- and software cuts.

polarization in the magnetized foil.

Actually, both effects are needed to get the full transverse polarization vector:

- From the polarization dependence of the annihilation cross section one derives the absolute value of $\boldsymbol{P}_T$,

- and from the residual μSR effect one determines time zero, i.e. the position of the precessing polarization vector $\boldsymbol{P}_\mu$ of the muon. Since P_{T_1} and P_{T_2} are defined relative to $\boldsymbol{P}_\mu$, this finally allows to determine the two components separately.

Fig. 4 shows the time distribution of annihilation events obtained for four different azimuthal angular regions. Their time dependence is

$$f_\nu(t) \sim \{1 + a_\nu \cos \omega t + b_\nu \sin \omega t\} \qquad (\nu = 1, \ldots, 4) \tag{5}$$

The Fourier coefficients of the time distributions as a function of φ are shown in Fig. 5, together with the averaged amplitude (circle) and their vectorial sum,

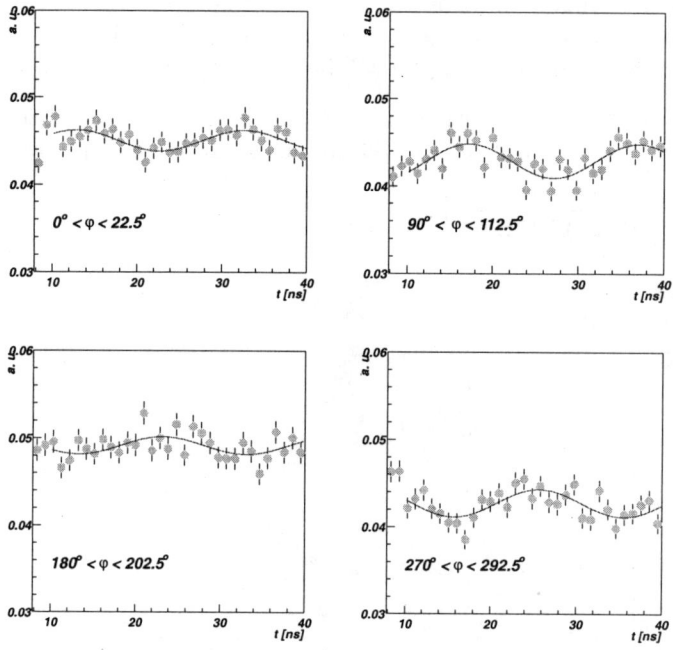

FIGURE 4. Time dependence of annihilation events for four different regions of the azimuthal emission angle of the decay positron. The oscillations are due to a small residual μSR effect. This allows for the determination of the muon polarization vector as a function of time.

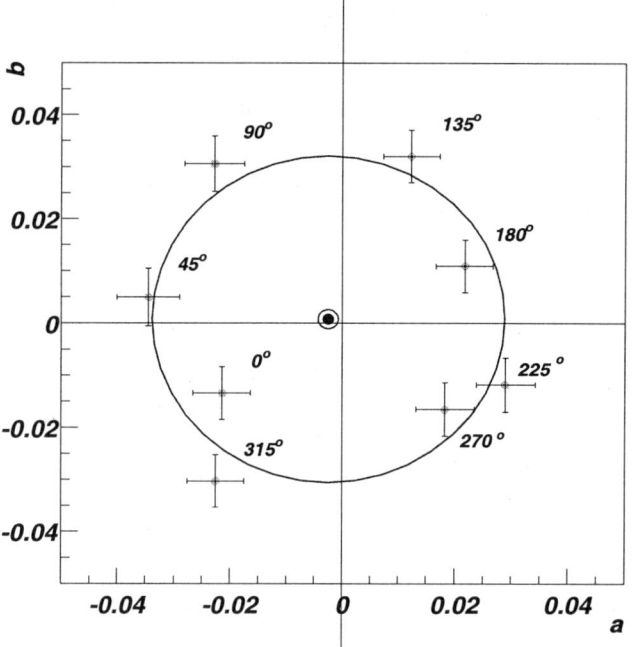

FIGURE 5. Fourier amplitudes of the residual μSR effect for different azimuthal emission angles φ of the decay positron. The circle is the average of the amplitudes, the small circle near to the origin is the vector sum of all values. It coincides, within statistics, with the origin and confirms the symmetry of the apparatus.

which coincicides, within statistics, with the origin of the diagram. This important test confirms the symmetry of the apparatus.

The properly normalized, time dependent annihilation cross section, after correcting for the μSR modulation, is given by:

$$\frac{1}{\sigma_0} \cdot \frac{d\sigma}{d\Omega} = 1 + \mathcal{A} \cdot S \cdot \cos(\omega t + \alpha) \qquad (6)$$

Here S is the electron polarization, and the amplitude $\mathcal{A}$ and the phase α are functions of the two transverse components of the positron polarization at the moment of annihilation, of the photon energies, and of the azimuthal angle of orientation ψ of the photon pair:

$$\mathcal{A} = \mathcal{A}(P_1, P_2, E_{\gamma_1}, E_{\gamma_2}, \psi)$$
$$\alpha = \alpha(P_1, P_2, E_{\gamma_1}, E_{\gamma_2}, \psi)$$

The values of $\mathcal{A}$ in this experiment vary between 0.1 and 0.95, with a maximum at 0.8; the electron polarization of 8 %, however, reduces possible effects by one order of magnitude. Fig. 6, finally, shows as a preliminary result of the evaluation of 10 % of the data, the energy dependence of the two transverse polarization components of the positrons from polarized muon decay at the moment of interaction. Both components are consistent with zero at the present precision.

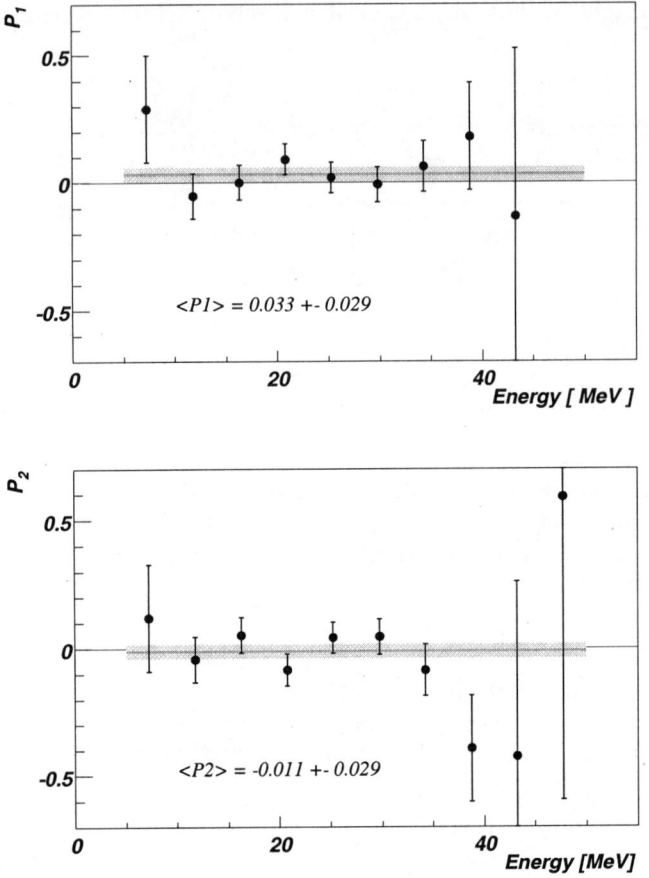

FIGURE 6. Energy dependence of the two transverse polarization components of the positrons from polarized muon decay at the moment of interaction. Preliminary result based on 10 % of the measured data.

Here the question arises: Could we have detected a nonzero signal if there is a nonzero transverse polarization?

The answer is yes:

In this experiment we not only make use of the time dependence of the annihilation and of the µSR distributions; in fact, we are able to simultaneously perform a third, independent experiment measuring the longitudinal polarization of the positrons from muon decay. We make use of the fact that positrons hitting the magnetized foil off the symmetry axis have a component of the longitudinal polarization in the direction of the electron polarization. If this projection is parallel for positrons to the right of the center, then it is antiparallel for positrons emitted to the left, and vice versa (see Fig. 7).

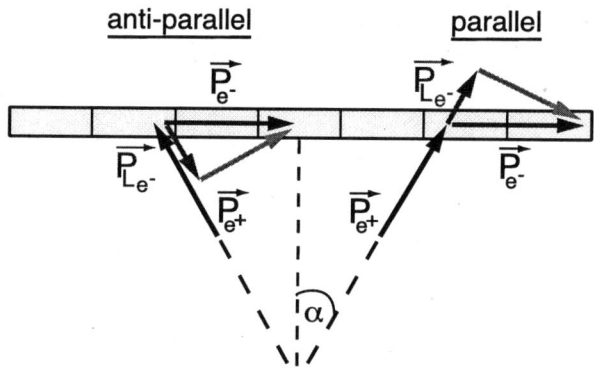

FIGURE 7. Principle of the measurement of the longitudinal polarization P_L of positrons from muon decay. Depending on which side the positron is emitted, P_L has a component parallel or antiparallel to the electron polarization in the magnetized foil. By reversing the electron polarization one gets a change in the rate of annihilation events for a given area on the foil, which allows one to deduce the longitudinal polarization.

The cross section for annihilation depends also on the longitudinal polarization. Thus, by dividing the fiducial area of the magnetized foils in seven vertical strips and calculating the asymmetry of the annihilation rate for each strip resulting from reversing the foil magnetization, one obtains the expected behaviour (see Fig. 8). By dividing the obtained asymmetries with the tangent of the polar angle, the calculated analyzing power and the electron polarization, as well as by correcting for background events, we obtain, in an *absolute* measurement, the longitudinal polarization of the positrons from muon decay. This confirms that our apparatus *is* sensitive to positron polarization and, at the same time, presents a novel way of measuring the longitudinal polarization with a setup of higher internal symmetry than used previously. The preliminary result, $P_L = 0.92 \pm 0.11$, agrees with the predictions of the standard model.

With an additional run in 2000 we expect to achieve our goal of reaching the proposed precision of $\Delta P_{T_1} = \Delta P_{T_1} = 3 \times 10^{-3}$. This will be an improvement by almost an order in magnitude.

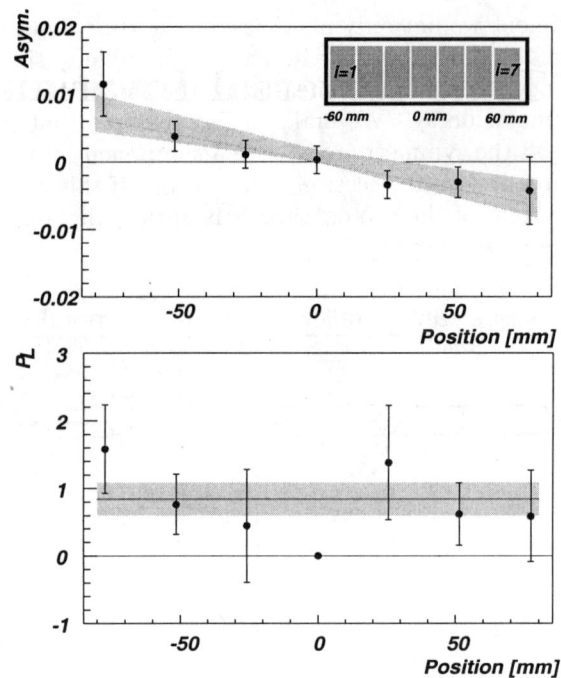

FIGURE 8. Rate asymmetries obtained in seven different regions of the magnetized target foil. The asymmetries are obtained by reversing the electron polarization, while the projection of the longitudinal polarization of the positron remains the same for each stripe. This leads to the observed change in sign between stripes on the left and on the right side of the symmetry axis. From this position-sensitive effect the longitudinal polarization of the positron was determined *absolutely* (lower figure). The horizontal line and the band show the average value and the error of P_L, respectively.

REFERENCES

1. I. Barnett et al., PSI proposal R-94-10.1 (1995).
2. W. Fetscher, H.-J. Gerber and K.F. Johnson, Phys. Lett. **173B** 102 (1986).
3. H. Burkard et al., Phys. Lett. **160 B** (1985) 343.
4. W. Fetscher and H.-J. Gerber, in *Precision Tests of the Standard Electroweak Model*, ed. P. Langacker, World Scientific, Singapore, 1995

Test of Time Reversal Invariance in $K_{\mu 3}$ Decay

Yoshitaka Kuno

Institute of Particle and Nuclear Studies (IPNS),
High Energy Accelerator Research Organization (KEK),
Oho 1-1, Tsukuba, Ibaraki, Japan 305-0801

(representing KEK-E246 Collaboration) [1]

Abstract. An experiment E246 at the 12-GeV proton synchrotron at KEK (KEK-PS), which is searching for the T-violating transverse muon polarization ($P_\mu^\perp$) in $K^+ \to \pi^0 \mu^+ \nu$ decay, is described. The search is sensitive to new mechanisms of T or CP violation beyond the Standard Model. The initial results of $P_\mu^\perp = -0.0042 \pm 0.0049(stat.) \pm 0.0009(syst.)$ and the T-violating parameter Im$(\xi) = -0.013 \pm 0.016(stat.) \pm 0.003(syst.)$ were determined from the data taken in 1996 and 1997.

INTRODUCTION

The triple-vector correlation is odd under the time-reversal operation and measurement of such a correlation would provide a promising testing ground to scrutinize the time-reversal invariance. One such example concerning K decay is the transverse muon spin polarization $P_\mu^\perp$ in $K^+ \to \pi^0 \mu^+ \nu$ ($K_{\mu 3}^+$) decay [1], where $P_\mu^\perp$ is defined as the component of muon spin polarization normal to the decay plane, determined by the μ^+ and π^0 momentum vectors. It is given by

$$P_\mu^\perp \equiv \frac{\vec{s}_{\mu^+} \cdot (\vec{p}_{\pi^0} \times \vec{p}_{\mu^+})}{|\vec{p}_{\pi^0} \times \vec{p}_{\mu^+}|}, \tag{1}$$

where $\vec{s}_{\mu^+}$ is the muon spin vector and $\vec{p}_{\mu^+}$ and $\vec{p}_{\pi^0}$ are the momentum vectors of the muon and neutral pion, respectively. Since the T-reversal operation changes

---

[1] The E246 collaboration is M. Abe, M. Aoki, I. Arai, Y. Asano, T. Baker, M. Blecher, M.D. Chapman, D.V. Dementyev, P. Depommier, M.P. Grigorjev, P. Gumplinger, M. Hasinoff, R. Henderson, K. Horie, W.S. Hou, H.C. Huang, Y. Igarashi, T. Ikeda, J. Imazato, A.P. Ivashkin, J.H. Kang, W. Keil, M.M. Khabibullin, A.N. Khotjantsev, Y.G. Kudenko, Y. Kuno, J.-M. Lee, K.S. Lee, G.Y. Lim, J.A. Macdonald, D.R. Marlow, C.R. Mindas, O.V. Mineev, C. Rangacharyulu, S.K. Sahu, S. Sekikawa, H.M. Shimizu, S. Shimizu, Y.-H. Shin, Y.-M. Shin, K.S. Sim, A. Suzuki, T. Tashiro, A. Watanabe, D.H. Wright, and T. Yokoi.

the sign of $P_\mu^\perp$, a non-zero value of $P_\mu^\perp$ would signal T-violation. If CPT invariance is obeyed, T-violation implies CP-violation.

A search for $P_\mu^\perp$ in $K_{\mu3}^+$ decay is sensitive to new mechanisms of CP violation beyond the Standard Model. The physics motivation of new CP violation arises from the observed baryon asymmetry in the universe, which cannot be explained by the CP violation in the Standard Model alone [2]. Therefore there must be new additional sources of CP violation. Furthermore, recent theoretical progress of electroweak baryogenesis suggests that new CP violation sources might exist at the electroweak scale which can be accessible experimentally.

$P_\mu^\perp$ IN $K_{\mu3}^+$ DECAY

As a clean search for T-violating phenomena, the measurement of $P_\mu^\perp$ in $K_{\mu3}^+$ decay has several striking advantages. First, the final-state electromagnetic interaction (FSI), which would otherwise mimic a fake T-odd effect, is negligible, of the order of 10^{-6}, in $K_{\mu3}^+$ decay [3]. This is due to the fact that only one charged particle exists in the final state. However, this small FSI is not always the case; for instance, in triple correlations in nuclear β decays and $P_\mu^\perp$ in $K_L^0 \to \pi^- \mu^+ \nu$ ($K_{\mu3}^0$) decay, the FSI is predicted [4] to be as large as 10^{-3}. This advantage in $K_{\mu3}^+$ decay allows a wider window to search for T-violation mechanisms, free of FSI-induced background.

Second, $P_\mu^\perp$ in $K_{\mu3}^+$ decay has no contribution from the CKM phase in the minimal Standard Model at the tree level, and higher-order contributions are extremely small ($\sim 10^{-6}$). This implies that the observation of a non-zero $P_\mu^\perp$ value would be a definite signature of new physics beyond the minimal Standard Model.

In $K_{\mu3}^+$ decay, the hadronic matrix element can be described as

$$< \pi|J|K > = f_+^K(q^2)(\tilde{p}_K + \tilde{p}_\pi) + f_-^K(q^2)(\tilde{p}_K - \tilde{p}_\pi) \qquad (2)$$

where $\tilde{p}_K$ and $\tilde{p}_\pi$ are the four momenta of the kaon and pion, respectively. $f_+^K(q^2)$ and $f_-^K(q^2)$ are the form factors of the hadronic matrix elements as a function of momentum transfer squared (q^2). Time-reversal invariance requires that the phases of f_+^K and f_-^K are relatively the same; in another words, if the parameter $\xi(q^2) \equiv f_-^K(q^2)/f_+^K(q^2)$ is defined, ξ should be a real number. Conversely, a non-zero value of Imξ would indicate T-violation. Based on eq.(2), $P_\mu^\perp$ can be calculated as a function of the energies of the muon (E_μ) and neutral pion (E_{π^0}) [5]. Then, $P_\mu^\perp$ is given by [6]

$$P_\mu^\perp \cong \text{Im}\xi \left(\frac{m_\mu}{m_K}\right) \frac{|\vec{p}_\mu|}{[E_\mu + |\vec{p}_\mu|\vec{n}_\mu \cdot \vec{n}_\nu - m_\mu^2/m_K]} \qquad (3)$$

in the K^+ rest frame. $P_\mu^\perp$ is proportional to Imξ with the kinematic factor which depends on the phase space of $K_{\mu3}$ decay sampled. It is $P_\mu^\perp \sim 0.3 \times$ Imξ for E246.

The present experimental upper limit on $P_\mu^\perp$ in $K_{\mu3}^+$ decay, which was obtained from the previous experiment using in-flight K^+ decays at Brookhaven National Laboratory (BNL) [7] is $P_\mu^\perp = (-4.2 \pm 6.7) \times 10^{-3}$ in the K^+ rest frame. This yielded $\text{Im}\xi = -0.016 \pm 0.025$, or $|\text{Im}\xi| \leq 0.049$ at 90 % confidence level.[2]

THEORETICAL PREDICTIONS OF $P_\mu^\perp$ IN $K_{\mu3}^+$ DECAY

In general, neither an effective vector (V) nor axial-vector (A) interaction introduce $P_\mu^\perp$, but only effective scalar (S) or pseudoscalar (P) interactions give a non-zero $P_\mu^\perp$ [8]. Some examples of theoretical models are given below.

Three Higgs Doublet Models

$P_\mu^\perp$ is sensitive to the extension of the Higgs sectors, such as multi Higgs-doublet models, which need at least three Higgs-doublets to generate CP violation [9]. The three Higgs-doublet model (3HDM) has two charged Higgs particles H_i^+ (i=1,2). $P_\mu^\perp$ in $K_{\mu3}^+$ decay arises from the interference between the W^+-exchange and H^+-exchange diagrams. The former contributes mostly to $f_+(q^2)$ and the latter only to $f_-(q^2)$. $\text{Im}\xi$ in $K_{\mu3}^+$ decay in the 3HDM is given by [6]

$$\text{Im}\xi = \text{Im}(\alpha_1\beta_1^*) \cdot \left(\frac{v_2}{v_3}\right)^2 \cdot \left(\frac{m_K}{m_{H_1^+}}\right)^2, \qquad (4)$$

where $\text{Im}(\alpha_1\beta_1^*)$ is a measure of the magnitude of CP-violation in 3HDM. m_K and m_{H^+} are the masses of the kaon and charged Higgs, respectively. v_i ($i = 1, 2, 3$) are the three Higgs vacuum-expectation-values (V.E.V.). Although there have been no estimations of the magnitudes of the V.E.V.s, one possible scenario [10] is that the three V.E.V.s are proportional to the fermion masses which they couple to: i.e. $v_1 : v_2 : v_3 \sim m_b : m_t : m_\tau$. This scenario gives a large $P_\mu^\perp$ in semileptonic K decays because of a large value of v_2/v_3 ($\sim m_t/m_\tau$) of about 100. Then, $P_\mu^\perp$ in $K_{\mu3}^+$ decay is the most sensitive among the other constraints (from the neutron electric dipole moment, $b \to s\gamma$ decay, and $B \to \tau\nu_\tau X$). The details are described elsewhere [11].

Supersymmetric models with squark family mixing

The supersymmetric (SUSY) contribution to $P_\mu^\perp$ in $K_{\mu3}^+$ decay could be significant when the squark family mixing at the quark-squark-gluino couplings is large [12]. In general, there exist four squark-flavor mixing matrices denoting the couplings

---

[2] Since the $P_\mu^\perp$ measurements in $K_{\mu3}^0$ decay might have FSI effects, only the $K_{\mu3}^+$ data is quoted instead of the combined result.

of quarks and their corresponding squarks ($\tilde{u}_L$, $\tilde{u}_R$, $\tilde{d}_L$, $\tilde{d}_R$). They contribute to a non-zero $P_\mu^\perp$ in $K_{\mu 3}^+$ decay if their relative phase is not zero. Considering these experimental constraints, they found that $P_\mu^\perp$ can be as high as 7×10^{-3} when the maximally-allowed values are taken. It should be noted that the minimum SUSY model [13] without large squark family mixing predicts $P_\mu^\perp$ of 10^{-6}.

Leptoquark models or Supersymmetric model with R-parity breaking

Leptoquark models are an attractive candidate for new physics beyond the Standard Model [14]. One of the manifestations of leptoquark models is a supersymmetric extension with R-parity breaking. In these models, $P_\mu^\perp$ in $K_{\mu 3}$ decay can be introduced by slepton ($\tilde{l}_i$) or down-type squark ($\tilde{d}_i$) exchange diagrams. For the squark exchange case, Imξ can be given by

$$\mathrm{Im}\xi = \sum_k \frac{\mathrm{Im}[\lambda'_{21k}(\lambda'_{22k})^*]}{4\sqrt{2}G_F \sin\theta_c (m_{\tilde{d}_k})^2} \cdot \frac{m_K^2}{m_\mu m_s} \quad (5)$$

where λ'_{ijk} are the coupling constants in the R-parity breaking superpotential (with i, j, k being family indices), G_F is the Fermi coupling constant, θ_c is the Cabbibo angle, and m_s is a strange quark mass. In this case, $\mathrm{Im}[\lambda'_{21k}(\lambda'_{22k})^*]$ is a measure of CP violation in this model.[3] Since the upper limits on $\mathrm{Im}[\lambda'_{21k}(\lambda'_{22k})^*]$ are determined by the previous $P_\mu^\perp$ measurement, $P_\mu^\perp$ could appear just below the limit [15].

KEK EXPERIMENT E246 DETECTOR

KEK-PS E246 employs K decays at rest, in contrast to the previous experiments which used in-flight K decays. There are several significant advantages of the use of the K decays at rest. For instance, it allows clean and precise determination of K-decay kinematics. This, together with the carefully-designed detector, will reduce systematic errors.

E246 Detector

Schematic side and end views of the E246 detector are shown in Fig.1. Incident K^+s of 650 MeV/c, which are produced by a 12-GeV proton beam from the KEK-PS, are slowed down in the degrader, and stopped in a target located at the center of the detector. The K^+-stopping target consists of an array of 256 plastic-scintillating fibers with 5×5 mm^2 square cross-section*. After exiting

---

[3] For the slepton exchange case, it can be replaced by $\mathrm{Im}[\lambda'_{i12}(\lambda_{2i2})^*]$.

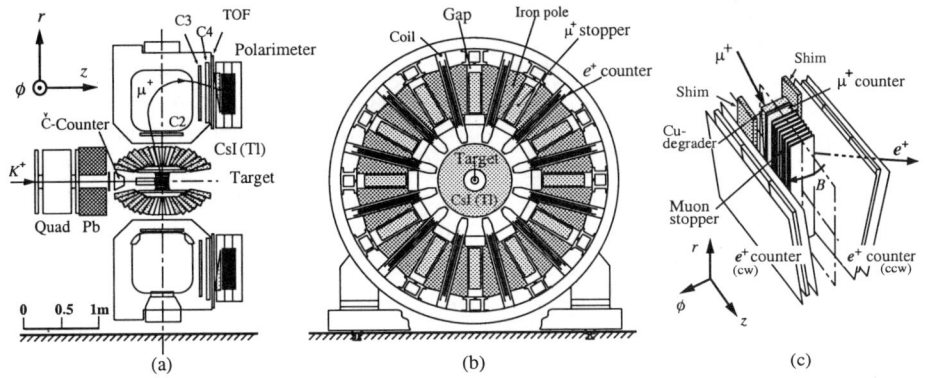

FIGURE 1. E246 detector; (a) side view, (b) end view, and 8c) one sector of polarimeter.

the target, the muons from K^+ decays enter the spectrometer (a superconducting toroidal magnet) where their momenta are analyzed. The magnet has 12 identical magnet-gaps with accurate 30° rotational symmetry. To track the muons, four sets of wire chambers are installed, one of which is located at the entrance and two at the exit of each of the 12 magnet gaps. The fourth one is a cylindrical drift chamber surrounding the target. In addition to these, there is an array of 32 thin ring-shaped plastic scintillator (ring counter) which surrounds the target to provide an additional coordinate of the track along the beam axis.

Photons from the π^0 decays are detected with a highly-segmented photon detector, consisting of an array of 768 thallium-doped CsI crystals with silicon PIN photodiode readout [16] in which each crystal points to the K^+ stopping target. The assembly covers about 75 % of 4π and has a beam hole for allowing the K^+s into the target, and 12 holes for μ^+s going out into the magnetic spectrometer. In addition to the conventional ADC/TDC readout, transient digitizers based on switched capacitor arrays are instrumented for each crystal [17].

The μ^+s exiting the spectrometer are stopped in a muon polarimeter installed at the exit of each magnet gap. The muon polarimeter consists of a muon stopper made of pure aluminum, and e^+ counters located at clockwise (cw) and counter-clockwise (ccw) sides of the stopper. The e^+ counters detect e^+s in $\mu^+ \to e^+\nu\bar{\nu}$ decay ($\tau_\mu = 2.2$ μsec). The muon polarization is deduced from the asymmetric angular distribution of e^+s, which are emitted preferentially along the μ^+ spin direction. The direction of the fringing magnetic field is parallel to $P_\mu^\perp$ so as to hold $P_\mu^\perp$, and at the same time it precesses the in-plane polarization components to average out.

$P_\mu^\perp$ Measurement and Systematic Cancelation

In E246, the $K_{\mu3}^+$ events which have the π^0 moving along the detector axis, either forward or backward from the target, will be accepted as "gold-plated events" For those events, the decay plane can be set radially from the detector axis. $P_\mu^\perp$ is directed azimuthally in a *screw-sense* around the detector axis, as in Fig.2. $P_\mu^\perp$ would then manifest itself as a difference in the e^+ counts between the cw and ccw counters from the muon stopper. By summing the cw- and ccw- counts of all the 12 sectors, $P_\mu^\perp$ is given by

$$\frac{\sum_{i=1}^{12} N_i(cw)}{\sum_{i=1}^{12} N_i(ccw)} \cong 1 \pm 2\alpha P_\mu^\perp, \qquad (6)$$

where $N_i(cw)$ and $N_i(ccw)$ are total e^+ counts at the cw and ccw counters at the ith sector, respectively. α is the effective analyzing power (defined as the ratio of the observed asymmetry to the muon polarization).

There are two important techniques for suppressing bias asymmetries in E246. First, summing over all magnet sectors would cancel any non-screw type biases. For instance, a fake asymmetry from the e^+-counter inefficiency, which would have opposite signs in its two adjacent sectors since the same e^+ detector acts as the cw counter in one sector and the ccw counter in the neighboring sector, would be canceled by this summing. A shift of the K^+ stopping distribution at the target would introduce an asymmetric muon stopping distribution within the muon stopper, resulting in spurious asymmetries with the same sign in one hemisphere but with the opposite sign in the other hemisphere. Again, this can be canceled after summing over all sectors.

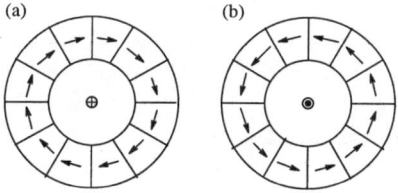

FIGURE 2. Schematic asymmetry directions in the 12 magnet sectors for (a) forward π^0 events and (b) backward π^0 events.

The second is the comparison of the $K_{\mu3}^+$ samples with forward π^0 and backward π^0 directions. Since $P_\mu^\perp$ is of the same magnitude but opposite in sign between these two samples as shown in Fig.2 and any biased asymmetries are likely to be independent of π^0 directions, their comparison would enable us to reduce the systematic errors significantly. A double ratio can be formed by these two samples as follows:

$$\frac{[\sum_{i=1}^{12} N_i(cw)/\sum_{i=1}^{12} N_i(ccw)]_{fwd}}{[\sum_{i=1}^{12} N_i(cw)/\sum_{i=1}^{12} N_i(ccw)]_{bwd}} \cong 1 + 4\alpha P_\mu^\perp. \tag{7}$$

It increases the asymmetry by a factor of two. In addition, by using the $K_{\mu 3}^+$ events with the π^0s moving transverse to the detector axis, we will be able to examine the zero-asymmetry level.

E246 ANALYSIS

The result presented here is based on the data taken in 1996 and 1997. Two independent off-line analyses were performed. They provided a consistency check and the estimation of systematic errors associated with the analysis. It was found that the two analyses were consistent one another. The final results shown here are obtained from the combination of the two analyses [18].

$K_{\mu 3}^+$ Event Selection

The $K_{\mu 3}^+$ event selection consists of identification of K^+ decays at rest in the K^+ stopping target, momentum selection (100 MeV/c < P_μ < 190 MeV/c), muon identification by time-of-flight, muon trajectory matching with hits in the target and the ring counter, π^0 mass cut, and a reconstructed kinematics consistent with $K_{\mu 3}^+$ decay. A 8-μsec offline time window of e^+s from μ^+ decay was set 15 nsec after the K^+ decay timing to eliminate K_{e3} decay. After further selection of the forward and backward π^0s, a total number of gold-plated $K_{\mu 3}^+$ samples of 2.1 M events is obtained. Also $K_{\mu 3}$ samples with only one photon from π^0 decay detected (E_γ >70 MeV) are also accumulated to increase the total statistics, since the high energy photon in asymmetric π^0 decay carries most of the information of the original π^0 direction. This one-photon sample contains about 1.8 M events. A total of 3.9 M good $K_{\mu 3}^+$ events have been collected. In Fig.3 shows a typical time spectrum of all good events (sum of 2γ and 1γ samples).

Major physics backgrounds are $K^+ \to \pi^+\pi^0$ with π^0 decay in flight (DIF), $K^+ \to \pi^0 e^+ \nu$, and $K^+ \to \pi^+\pi^0\pi^0$ which are estimated to be 4 %, < 0.1 %, and < 0.5%, respectively, in these event samples. These background decays only dilute the asymmetry, but do not introduce any spurious asymmetries.

The analyzing power α was experimentally obtained from the $K_{\mu 3}^+$ samples with the π^0 moving perpendicular to the detector axis. For those events, the e^+ asymmetry associated with the inplane muon polarization (P_μ^N) can be measured by the polarimeter. It is then compared with the expected P_μ^N which was calculated with the known value of Reξ. Then $\alpha = 0.198$ was determined. Since the stopping distributions of muons at the polarimeter were determined to be almost identical for the samples for P_μ^N and $P_\mu^\perp$, the same α value was used for the $P_\mu^\perp$ measurement.

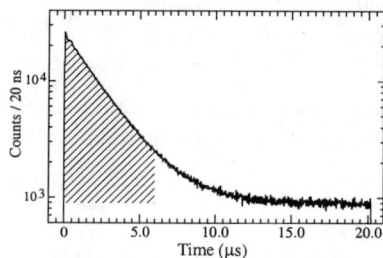

FIGURE 3. Positron time spectra of all good events. The hatched region is the analyzed signal region after subtraction of the constant background deduced from fitting in the region from 6.0 to 19.6 µs.

Studies of Systematic Errors

Extensive studies of the possible systematic errors was done before examining $P_\mu^\perp$, based on real data as much as possible. The summary of the studies of systematic errors is given in Table 1.

Among many studies, two examples are given. One is the measurement of the e^+ asymmetry without any π^0 direction tagging. As mentioned earlier, a vanishing e^+ asymmetry is expected after summing over all sectors. It was confirmed at a level of $< 10^{-3}$, which is further reduced by the cancelation by forward and backward π^0s.[4] Another study is to examine distributions of the angle between the $K_{\mu 3}^+$ decay plane and the mid-plane of the magnet gap, where any deviation of the distribution from the expected would introduce bias asymmetry from the inplane muon polarization. A possible contribution to $P_\mu^\perp$ of $< 10^{-3}$ was determined. Further cancelation by the comparison of forward and backward π^0 events are estimated for each bias source separately. As a result of all the examinations, a combined systematic error of less than 1.0×10^{-3} is obtained.

Results

The E246 initial result of $P_\mu^\perp$ in $K^+ \to \pi^0 \mu^+ \nu$ decay from the data collected in 1996 and 1997 [18] is

$$P_\mu^\perp = -0.0042 \pm 0.0049(stat) \pm 0.0009(sys). \tag{8}$$

which yields

$$\mathrm{Im}\xi = -0.013 \pm 0.016(stat) \pm 0.003(syst). \tag{9}$$

The 90% confidence limits are given as $|P_\mu^{perp}| < 0.011$ and $|\mathrm{Im}\xi| < 0.033$. At this moment, the statistical error dominates the systematic error.

---

[4] This cancelation factor is about a few tens, estimated from real data.

TABLE 1. Summary of systematic errors. $\sum_{12}$ and fwd/bwd denote the cancelation capabilities by the azimuthal symmetry of 12 sectors, and by the comparison of π^0 directions, respectively.

Source	Cancelation by Σ_{12}	fwd/bwd	$\delta P_T \times 10^5$
e^+ counter r-rotation	no	yes	5
e^+ counter z-rotation	no	yes	2
e^+ counter ϕ-offset	no	yes	22
e^+ counter r-offset	yes	yes	< 1
e^+ counter z-offset	yes	yes	< 1
μ^+ counter ϕ-offset	no	yes	< 1
MWPC ϕ-offset (C4)	no	yes	25
CsI(Tl) misalignment	yes	yes	16
$\vec{B}$ offset (ϵ)	no	yes	30
$\vec{B}$ rotation (δ_r)	no	yes	3.7
$\vec{B}$ rotation (δ_z)	no	no	53
K^+ stopping distribution	yes	yes	< 30
Decay plane angle (θ_r)	no	yes	20
Decay plane angle (θ_z)	no	no	9
$K_{\pi 2}$ DIF background	no	yes	6
K^+ DIF background	yes	no	< 19
e^+ spectrum background	yes	yes	8
Analysis	-	-	38
Total			92

$P_\mu^\perp$ in $K^+ \to \mu^+ \nu \gamma$ decay, in which the decay plane is determined by the muon and photon momentum vectors, is also being investigated at E246.

SUMMARY

E246 completed its initial result from the data taken in 1996 and 1997. E246 has taken more data since then, and the total data are more than twice the data shown here. A naive extrapolation of the final statistical sensitivity of Imξ to the whole data is about 1×10^{-2}, which would be about 2.5 times better than the previous BNL K^+ experiment. A request for a beam-time extension has been approved at KEK-PS, and the ultimate sensitivity of in the range of 0.7×10^{-2} is anticipated.

ACKNOWLEDGMENT

The author is grateful to all of the KEK-E246 collaboration members.

REFERENCES

1. Sakurai J.J., *Phys. Rev.* **109**, 980 (1958).
2. Mclerran L., Shaposhnikov M., Turok N., and Voloshin M., Phys. Lett. **B 25 6**, 451 (1991); Turok N., and Voloshin M., Phys. Lett. **B 256**, 451 (1991); Turok N., and Zadrozny J., Nucl. Phys. **B 358**, 471 (1991); Dine M., Huet P., Singleton R., and Susskind L., Phys. Lett. **B 257**, 351 (1991).
3. Zhitnitskii A.R., *Yad. Fiz.* **31**, 1014 (1980) [*Sov. J. Nucl. Phys*, **31**, 529 (1980)].
4. Adkins G.S., *Phys. Rev.* D **28**, 2885 (1983) and references therein.
5. Cabbibo N., and Maksymowicz A., *Phys. Lett.* **9**, 352 (1964); *Phys. Lett.* **11**, 360 (1964); *Phys. Lett.* **14**, 2 (1966); MacDowell S.W., *Nuovo Cimento* **9**, 258 (1958).
6. Bélanger G., and Geng C.Q., *Phys. Rev.* D **44**, 2789 (1991).
7. Campbell M.K., *et al.*, *Phys. Rev. Lett.* **47**, 1032 (1981); Blatt, S.R., *et al.*, *Phys. Rev.* D **27**, 1056 (1983).
8. Leurer M., *Phys. Rev. Lett.* **62**, 1967 (1989); Castoldi P., Frère J.-M., and Kane G.L.,*Phys. Rev.* D **39**, 2633 (1989).
9. Weinberg S., *Phys. Rev. Lett.* **37**, 657 (1976) and as a recent review, Cheng H.-Y., *Int. J. Mod. Phys.* A **7**, 1059 (1992) and references therein.
10. Garisto R., and Kane G., *Phys. Rev.* D **44**, 2038 (1991).
11. Kuno Y., *Nucl. Phys. B (Proc. Suppl.)* **37A**, 87 (1994); Kuno Y., *Chinese Journal of Physics* **32**, 1015 (1994).
12. Wu G.-H. and Ng J.N., *Phys. Lett.* B **392**, 93 (1997).
13. Christova E. and Fabbrichesi M., *Phys. Lett.* B **315**, 113 (1993).
14. Altarelli G., Ellis J., Giudice G.F., Lola S. and Mangano M.L., *Nucl. Phys.* B **506**, 3 (1997).
15. Fabbrichesi M. and Vissani F., *Phys. Rev.* D **55**, 5334 (1997).
16. Dementyev D.V., *et al.*, *Nucl. Instrum. Methods* A **379**, 499 (1996).
17. Wixted R.L., *et al.*, *Nucl. Instrum. Methods* A **386**, 483 (1997).
18. Abe M., *et al.*, *Phys. Rev. Lett.* **83**, 4253 (1999).

T Violation and CPT Tests at CPLEAR

Lukas A. Schaller

*University of Fribourg, Switzerland,
for the CPLEAR Collaboration*[1]

Abstract. The CPLEAR experiment at LEAR/CERN employed the high flux of antiprotons stopped in hydrogen to produce, via strong $p\bar{p}$ interaction, neutral kaons strangeness-tagged at the production time $t = 0$ by the charge of the associated charged kaon. At the decay time $t = \tau$, in the semileptonic decay, the strangeness of the neutral kaon is tagged by the lepton charge. In this way, the probabilities of a K^0 transforming into a $\overline{K}^0$ and vice versa as a function of the neutral-kaon proper time can be determined, yielding a direct measurement of T violation. Other semileptonic decay asymmetries allow to set lower limits on CPT violating parameters. Together with the Bell-Steinberger unitarity relation, it can be shown that the directly measured violation of T invariance in the mixing cannot be mimicked by CPT violation. Finally, very low limits are established for possible $K^0 - \overline{K}^0$ mass and decay-width differences.

INTRODUCTION

LEAR, the Low Energy Antiproton Ring at CERN, offered unique possibilities to study the symmetries between matter and antimatter. CPLEAR, taking advantage of tagging the strangeness of the neutral kaons both at production and decay time, has studied many parameters related to CP, T and CPT symmetries with higher precision than was possible before. Here, the CPLEAR results are presented regarding the direct measurement of T violation as well as lower limits for CPT violating parameters. First, we review the neutral kaon system and the relevant parameters to be determined from the semileptonic decay asymmetries. Secondly, experimental setup and analysis are discussed. Third, the direct measurement of the T-violation parameter Re(ϵ), including a possible violation of the $\Delta S = \Delta Q$ rule, is described. Next, limits are given for the CPT-violating parameters Re(δ), Im(δ), Re(y), Re(x_-) and Im(x_+). Employing in addition the Bell-Steinberger relation based on unitarity, it turns out that the measured T

---

[1] University of Athens, Greece, University of Basel, Switzerland, Boston University, USA, CERN, Geneva, Switzerland, LIP and University of Coimbra, Portugal, Delft University of Technology, Netherlands, University of Fribourg, Switzerland, University of Ioannina, Greece, University of Liverpool, UK, J. Stefan Institute, and Phys. Dep., University of Ljubljana, Slovenia, CPPM, IN2P3-CNRS et Université d'Aix-Marseille II, France, CSNSM, IN2P3-CNRS, Orsay, France, Paul Scherrer Institute (PSI), Switzerland, CEA, DSM/DAPNIA, CE-Saclay, France, Royal Institute of Technology, Stockholm, Sweden, University of Thessaloniki, Greece, ETH-IPP Zürich, Switzerland.

violation in the neutral-kaon mixing matrix cannot be mimicked by a possible CPT violation in the decay. From the parameter δ, stringent limits can also be obtained for the $K^0 - \overline{K}^0$ decay-width and mass differences. Finally, the results are summarized.

PHENOMENOLOGY

The neutral kaon system

The states K^0 and $\overline{K}^0$ describing the neutral kaon and its antiparticle are states of opposite strangeness, a quantity which is conserved in strong and electromagnetic interactions. The weak interaction however allows the K^0 to change into a $\overline{K}^0$ and vice versa ($\Delta S = 2$ processes) or to decay into non-strange final states ($\Delta S = 1$ processes). The time evolution of the neutral kaons can be expressed as follows:

$$|\Psi(t)\rangle = \alpha(t)|K^0\rangle + \beta(t)|\overline{K}^0\rangle, \quad \frac{d}{dt}\Psi = -i\Lambda\Psi, \tag{1}$$

Here, Λ is the mixing matrix, which can be written in terms of the mass matrix M and the decay matrix Γ:

$$\Lambda \equiv \begin{pmatrix} \Lambda_{K^0 K^0} & \Lambda_{K^0 \overline{K}^0} \\ \Lambda_{\overline{K}^0 K^0} & \Lambda_{\overline{K}^0 \overline{K}^0} \end{pmatrix} \equiv M - \frac{i}{2}\Gamma \equiv \begin{pmatrix} M_{K^0 K^0} & M_{K^0 \overline{K}^0} \\ M_{\overline{K}^0 K^0} & M_{\overline{K}^0 \overline{K}^0} \end{pmatrix} - \frac{i}{2} \begin{pmatrix} \Gamma_{K^0 K^0} & \Gamma_{K^0 \overline{K}^0} \\ \Gamma_{\overline{K}^0 K^0} & \Gamma_{\overline{K}^0 \overline{K}^0} \end{pmatrix}. \tag{2}$$

The diagonal elements of M and Γ signify the masses and decay widths of K^0 and $\overline{K}^0$, respectively. The off-diagonal elements are complex conjugates. The eigenvalues

$$\lambda_{L,S} = m_{L,S} - i/2\Gamma_{L,S} \tag{3}$$

correspond to the physical states K_L and K_S, with $\Delta m \equiv m_L - m_S = 2|M_{K^0\overline{K}^0}|$ and $\Delta\Gamma \equiv \Gamma_S - \Gamma_L = 2|\Gamma_{K^0\overline{K}^0}|$. The symmetry properties of the matrix elements are shown in Table 1, together with the parameters ϵ and δ commonly used to describe the breaking of the symmetries CPT, T and CP [1,2]:

TABLE 1. Properties of the Λ matrix elements under the assumption of CPT, T and CP invariance and the parameters which describe the breaking of these symmetries.

Symmetry	Λ-matrix properties	Parameters
CPT	$\Lambda_{K^0 K^0} = \Lambda_{\overline{K}^0 \overline{K}^0}$	δ
T	$|\Lambda_{K^0 \overline{K}^0}| = |\Lambda_{\overline{K}^0 K^0}|$	ϵ
CP	$\Lambda_{K^0 K^0} = \Lambda_{\overline{K}^0 \overline{K}^0}$ and $|\Lambda_{K^0 \overline{K}^0}| = |\Lambda_{\overline{K}^0 K^0}|$	$\epsilon_L = \epsilon - \delta$, $\epsilon_S = \epsilon + \delta$

The T and CPT violating parameters ϵ and δ are assumed to have absolute value $\ll 1$ and can be expressed as

$$\epsilon = \frac{\Lambda_{\overline{K}^0 K^0} - \Lambda_{K^0 \overline{K}^0}}{2(\lambda_L - \lambda_S)}, \quad \delta = \frac{\Lambda_{\overline{K}^0 \overline{K}^0} - \Lambda_{K^0 K^0}}{2(\lambda_L - \lambda_S)}. \quad (4)$$

It should be stressed that the study of the time evolution of the neutral kaons and the direct measurement of the three parameters $\mathrm{Re}(\epsilon)$, $\mathrm{Re}(\delta)$ and $\mathrm{Im}(\delta)$ is the most direct way for testing T and possible CPT violation in the neutral-kaon mixing matrix. On the other hand, CP-violating effects are described via $\epsilon_L = \epsilon - \delta$ and $\epsilon_S = \epsilon + \delta$ respectively.

Semileptonic decay asymmetries

The experimental method of CPLEAR consisted of measuring asymmetries between the decay rates of K^0 and $\overline{K}^0$ into various final states f ($f = \pi^+\pi^-$, $\pi^0\pi^0$, $\pi^+\pi^-\pi^0$, $\pi^0\pi^0\pi^0$, $e\pi\nu$) as functions of the proper decay time τ:

$$A_f(\tau) = \frac{R(\overline{K}^0_{t=0} \to f_{t=\tau}) - R(K^0_{t=0} \to f_{t=\tau})}{R(\overline{K}^0_{t=0} \to f_{t=\tau}) + R(K^0_{t=0} \to f_{t=\tau})}. \quad (5)$$

Such asymmetries have the advantage that many acceptances which would otherwise be present in absolute measurements cancel. If, as is the case for this talk, we restrict ourselves to semileptonic ($e\pi\nu$) decays, the following four decay rates, labelled by kaon strangeness and electron charge, can been measured:

$$R_+(\tau) = R(K^0_{t=0} \to e^+\pi^-\nu_{t=\tau}), \quad \overline{R}_-(\tau) = R(\overline{K}^0_{t=0} \to e^-\pi^+\bar\nu_{t=\tau}), \quad (6)$$
$$R_-(\tau) = R(K^0_{t=0} \to e^-\pi^+\bar\nu_{t=\tau}), \quad \overline{R}_+(\tau) = R(\overline{K}^0_{t=0} \to e^+\pi^-\nu_{t=\tau}).$$

Theoretically, these rates correspond to four decay amplitudes which can be described by a combination of the following 4 parameters, with T denoting the corresponding transition matrix:

$$\langle l^+\pi^-\nu|T|K^0\rangle = a+b, \quad \langle l^-\pi^+\bar\nu|T|\overline{K}^0\rangle = a^* - b^*,$$
$$\langle l^-\pi^+\bar\nu|T|K^0\rangle = c+d, \quad \langle l^+\pi^-\nu|T|\overline{K}^0\rangle = c^* - d^*. \quad (7)$$

Here, the charged leptons l^{+-} are either e^{+-} or μ^{+-}. $\mathrm{Re}(a)$ is CPT, T and CP symmetric, while the imaginary parts of all 4 parameters a, b, c and d are T violating. The parameters b and d are CPT violating, while c and d violate the $\Delta S = \Delta Q$ rule. For convenience in fitting the respective asymmetries, the following parameters are also introduced:

$$y = -b/a, \quad x = (c^* - d^*)/(a+b), \quad \bar{x} = (c^* + d^*)/(a-b), \quad (8)$$
$$x_+ = (x+\bar{x})/2, \quad x_- = (x-\bar{x})/2.$$

The quantity y describes CPT violation in semileptonic decays if the $\Delta S = \Delta Q$ rule holds. The quantity x_+ describes the violation of the $\Delta S = \Delta Q$ rule if CPT holds. In this case, $x_+ = c^*/a = x = \bar{x}$, while $x_- = 0$. Finally, if both CPT and the $\Delta S = \Delta Q$ rule are broken, all 5 expressions in (8) are different.

EXPERIMENT

The detector

The CPLEAR detector had a cylindrical geometry and was mounted inside a solenoid magnet with a field of 0.44 T. It allowed the detection of neutral-kaon decays in the decay-time range $0 < \tau < 20\ \tau_S$, where τ_S is the K_S mean life. The 200 MeV/c antiprotons from LEAR/CERN stopped and annihilated inside a gaseous, pressurized hydrogen target in the centre of the detector. The K^0 and $\overline{K}^0$ were produced via the $p\bar{p}$ annihilation channels

$$p\bar{p} \to K^-\pi^+ K^0 \text{ and } p\bar{p} \to K^+\pi^- \overline{K}^0. \tag{9}$$

Strangeness conservation enabled the initial strangeness of the neutral kaon to be tagged by the charge of the accompanying charged kaon. Going radially outwards from the target, the detector consisted of a tracking device (two multiwire proportional chambers, six drift chambers and two layers of streamer tubes), a scintillator - Cherenkov - scintillator 'sandwich' for particle identification and a lead/gas-sampling electromagnetic calorimeter. To cope with the low branching ratio for reactions (9) (2×10^{-3} each), a multilevel trigger system was specially designed to provide full event reconstruction and selection in a few μs. More details regarding the CPLEAR detector can be found in Ref. [3].

Selection and analysis of semileptonic events

The desired $p\bar{p}$-annihilations followed by the decay of the neutral kaon into $e\pi\nu$ were first selected by requiring that the events had four charged tracks and zero total charge, and by identifying one of the decay tracks as an electron or a positron. The lepton identification was performed with a neural network algorithm [4]. No attempt was made to identify muons – the probability to identify a muon as an electron was about 15%. To further reduce the background, kinematically constrained fits were used. A total of 1.3×10^6 $e\pi\nu$-events, with measured decay times $\tau > \tau_S$ survived the above selection criteria. The background-to-signal ratios for the different background channels as a function of the decay time were obtained by a Monte Carlo simulation (Fig. 1a). There is excellent agreement between real and simulated data, as can be seen from Fig. 1b. A conservative estimate of 10% uncertainty was used for the systematic errors on the background levels.

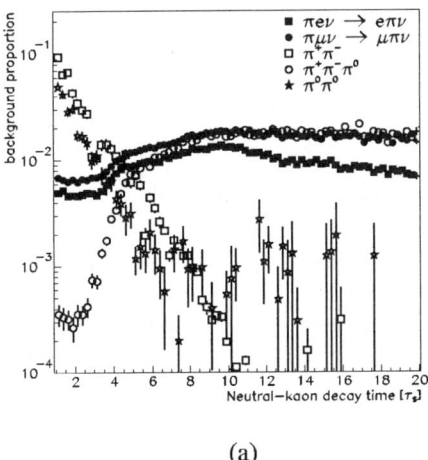 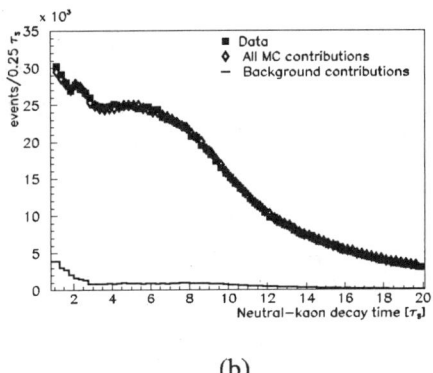

FIGURE 1. a) Background-to-signal ratios for different background channels as indicated in the figure. b) Decay time distribution for real and simulated data.

After background and regeneration correction – for the latter, a dedicated run was performed at CPLEAR in 1996 [5] – the numbers N of events, measured and labelled by the initial kaon strangeness and the decay electron charge, enter the various asymmetries. The detection efficiencies common to the two processes being compared cancel. Differences in the geometrical acceptances are compensated to first order by frequently reversing the magnetic field. However, different detection probabilities for the charged $(K\pi)$ and $(e\pi)$ pairs used for the strangeness tagging lead to different corrections for each event sample. These corrections were performed on an event-by-event basis using two normalization factors, namely ξ and η:

$$\xi = \epsilon(K^+\pi^-)/\epsilon(K^-\pi^+) = \text{efficiency ratio of the charged particles} \quad (10)$$
$$\text{at the primary or production vertex,}$$

$$\eta = \epsilon(\pi^+e^-)/\epsilon(\pi^-e^+) = \text{efficiency ratio of the charged particles} \quad (11)$$
$$\text{at the secondary or decay vertex.}$$

Since ξ does not depend on the decay mode, it can be obtained from the data set of $\pi^+\pi^-$ decays between 1 and 4 τ_S for which statistics is high and background very small [6]. In this time interval, the following formula holds

$$\frac{\xi N(K^0_{t=0} \to \pi^+\pi^-_{t=\tau})}{N(\overline{K}^0_{t=0} \to \pi^+\pi^-_{t=\tau})} = (1 - 4\text{Re}(\epsilon_L)) \times (1 + 4|\eta_{+-}|\cos(\Delta m\tau - \phi_{+-})e^{1/2\Gamma_S\tau}). \quad (12)$$

The oscillating factor on the right side remains smaller than 4% and is known to a precision of 10^{-4}. However, from the $\pi^+\pi^-$ samples, only the factor $\alpha_{2\pi} = \xi(1 + 4\text{Re}(\epsilon_L))$ can be evaluated, with a statistical error of 4.3×10^{-4}. If ξ alone is needed, as was the case for the direct determination of T violation, the quantity $\text{Re}(\epsilon_L)$ is taken from the semileptonic charge asymmetry parameter δ_l,

$$\delta_l = \frac{\Gamma(K_L \to l^+\pi^-\nu) - \Gamma(K_L \to l^-\pi^+\bar{\nu})}{\Gamma(K_L \to l^+\pi^-\nu) + \Gamma(K_L \to l^-\pi^+\bar{\nu})} = 2\,\text{Re}(\epsilon_L) - 2(\text{Re}(x_-) + \text{Re}(y)) \quad (13)$$

experimentally given as $\delta_l = 3.27(12) \times 10^{-3}$ [7]. In the limit of CPT-invariant semileptonic decays, $\delta_l = 2\,\text{Re}(\epsilon_L)$, and $\langle\xi\rangle = 1.12013(43)$.

As far as the efficiency factor η at the decay vertex is concerned, this quantity was measured as a function of the momenta of the decay pions (p_π) and electrons (p_e). Specifically, a sample of π^+ and π^- tracks was selected from minimum bias data, namely from $p\bar{p} \to 2\pi^+\pi^-$, and a sample of e^+e^- pairs was selected from $2\pi^0$ decays of neutral kaons, with $\pi^0 \to 2\gamma$ and $\gamma \to e^+e^-$. The value of η, averaged over the particle momenta, is $\langle\eta\rangle = 1.014(2)$.

DIRECT MEASUREMENT OF T VIOLATION

If time-reversal invariance holds, the probability P, that a $\overline{K}^0$ at $t = 0$ is observed as a K^0 at time $t = \tau$ should be equal to the probability that a K^0 at time $t = 0$ is observed as a $\overline{K}^0$ at $t = \tau$. Any difference between these two probabilities is a clear signal of T violation and can be measured via the time-reversal asymmetry

$$A_T(\tau) = \frac{P(\overline{K}^0_{t=0} \to K^0_{t=\tau}) - P(K^0_{t=0} \to \overline{K}^0_{t=\tau})}{P(\overline{K}^0_{t=0} \to K^0_{t=\tau}) + P(K^0_{t=0} \to \overline{K}^0_{t=\tau})}. \quad (14)$$

Experimentally, this requires the knowledge of the strangeness of the neutral kaons at $t = 0$ and $t = \tau$. As already outlined above, CPLEAR can tag both the strangeness at the primary vertex by measuring the charge of the accompanying kaon and at the secondary vertex by using semileptonic decays and measuring the charge of the accompanying lepton, assuming the validity of the $\Delta S = \Delta Q$ rule. Specifically, we measured, as a function of time, the decay rate asymmetry

$$A_T^l(\tau) = \frac{R(\overline{K}^0_{t=0} \to e^+\pi^-\nu_{t=\tau}) - R(K^0_{t=0} \to e^-\pi^+\bar{\nu}_{t=\tau})}{R(\overline{K}^0_{t=0} \to e^+\pi^-\nu_{t=\tau}) + R(K^0_{t=0} \to e^-\pi^+\bar{\nu}_{t=\tau})}. \quad (15)$$

In the limit of CPT symmetry in the semileptonic decays and the validity of the $\Delta S = \Delta Q$ rule, the asymmetry (15) is identical with (14) and amounts, at long decay times, to $4\,\text{Re}(\epsilon)$. Including the normalization factors at the production and decay vertices as discussed in the preceeding section, CPLEAR [8] has measured the asymmetry

$$A_T^{\text{exp}}(\tau) = \frac{\eta N(\overline{K}^0_{t=0} \to e^+\pi^-\nu_{t=\tau}) - \xi N(K^0_{t=0} \to e^-\pi^+\bar{\nu}_{t=\tau})}{\eta N(\overline{K}^0_{t=0} \to e^+\pi^-\nu_{t=\tau}) + \xi N(K^0_{t=0} \to e^-\pi^+\bar{\nu}_{t=\tau})}. \quad (16)$$

This asymmetry is shown in Fig. 2a, with the systematic errors given in Fig. 2b:

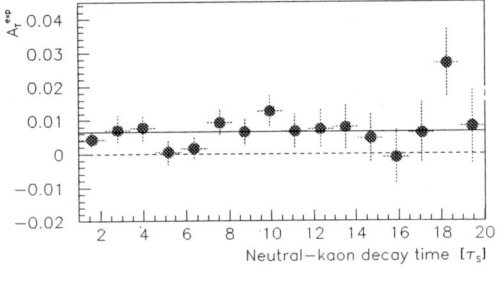

FIGURE 2. a) The experimental asymmetry $A_T^{\text{exp}}(\tau)$. b) Summary of systematic errors.

There is a clear offset from zero, with the average value being

$$\langle A_T^{\text{exp}} \rangle_{(1-20)\tau_S} = (6.6 \pm 1.3 \pm 1.0) \times 10^{-3}. \quad (17)$$

The first error is statistical, the second systematic. The χ^2 value per degree of freedom is 0.84.

Allowing for both CPT and $\Delta S = \Delta Q$ violation, the theoretical expression for $A_T^l(\tau)$ becomes

$$A_T^l(\tau) = 4\text{Re}(\epsilon) - 2(\text{Re}(x_-) + \text{Re}(y)) \quad (18)$$
$$+ 2\frac{\text{Re}(x_-)(e^{-1/2\Delta\Gamma\tau} - \cos(\Delta m\tau)) + \text{Im}(x_+)\sin(\Delta m\tau)}{\cosh(1/2\Delta\Gamma\tau) - \cos(\Delta m\tau)}$$
$$\xrightarrow[\tau \gg \tau_S]{} 4\text{Re}(\epsilon) - 2(\text{Re}(x_-) + \text{Re}(y)). \quad (19)$$

If CPT holds, the experimental value (17) for decay times $\tau \gg \tau_S$ is just $4\text{Re}(\epsilon)$, in perfect agreement with the value expected from η_{+-} [6]. If CPT holds but the $\Delta S = \Delta Q$ rule is broken, the parameters x_- and y disappear, but x_+ is different from zero. In this case, the fit to Equation (18) yields $4\text{Re}(\epsilon) = (6.2 \pm 1.4 \pm 1.0) \times 10^{-3}$ and $\text{Im}(x_+) = (1.2 \pm 1.9 \pm 0.9) \times 10^{-3}$. Again, the first error in the brackets is statistical, the second systematic. Note that $\text{Im}(x_+)$ is T violating anyway!

If CPT is violated, we can use the approximation (19) for long decay times. However, because in the normalization factor $\alpha_{2\pi} = \xi(1 + 4\text{Re}(\epsilon_L))$ we replaced the quantity $2\text{Re}(\epsilon_L)$ by the experimental value of δ_l, we have to correct for the CPT violating terms $2\text{Re}(x_- + y)$. Hence, an additional term $- 2\text{Re}(x_- + y)$ enters the expression $A_T^{\text{exp}}(\tau)$. For $\tau \gg \tau_S$, we then have

$$A_T^{\text{exp}}(\tau) \xrightarrow[\tau \gg \tau_S]{} 4\text{Re}(\epsilon) - 4\text{Re}(x_- + y). \quad (20)$$

After the next section, we shall return to the significance of the term $\text{Re}(x_- + y)$ showing that this term can be safely neglected.

LIMITS ON CPT VIOLATING PARAMETERS

Re(δ), Im(δ), Re(x_-) and Im(x_+)

Here, the relevant asymmetry – with P = respective probabilities – is of the form

$$A_\delta = \frac{P(\overline{K}^0_{t=0} \to \overline{K}^0_{t=\tau}) - P(K^0_{t=0} \to K^0_{t=\tau})}{P(\overline{K}^0_{t=0} \to \overline{K}^0_{t=\tau}) + P(K^0_{t=0} \to K^0_{t=\tau})}. \tag{21}$$

Experimentally, the double asymmetry

$$A_\delta^{\exp}(\tau) = \frac{\eta N(\overline{K}^0_{t=0} \to e^+\pi^-\nu_{t=\tau}) - \alpha_{2\pi} N(K^0_{t=0} \to e^-\pi^+\bar\nu_{t=\tau})}{\eta N(\overline{K}^0_{t=0} \to e^+\pi^-\nu_{t=\tau}) + \alpha_{2\pi} N(K^0_{t=0} \to e^-\pi^+\bar\nu_{t=\tau})}$$
$$+ \frac{N(\overline{K}^0_{t=0} \to e^-\pi^+\bar\nu_{t=\tau}) - \eta\alpha_{2\pi} N(K^0_{t=0} \to e^+\pi^-\nu_{t=\tau})}{N(\overline{K}^0_{t=0} \to e^-\pi^+\bar\nu_{t=\tau}) + \eta\alpha_{2\pi} N(K^0_{t=0} \to e^+\pi^-\nu_{t=\tau})} \tag{22}$$

has been measured. In the long decay time limit this expression becomes

$$A_\delta(\tau) \xrightarrow[\tau \gg \tau_S]{} 8\,\mathrm{Re}(\delta). \tag{23}$$

Fitting the $A_\delta^{\exp}(\tau)$ data with the complete phenomenological expression which depends, in addition to the normalization factor $\alpha_{2\pi} = \xi(1 + 4\,\mathrm{Re}(\epsilon_L))$, on the parameters Re($\delta$), Im($\delta$), Re($x_-$) and Im($x_+$), yields the values [9]

$$\mathrm{Re}(\delta) = (3.0 \pm 3.3 \pm 0.6) \times 10^{-4},\ \mathrm{Im}(\delta) = (1.5 \pm 2.3 \pm 0.3) \times 10^{-2}, \tag{24}$$
$$\mathrm{Re}(x_-) = (0.2 \pm 1.3 \pm 0.3) \times 10^{-2},\ \mathrm{Im}(x_+) = (1.2 \pm 2.2 \pm 0.3) \times 10^{-2}.$$

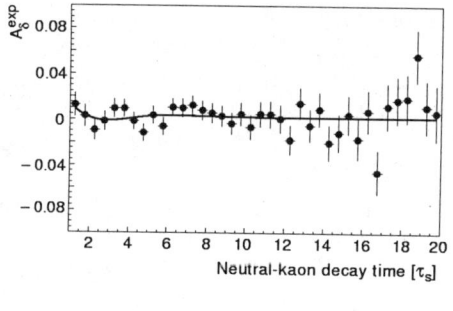

Source	Re(δ) [10^{-4}]
Bckg level	± 0.01
Bckg asym.	± 0.02
ξ	± 0.5
η	± 0.02
τ resolution	negligible
Regeneration	± 0.25
Total syst.	± 0.6

(a)   (b)

FIGURE 3. (a) The experimental asymmetry $A_\delta^{\exp}(\tau)$. (b) Summary of systematic errors.

Figure 3 shows the fit including a summary of systematic errors. Re(δ) is compatible with zero and about 50 times more accurate than in previous measurements. It does neither depend on an external measurement of Re(ϵ_L) nor on y.

Limits including unitarity

The above obtained limits for the CPT violating parameters can be improved by performing a global fit of the neutral-kaon data using the Bell-Steinberger or unitarity relation [10]. Essentially, this amounts to include all known decay channels of the neutral kaons in the analysis. If A_{fL} and A_{fS} are the K_L and K_S decay amplitudes to a given final state f, one obtains with Bell-Steinberger the constraint

$$\text{Re}(\epsilon) - i\text{Im}(\delta) = \frac{1}{2(i\Delta m + \frac{\Gamma_s + \Gamma_L}{2})} \times \sum_f A_{fL} A_{fS}^* . \quad (25)$$

The analysis yields [11]

$$\begin{aligned}
&\text{Re}(\delta) = (2.4 \pm 2.8) \times 10^{-4}, \ \text{Im}(\delta) = (2.4 \pm 5.0) \times 10^{-5}, \\
&\text{Re}(\epsilon) = (164.9 \pm 2.5) \times 10^{-3}, \ \text{Re}(y) = (0.3 \pm 3.1) \times 10^{-3}, \\
&\text{Re}(x_-) = (-0.5 \pm 3.0) \times 10^{-3}, \ \text{Im}(x_+) = (-2.0 \pm 2.7) \times 10^{-3},
\end{aligned} \quad (26)$$

with a χ^2/d.o.f. of 1.09. While the accuracy of $\text{Re}(\delta)$ is only slighty better than in the direct measurement [see (24)], the error of $\text{Im}(\delta)$ is 500 times smaller. In addition, the two parameters $\text{Re}(x_-)$ and $\text{Re}(y)$ have a strong negative correlation, and their sum can thus be given with a considerably smaller error than the individual terms:

$$\text{Re}(x_- + y) = (-0.2 \pm 0.3) \times 10^{-3} . \quad (27)$$

This term enters into equation (19). It's smallness excludes the possibility that the measured direct time violation may eventually be due to CPT violation in the decay (see also conclusions).

K^0 and $\overline{K}^0$ mass and decay width differences

With the values of $\text{Re}(\delta)$ and $\text{Im}(\delta)$ given in (26), together with the superweak phase $\phi_{SW} = (43.5 \pm 0.1)°$ [7], the projections of δ along the ϕ_{SW} axis become

$$\begin{aligned}
\delta_\| &= \text{Re}(\delta)\cos(\phi_{SW}) + \text{Im}(\delta)\sin(\phi_{SW}), \\
\delta_\perp &= -\text{Re}(\delta)\sin(\phi_{SW}) + \text{Im}(\delta)\cos(\phi_{SW}) .
\end{aligned} \quad (28)$$

These projections are directly proportional to the K^0 and $\overline{K}^0$ mass and decay width differences

$$\delta_\| = \frac{1}{4} \frac{\Gamma_{K^0 K^0} - \Gamma_{\overline{K}^0 \overline{K}^0}}{\sqrt{\Delta m^2 + (\frac{\Delta\Gamma}{2})^2}} \text{ and } \delta_\perp = \frac{1}{2} \frac{M_{K^0 K^0} - M_{\overline{K}^0 \overline{K}^0}}{\sqrt{\Delta m^2 + (\frac{\Delta\Gamma}{2})^2}} . \quad (29)$$

Using $\Delta m = (3.49 \pm 0.01) \times 10^{-15}$ GeV/c^2 and $\Delta\Gamma = (7.355 \pm 0.007) \times 10^{-15}$ GeV [7], we then obtain [12]:

$$\Gamma_{K^0 K^0} - \Gamma_{\overline{K}^0 \overline{K}^0} = (3.9 \pm 4.2) \times 10^{-18} \text{ GeV}, \tag{30}$$
$$M_{K^0 K^0} - M_{\overline{K}^0 \overline{K}^0} = (-1.5 \pm 2.0) \times 10^{-18} \text{ GeV}/c^2.$$

If we assume that only the mass matrix violates CPT, the limit for a possible difference of its diagonal terms becomes $< 12.7 \times 10^{-19}$ GeV/c^2. If only the $\pi\pi$ channel contributes to the unitarity relation, this limit falls to $< 4.4 \times 10^{-19}$ GeV/c^2. Both limits are given at 90% confidence level.

CONCLUSIONS

CPLEAR has established lower limits for many relevant CPT violating parameters. In particular, by using the Bell-Steinberger relation, the limit for Im(δ) was lowered by more than two orders of magnitude. CPLEAR has also obtained the same masses and decay widths for K^0 and $\overline{K}^0$ down to 10^{-18} GeV. The fact that the phases ϕ_{+-}, ϕ_{00} and ϕ_{SW} are all the same within less than 0.3^0 [6, 7] is a further confirmation that CPT still remains a good symmetry.

As far as microscopic reversibility is concerned, CPLEAR has, for the first time, found direct evidence for T violation in the neutral-kaon mixing matrix. Due to the low limits for the CPT violating parameters Re($x_- + y$) which enter in the expression of the experimentally measured time reversal asymmetry $A_T^{\exp}(\tau)$, the hypothesis of T conservation in the mixing and CPT violation in the decay can be excluded. Indeed, J. Ellis et al. have recently shown that Re($x_- + y$) is at least one order of magnitude too small to mimic T violation in the mixing by CPT violation in the decay [13].

REFERENCES

1. C.D. Buchanan et al., *Phys. Rev.* **D45**, 4088 (1992).
2. L. Maiani, in *2nd DAΦNE Physics Handbook*, INFN, Frascati, 1995, pp. 3-26.
3. R. Adler et al., CPLEAR Collaboration, *Nucl. Instrum. Meths.* **A309**, 76 (1996).
4. CPLEAR/DET/95–02 (1995), Internal Report.
5. A. Angelopoulos et al., CPLEAR Collaboration, *Phys. Lett.* **B413**, 422 (1997).
6. A. Apostolakis et al., CPLEAR Collaboration, Phys. Lett. **B458**, 545 (1999).
7. C. Caso et al., Particle Data Group, *Eur. Phys. J.* **C3**, 1 (1998).
8. A. Angelopoulos et al., CPLEAR Collaboration, *Phys. Lett.* **B444**, 43 (1998).
9. A. Angelopoulos et al., CPLEAR Collaboration, *Phys. Lett.* **B444**, 52 (1998).
10. J.S. Bell, J. Steinberger, in *Proc. of the Oxford Int. Conf. on Elementary Particles*, Rutherford Laboratory, Chilton, England, 1965, pp. 195-222.
11. A. Apostolakis et al., CPLEAR Collaboration, *Phys. Lett.* **B456**, 297 (1999).
12. A. Angelopoulos et al., CPLEAR Collaboration, *Phys. Lett.* **B471**, 332 (1999).
13. J. Ellis and N.E. Mavromatos, *Phys. Rep.* **320**, 341 (1999).

Latest Results from BaBar

Brad Abbott
For the BaBar Collaboration

Lawrence Berkeley National Laboratory
1 Cyclotron Road, MS 50A-2160, Berkeley, CA 94720
E-mail: bkabbott@lbl.gov

Abstract. The BaBar experiment at SLAC recorded its first collisions on May 26, 1999 and has collected nearly 3.5 fb^{-1} by the middle of March 2000. Detector performance studies and preliminary results on physics are presented, emphasizing the components of the analysis on the CP violating parameter $\sin(2\beta)$ using $B^\circ \to J/\Psi K^\circ_s$ decays.

INTRODUCTION

The BaBar experiment [1] located at SLAC is primarily designed to measure CP violation in B° decays. As of March, 2000 the BaBar detector has accumulated nearly 3.5 fb^{-1} of data since its initial startup on May 26, 1999. The primary focus has been to understand the detector performance and preliminary results from each detector subsystem are presented. Also presented are preliminary studies focusing on the measurement of $\sin(2\beta)$ in the final state $B^\circ \to J/\Psi K^\circ_s$.

The relatively large branching ratio (10^{-5}) into a CP eigenstate, clean decay ($J/\Psi \to l^+l^-$, $K^\circ_s \to \pi^+\pi^-$) and theoretical ease of equating the measured asymmetry to $\sin(2\beta)$ make $B^\circ \to J/\Psi K^\circ_s$ particularly interesting. The essential ingredients necessary to make this measurement are: reconstruction of one B into J/Ψ and K°_s, tagging the flavor of the other B and measuring the differences between the z vertex of the two B decays(Δz). By using the approximation that $\Delta z = \beta\gamma c \Delta t$, the time between the two B decays (Δt) can be determined. The time dependent asymmetry for any CP eigenstate f can be parameterized as

$$\frac{\Gamma(B^\circ(t) \to f) - \Gamma(\bar{B}^\circ(t) \to f)}{\Gamma(B^\circ(t) \to f) + \Gamma(\bar{B}^\circ(t) \to f)} = C_f \cos(\Delta mt) + S_f \sin(\Delta mt) \qquad (1)$$

The parameters C_f and S_f are related to the phase angles in the CKM unitarity triangle. For the decay $B^\circ \to J/\Psi K^\circ_s$, $C_f = 0$ and $S_f \equiv \sin(2\beta)$.

PEP-II

Producing a large enough sample of B decays in order to make a statistically significant measurement of $\sin(2\beta)$ requires unprecendented luminosity. PEP-II is an asymmetric e^+e^- collider which collides 3.1 GeV positrons with 9 GeV electrons. The $B\bar{B}$ pairs are produced from $\Upsilon(4S)$ decays with a boost $\beta\gamma$ of 0.56 in the z direction. This large boost gives an average distance between B decays of 250 μm, which can be measured using modern silicon detectors. The PEP-II design goal is to reach a luminosity of 3×10^{33}cm^{-2}s^{-1}. The accelerator is rapidly approaching this goal and has already achieved a peak luminosity of 1.7×10^{33}cm^{-2}s^{-1}.

A scan was performed on June 15-17, 1999 by varying the electron beam energy to scan the $\Upsilon(4S)$ resonance. The ratio of the number of multihadron events to Bhabha events was determined for each center of mass energy. The data were fit to a model including the $\Upsilon(4S)$ lineshape and the known spread in the PEP-II center of mass energy. The data with the fit superimposed is shown in Figure 1.

The results of the fit yield

$$M_{\Upsilon(4S)} = 10.5841 \pm 0.0007 \text{ GeV}, \quad \Gamma_{\Upsilon(4S)} = 11.1 \pm 3.4 \text{ MeV}$$

where the fitted mass is measured on the PEP-II energy scale.

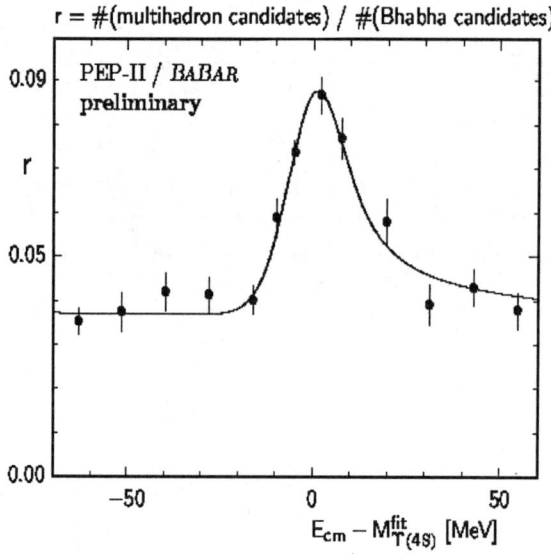

FIGURE 1. The ratio of multihadron to Bhabha candidates selected from samples recorded by varying the center of mass energy. The curve shows a fit to the data.

DETECTOR PERFORMANCE

The detector closest to the beam pipe is the silicon vertex tracker (SVT). It consists of five layers of double-sided, AC-coupled 300 μm thick silicon. The angular acceptance is limited by the machine components and extends from -0.87 $\leq \cos\Theta_{lab}$ $\leq$ 0.96. The SVT's primary function is to provide the vertex information, but it must also provide independent tracking for charged particles with low transverse momentum between approximately 40 MeV/c and 100 MeV/c. The alignment of the SVT is critical to achieve optimal vertex resolution. The spatial resolution achieved for layer 1 which is the most important layer for determining the B decay vertex is shown in Figure 2. Also shown in Figure 2 is a comparison to Monte Carlo expectations with perfect wafer alignment. The agreement to Monte Carlo expectations is quite good.

The determination of the track momentum is primarily measured in the drift chamber. The drift chamber consists of 7104 hexagonal drift cells with axial wires running parallel to the detector axis and u-v stereo layers with angles between 42 and 70 mrad. The gas is a 80%/20% He/C_4H_{10} mix to reduce multiple scattering. The drift chamber resides inside a 1.5T superconducting magnet. The momentum resolution for the drift chamber is shown in figure 3. The drift chamber was designed to provide an average single hit resolution of 140 μm; 125 μm has been achieved. Good dE/dX resolution is also achieved with the drift chamber. The drift chamber

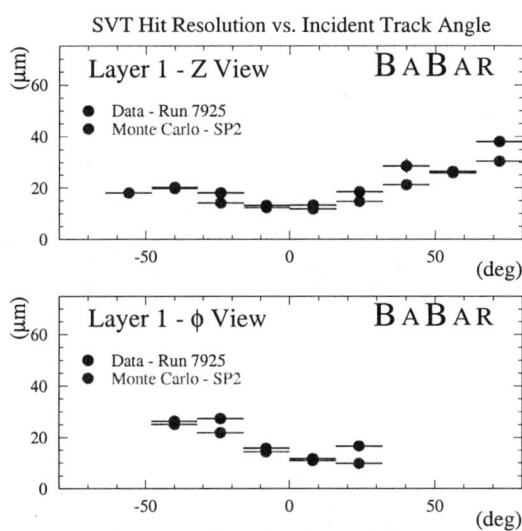

FIGURE 2. The measured hit resolution as a function of incident track angle for the inner layer of the SVT. Also shown is the Monte Carlo expectation for the hit resolution.

provides a dE/dX resolution of 7.5% for Bhabhas and provides a 3σ separation between kaons and pions up to 0.7 GeV.

Identification of kaons is the primary responsibility of the DIRC. The DIRC (Detection of Internally Reflected Cherenkov light) consists of 144 quartz bars 4.7 m in length arranged in a 12 sided polygon around the drift chamber. A charged particle traversing the quartz emits Cherenkov light which reflects within the bar until it reaches the end where it is transmitted via a water filled tank to an array of photomultiplier tubes. The measure of the Cherenkov angle along with momentum information from the drift chamber allows the particle type to be identified. Figure 4 shows the mass of D $\to$ Kπ candidates with no particle identification and for kaons identified in the DIRC. Within the acceptance, the DIRC provides a background rejection of approximately 5 while retaining a kaon efficiency of nearly 80%.

The primary goal of the calorimeter is to identify π^o mesons with high efficiency and good resolution, to measure photons between 20 MeV and 4 Gev and to aid in electron ID. The electromagnetic calorimeter consists of 6580 Tl-doped CsI crystals arranged in a barrel and endcap structure which are readout by photodiodes. To identify π^os, the invariant mass distribution of all neutral clusters(photon candidates) are formed. The invariant mass distribution for photons above 100 Mev with an energy sum above 1000 MeV is shown in Figure 5. The measured π^o mass is consistent with expectations while the resolution is slightly above the Monte Carlo value.

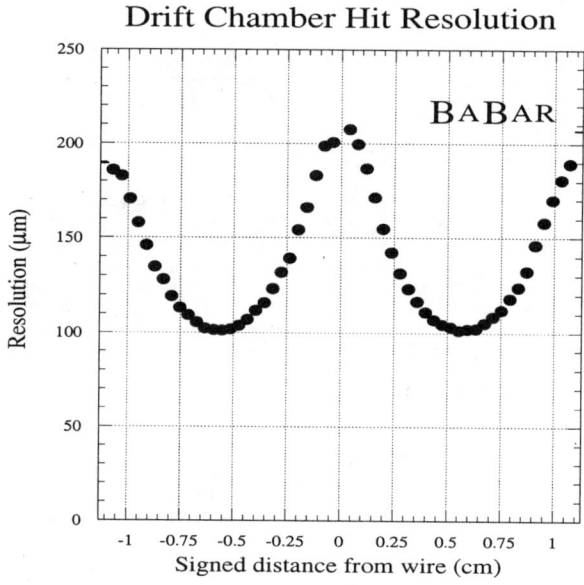

FIGURE 3. Measured spatial resolution in the drift chamber as a function of the distance from the signal wire.

The primary goal of the instrumented flux return (IFR) detector is to identify muons and K_L^o. The detector consists of 19 layers of resistive plate chambers which provide coverage for the 2500 m^2 area. Muons are identified within the IFR by using a tight selection which requires a minimum ionizing particle in the calorimeter, a significant number of layers hit in the IFR and a large number of interaction lengths traversed. The efficiency for identifying muons from ee$\mu\mu$ decays and the corresponding pion fake rate is shown in Figure 6.

PRELIMINARY PHYSICS RESULTS

D^* mesons are identified in the decay $D^* \to D^o \pi^+$, with the D^o decaying to Kπ. Using only kinematic cuts and no particle identification criteria, D^o candidates are found by combining two oppositely charged tracks coming from a common vertex. D^* candidates are formed by combining the D^o candidates with any other charged track. A mass constraint is applied requiring that the D^o and D^*-D^o mass is as expected. Requiring that the D^* originate from the beam spot is also used to improve the D^*-D^o mass resolution. The mass distributions are shown in Figure 7.

K_s^o decaying to $\pi^+\pi^-$ are found by combining two oppositely charged tracks from a common vertex. Additional cuts require the K_s^o to point back to the interaction point and require a minimum transverse momentum in the decay. The K_s^o mass

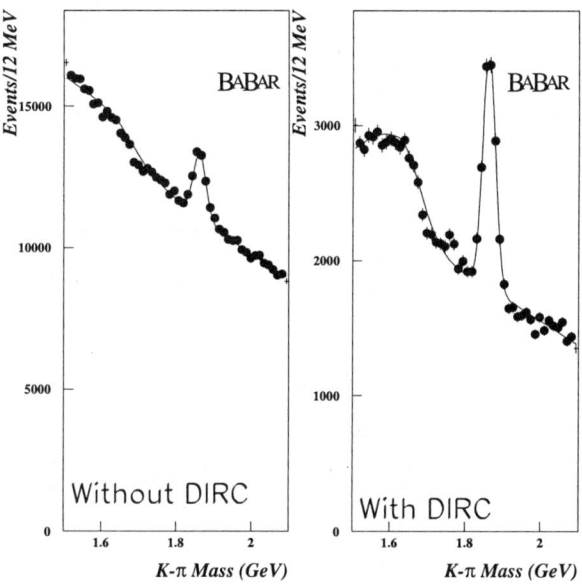

FIGURE 4. The D$\to$ Kπ mass distribution with and without DIRC kaon identification. Within the acceptance, the DIRC provides a background rejection of approximately 5 while retaining a kaon efficiency of nearly 80%.

distribution is shown in Figure 8. The narrow peak, containing 70% of the events, has a width of 2.8 MeV while the wide peak has a width of 11 MeV.

J/Ψ candidates are formed by combining e^+e^- and $\mu^+\mu^-$ coming from a common vertex. Electrons are identified by requiring a shower shape consistent with expectations and the ratio of energy measured in the calorimeter to the measured track momentum be close to unity. Muons are identified by requiring a minimum ionizing track within the calorimeter and hits in the IFR. The mass distribution for e^+e^- is shown in Figure 9. The long low mass tail in the e^+e^- mass distribution is due to bremsstrahlung. The mass resolution for both e^+e^- and $\mu^+\mu^-$ decays is approximately 15 MeV.

Candidate $B^\circ \to J/\Psi\ K^\circ_s$ decays are found by selecting events with a J/Ψ candidate and combining it with a K°_s candidate which form a B° with the expected mass. To better constrain the fit we use ΔE and the beam constrained mass M_b.

$$M_b = \sqrt{E(beam)^2 - p_B^2}, \Delta E = E_B - E_{beam} \qquad (2)$$

Shown in Figure 10 is the distribution of events in the ΔE - M_b plane. There are 12 events located within the signal region indicated by the rectangle with an expected background of 1.4 events. The event sample used for this analysis consists of a total luminosity of 600 pb^{-1}. A similar analysis was performed by studying $B^+ \to J/\Psi\ K^+$ decays. This analysis is an important control sample since it is very similar to $B^\circ \to J/\Psi\ K^\circ_s$, but with no expected asymmetry. Using 600 pb^{-1} of

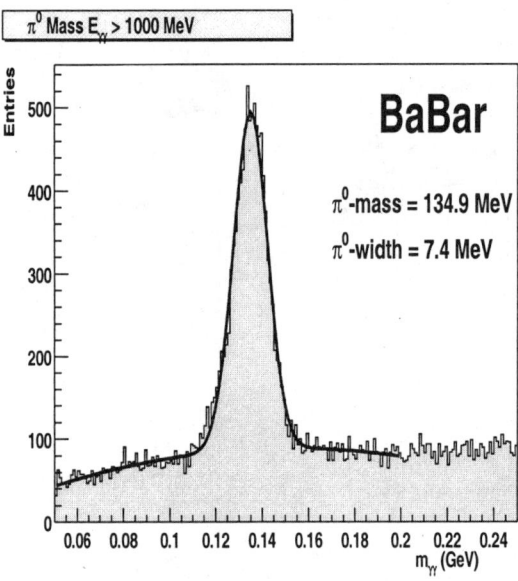

FIGURE 5. The π° mass determined by combining neutral energy clusters in the calorimeter with energy greater than 100 MeV and the energy sum above 1000 MeV.

data, 32 candidate $B^+ \to J/\Psi\ K^+$ events are found with an estimated background of 5.8 events.

The Δz distribution for $B^+ \to J/\Psi\ K^+$ and $B^\circ \to J/\Psi\ K^*$ candidates using 1 fb^{-1} of data is shown in Figure 11. Also shown is the Monte Carlo expectations. This sample provides a good sample to understand detector induced asymmetries. Comparisons to the Monte Carlo expectations show good agreement with no measureable asymmetries.

CONCLUSIONS

The BaBar experiment has collected nearly 3.5 fb^{-1} of data and initial studies of detector performance are underway. The preliminary results on detector performance show the detector is working well. Preliminary results on physics emphasizing the measurement of the CP violating parameter $\sin(2\beta)$ have been presented. The BaBar experiment expects to accumulate 10 fb^{-1} of data by summer of 2000.

REFERENCES

1. BaBar Collaboration, BaBar Technical Design Report, SLAC-R-95-457, March 1995.

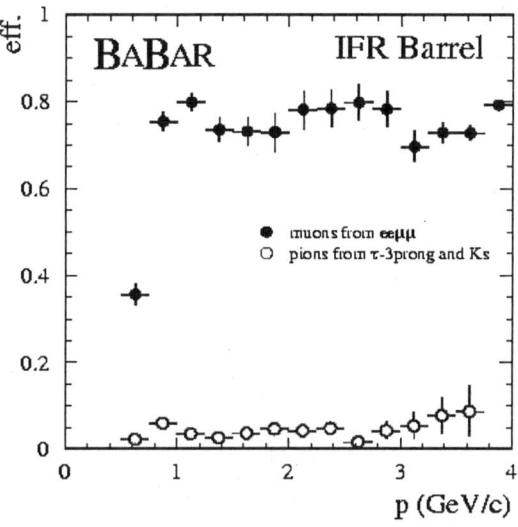

FIGURE 6. Efficiency for identifying muons in the IFR. The solid points are the efficiency for identifying muons from $ee\mu\mu$ events. The open points are the efficiency to identify a pion from $\tau \to 3$ prong decays and K_s° decays. The muon efficiency includes a 10% effect due to the geometry of the IFR.

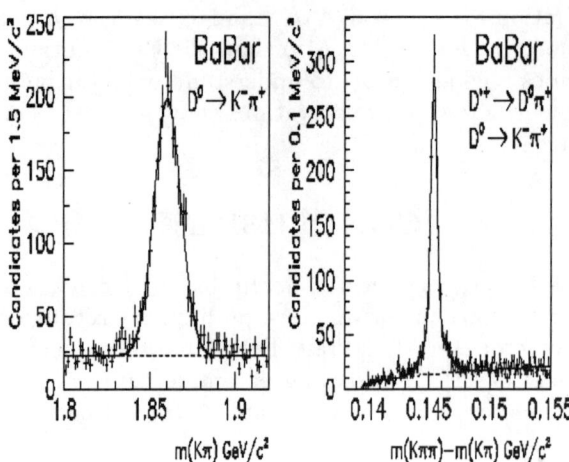

FIGURE 7. The D meson mass and D^*-D^o mass from data. The mass resolution for the D meson is 8 MeV. The D^*-D^o mass peak is fit to a double gaussian which yields a narrow gaussian width of 252 KeV with 59% of the data in the narrow gaussian peak.

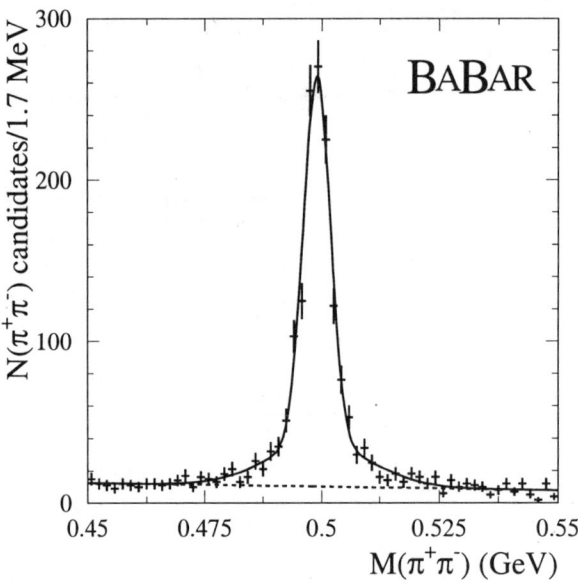

FIGURE 8. Mass distribution for $K_s^o \to \pi^+\pi^-$ The narrow peak, containing 70% of the events, has a width of 2.8 MeV while the wide peak has a width of 11 MeV.

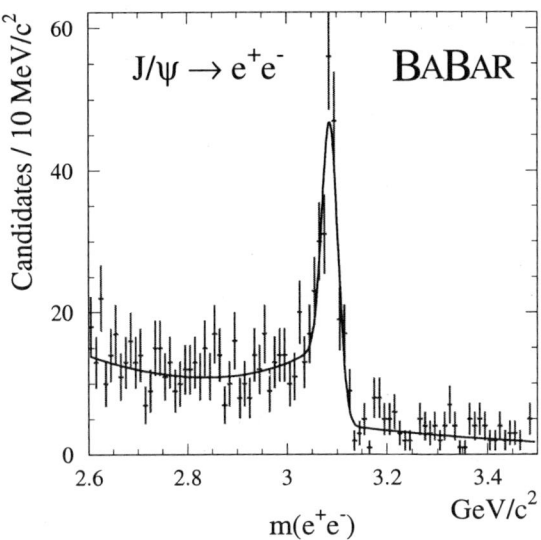

FIGURE 9. Mass distribution for $J/\Psi \to e^+e^-$. The low mass tail is due to bremsstrahlung.

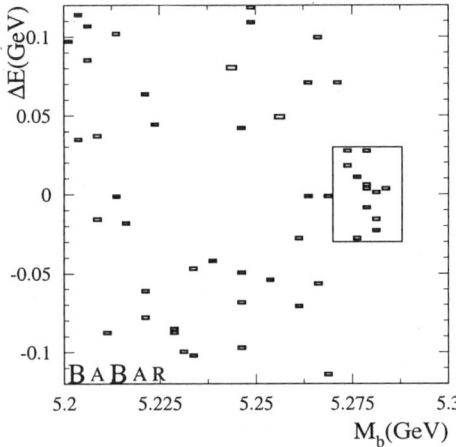

FIGURE 10. The distribution of events for $B \to J/\Psi \; K_s^o$ candidates in the Δ E- M_b plane using 600 pb^{-1}. There are 12 events found in the signal region indicated by the rectangle with an expected background of 1.4 events.

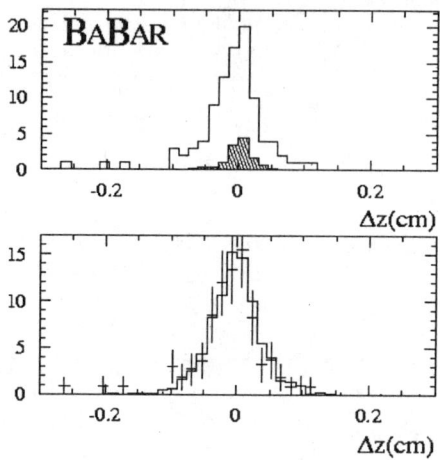

FIGURE 11. The Δz distribution for $B^+ \to J/\Psi\ K^+$ and $B^o \to J/\Psi\ K^*$ candidates. Top Figure: Open histogram is the Δz distribution for signal events. The shaded histogram is the Δz distribution for the expected background. Bottom Figure: Open histogram is the signal after background subtraction. The crosses are the Monte Carlo expectations.

Symmetries in Parton Distributions

J.T. Londergan

*Dept. of Physics and Nuclear Theory Center, Indiana University
Bloomington, IN 47405, USA;
E-mail: tlonderg@iucf.indiana.edu* [1]

Abstract. Significant violation of flavor symmetry in the proton sea has been revealed by comparing pp and pD Drell-Yan processes. We review this and discuss theoretical interpretation of these results. We argue that substantial flavor-dependent effects should occur in hyperon production, and review experimental evidence for such effects. We review the status of parton charge symmetry. Precise measurements of structure functions in muon and neutrino reactions initially suggested substantial charge symmetry violating (CSV) contributions, but re-analysis of this data shows small parton CSV effects.

The basic features of parton distributions have been well established through measurements of deep inelastic scattering [DIS], Drell-Yan processes and direct photon experiments. Precision tests of approximate symmetries allow us to understand the details of nucleon parton distributions. Here we will review the current status of flavor symmetry and charge symmetry of parton distributions.

Flavor Symmetry in Parton Distributions

Flavor symmetry refers to the expectation that the antiquark distributions for up, down and strange quarks should be equal. Although there is no symmetry principle which governs this notion, if antiquarks arose exclusively from gluon radiation and if all quarks had identical mass, then antiquarks of all flavors should be produced in equal amounts (ignoring Pauli principle effects). The first statistically significant evidence for flavor symmetry effects in the light quark sector came from the NMC experiment [1], which measured μp and μD DIS, and accurately tested the Gottfried Sum Rule S_G, which is given by

$$S_G = \int_0^1 [F_2^{\mu p}(x) - F_2^{\mu n}(x)] \frac{dx}{x} = \frac{1}{3} - \frac{2}{3} \int_0^1 [\bar{d}^p(x) - \bar{u}^p(x)] \, dx \qquad (1)$$

Assuming $\bar{d}^p(x) = \bar{u}^p(x)$ one predicts $S_G = 1/3$. Perturbative QCD (pQCD) predicts very small deviations from 1/3. The NMC result $S_G = 0.235 \pm 0.026$ was

---

[1] Research supported in part by NSF research contract PHY-9722706.

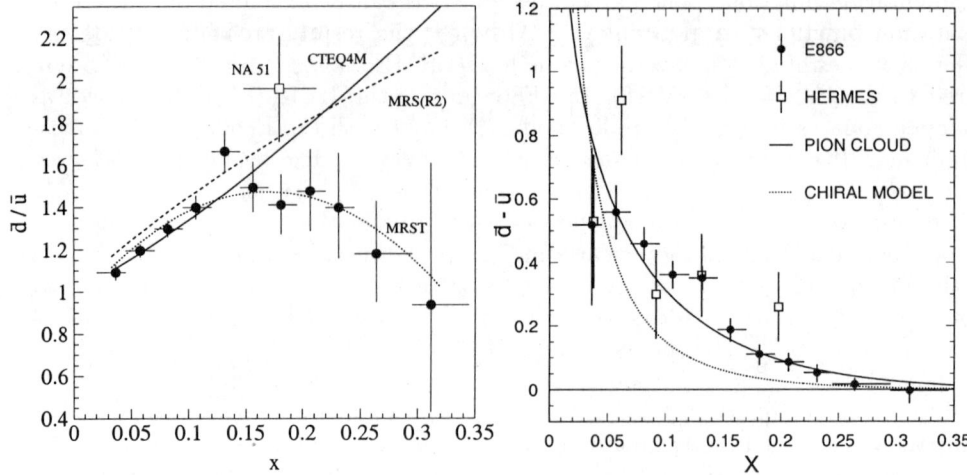

FIGURE 1. (a) The ratio R_{DY} of Eq. 2 as measured by E866 experiment. (b) The difference $\bar{d}^p(x) - \bar{u}^p(x)$ extracted from experiment. Solid circles: E866 data; open squares: HERMES data. Solid curve: meson-cloud model of Ref. [3]; dotted curve: chiral model of Ref. [4].

four standard deviations lower than the "naive" prediction, apparently indicating substantial flavor symmetry violation in the proton sea.

The E866 experiment at Fermilab then compared pp and pD Drell-Yan [DY] processes [2]. For large x_F the ratio of DY cross sections is given approximately by

$$R_{DY} \equiv \sigma^{pD}_{DY}/(2\sigma^{pp}_{DY}) \approx (1 + \bar{d}^p(x_t)/\bar{u}^p(x_t))/2 , \qquad (2)$$

where x_t is the momentum fraction of the target sea quarks. This ratio will be larger than one if $\bar{d}^p(x) > \bar{u}^p(x)$. The E866 results are shown in Fig. 1. Fig. 1a shows the ratio $\bar{d}^p(x)/\bar{u}^p(x)$ extracted from the E866 data. This clearly shows that $\bar{d}^p(x) > \bar{u}^p(x)$ for $x < 0.3$. If the averaged sea quark distributions are taken from other experiments, this data can be used to extract the difference $\bar{d}^p(x) - \bar{u}^p(x)$. This is shown in Fig. 1b. The data agrees rather well with pion-cloud calculations of Kumano [3]; a chiral model calculation of Eichten et al. [4] produces flavor asymmetries which are softer than experiment. The E866 group has estimated the integrated difference in light sea quarks as

$$< \bar{d}^p - \bar{u}^p > = 0.100 \pm 0.007(stat) \pm 0.017(syst) . \qquad (3)$$

This is about 2/3 as large as the result $< \bar{d}^p - \bar{u}^p > = 0.147 \pm 0.039$ extracted by the NMC group.

The HERMES collaboration at the HERA facility [5] has studied the flavor dependence of the nucleon sea by measuring semi-inclusive DIS for charged hadrons produced from scattering of positrons on p, D and ^{3}He targets. The semi-inclusive cross sections are proportional to the product of parton distributions times quark fragmentation functions. The parton distributions can be extracted once the fragmentation functions are determined. Although the results are obtained at very different values of Q^2 (the average value for HERMES is $<Q^2> = 2.3$ GeV^2, compared to $<Q^2> = 54$ GeV^2 for the E866 data), the HERMES data (shown as the open squares in Fig. 1b) are in good agreement with the E866 Drell-Yan measurements. These results are summarized in a review article by Peng and Garvey [6].

The observed flavor symmetry effects are far greater than can be accounted for by perturbative QCD, so they clearly indicate substantial non-perturbative effects in the proton sea. The most promising theoretical model to date is the "meson-cloud" picture, which includes a quark "core" for the nucleon plus a "cloud" of baryon-meson Fock components. The virtual photon scatters from any of these components, a process first proposed by Sullivan [7]. Melnitchouk et al. [8] obtained qualitative agreement with E866 data from a model including nucleon, pion and Δ components (see the talk by Melnitchouk in the proceedings of this conference). An interesting chiral soliton model has been proposed by Diakonov and collaborators [9]. This picture is taken in the large-N_c limit. This model predicts large flavor-dependent spin effects in the nucleon sea, i.e. that the quantity $\Delta \bar{u}^p(x) - \Delta \bar{d}^p(x)$ will be large and positive. Such effects could be measured either in semi-inclusive DIS for polarized $e-N$ reactions, or by measuring the $W^\pm$ asymmetry in polarized pp collisions at RHIC. Similar large spin-dependent flavor effects are predicted in a statistical model of parton distributions by Bhalerao [10], and large spin effects may also be present in calculations which take into account Pauli effects in the nucleon quark distributions [11].

Flavor Dependent Effects in Hyperon Production

The Adelaide group [12] have produced a method for extracting leading twist quark distributions from quark model wavefunctions,

$$q^{\uparrow\downarrow}(x) = 2P^+ \sum_n |\langle n; p_n | \psi_+^{\uparrow\downarrow}(0) | B; PS \rangle|^2 \delta(P^+(1-x) - p_n^+) \quad (4)$$

Eq. 4 gives the quark distributions inside a baryon B, for quarks with helicity parallel ($q^\uparrow(x)$) or antiparallel ($q^\downarrow(x)$) to the baryon's helicity. This equation includes a sum over a complete set of states n, with momentum $p_n^+ = \sqrt{M_n^2 + p_n^2} + p_{nz}$. These are the states in a baryon which remain following a hard collision which removes a single quark. The contribution to a quark distribution $q(x)$ from an intermediate state with mass M_n peaks at

$$x_{max} \approx 1 - M_n/M \quad . \quad (5)$$

The effective mass M_n in Eq. 4 depends on the helicity and flavor of the struck quark. This in turn produces a spin-flavor dependence in the resulting parton distributions.

Such spin-flavor effects can be very significant in the case of hyperon parton distributions, as was first pointed out by Alberg et al. [13]. For large x the dominant contributions come from two-quark intermediate states. The diquark effective masses for non-strange quarks can be estimated from fitting $N - \Delta$ splitting; one obtains a scalar effective mass $M_2(S = 0) \sim 600$ MeV, and a vector effective mass $M_2(S = 1) \sim 800$ MeV. The corresponding effective masses for two-quark states containing one s quark can be obtained from $\Lambda - \Sigma$ splitting, giving $M_2'(S = 0) \sim 890$ MeV and $M_2'(S = 1) \sim 1010$ MeV. The simplest quark model picture of the Λ couples an s quark to a ud pair with $S = 0$. Arguments based on $SU(3)$ symmetry generally assume equal distributions for all three quarks in the Λ. Inserting the relevant states into Eq. 5, we see that removing a strange quark from the Λ should produce a peak in the s distribution at $x_{max} \approx 0.48$. If a u quark in the Λ is struck, this leaves behind a sd pair which is most probably in an $S = 1$ state. The resulting vector diquark produces a peak in the u quark distribution at $x_{max} \approx 0.12$. From this argument, one expects the s quark distribution in the Λ hyperon to be *much harder* than the corresponding u or d distributions. Applying these arguments to spin distributions, one expects the Δu quark distribution in the Λ to be positive at large x, even if the average of this distribution over all x is zero [14].

Similar arguments have been applied to the fragmentation functions of quarks into hyperons, by Boros et al. [14]. Analogous to the method for obtaining parton distributions from quark model wavefunctions, one can generate fragmentation functions $D(z)$ through the relation

$$\frac{1}{z} D_{q\Lambda}^{\uparrow\downarrow}(z) = P^+ \sum_n |\langle 0|\psi_+^{\uparrow\downarrow}(0)|\Lambda(PS); np_n\rangle|^2 \delta(P^+(\frac{1}{z} - 1) - p_n^+) \qquad (6)$$

Eq. 6 describes the process by which a quark fragments into a Λ hyperon plus an n-quark final state. The contribution from a state with mass M_n to the fragmentation will produce a peak at a value $z_{max} \approx 1/(1 + M_n/M_\Lambda)$. Therefore, the fragmentation function at large z will be dominated by contributions from quark fragmentation leading to a Λ plus a pair of antiquarks ($n = 2$). Because of quark mass differences, and spin-flavor effects in quark interactions, one expects that at large z the fragmentation of s quarks into Λ hyperons will be much larger than that for u or d quarks. This differs significantly from models based on $SU(3)$ quark symmetry, which generally assume that u, d, and s quarks have identical fragmentation into Λ's.

The results of this flavor dependence are shown in Fig. 2, which shows Λ production arising from e^+e^- annihilation at the Z resonance. Fig. 2a shows the spin-averaged production cross sections, which are proportional to the fragmentation functions of quarks into a Λ particle. The dot-dashed curve is the contribution of the s quark fragmentation producing two-quark final states; this contribution, termed "valence" fragmentation, is calculated in the MIT bag model by Boros et al. [14] at the bag scale, and evolved to M_Z^2. The dotted curve is the corresponding contribution from u and d valence fragmentation. The dashed curve is the

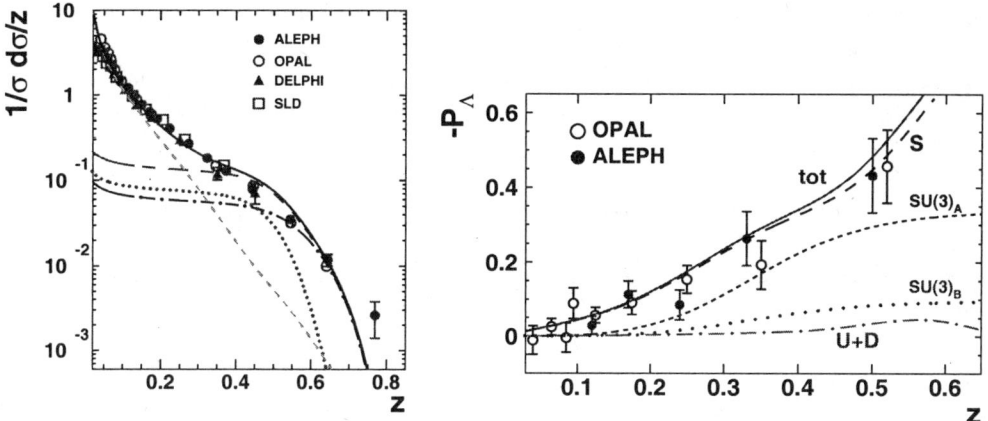

FIGURE 2. e^+e^- annihilation at the Z resonance. (a) Inclusive Λ production cross section. Dash-dot (dotted) curves: s quark (u, d quark) "valence" fragmentation into Λ. Dashed line: "sea" fragmentation contribution. (b) Λ polarization. Solid curve: full calculation; dashed curve: s quark contribution. Dashed, dotted curves: results of assuming $SU(3)$ symmetry for unpolarized quark fragmentation.

contribution from quark fragmentation leading to $n = 4$ and higher states. This contribution is fit to the data, which has been taken at both LEP and SLD facilities [15]. This simple model gives quite a good fit to the experiment; at large z note that the cross sections are dominated by s quark fragmentation.

The resulting longitudinal Λ polarization is shown in Fig. 2b. At large z the polarization is large and negative; it is predicted that $P_\Lambda \to -0.94$ in the limit $z \to 1$. The solid curve is the calculation of Boros et al., while the dotted and dashed curves are the results assuming $SU(3)$ symmetry for the spin-averaged fragmentation. Curve $SU(3)_A$ assumes the naive quark model result, $\Delta D_{s\Lambda} = D_{s\Lambda}; \Delta D_{u\Lambda} = \Delta D_{d\Lambda} = 0$. The curve $SU(3)_B$ results from using the parton spin distributions measured for nucleons, and converts these the the corresponding Λ parton distributions using $SU(3)$. This produces hyperon polarizations which differ radically from the data. De Florian et al. [16] obtain closer agreement with the polarized data using $SU(3)$-based arguments, however their spin-averaged fragmentation functions are freely varied to reproduce the hyperon production cross sections.

Λ polarization can also be measured in semi-inclusive DIS. Here one detects charged particles, such as mesons, which arise from DIS of leptons on nucleons. In such reactions Λ production at large z is dominated by u quark fragmentation, since u quark distributions are much larger than s quarks for nucleons in the valence region, and the up quark contribution is further enhanced by the coupling of the virtual photon, which is proportional to the squared charge of the contributing quark. For large z values, from our preceding arguments one expects the resulting Λ polarization in semi-inclusive DIS to be positive. This is in contradiction to SU(3)-based models in which the resulting polarization should be zero or negative. Recent preliminary results from the HERMES collaboration [17] produce a data point at a single z value. This result is slightly positive, but with significant error bars. Reasonably precise measurements at significantly larger z could distinguish between these models of semi-inclusive DIS.

Charge Symmetry in Parton Distributions

It is widely believed that parton distributions for light quarks should obey charge symmetry. For hadrons, the operation of charge symmetry interchanges protons and neutrons. The corresponding operation at the quark level interchanges up and down quarks, while changing proton and neutron labels. If parton distributions are invariant under charge symmetry, then we expect $d^n(x) = u^p(x)$, $u^n(x) = d^p(x)$ and $s^n(x) = s^p(x)$. An identical set of relations should hold for the antiquark distributions. At low energies, nuclear physics experiments generally show that charge symmetry is conserved at or below the 1% level in a variety of observables [18]. It is difficult to see why charge symmetry should be so well obeyed at low energies if it is violated at the quark level. As a result, all phenomenological parton distributions assume charge symmetry, which allows all neutron quark distributions to be replaced by the corresponding distributions in the proton. Ma [19] emphasized that experiments which test flavor symmetry rely on the assumption that charge symmetry is valid; he pointed out that the DY and NMC experiments could be reproduced, even if flavor symmetry was exact, by assuming sufficiently large charge symmetry violation (CSV) for partons.

Parton charge symmetry has been investigated recently by several groups [20–22]. If one defines the parton CSV distributions through

$$\delta d(x) = d^p(x) - u^n(x), \qquad \delta u(x) = u^p(x) - d^n(x) \tag{7}$$

then for valence distributions one found that $\delta d_v(x) \approx -\delta u_v(x)$. Since at large x $d^p \ll u^p$, the fractional CSV terms were expected to be much larger for the "minority" quark term $d_v(x)$ than for $u_v(x)$. Benesh and Londergan [21] derived an approximate relation which related parton CSV terms to the corresponding quark distributions

$$\delta q(x) \approx \frac{2m\,\delta m(1-x)}{M^2(1-x)^2 + m^2} \frac{dq(x)}{dx} \tag{8}$$

From this equation, we can estimate the magnitude of charge symmetry violation directly from phenomenological parton distributions without using quark models. The results obtained are in good agreement with direct quark model calculations of CSV effects.

Cao and Signal [23] calculate CSV effects arising from meson cloud models. These effects are significantly smaller than those calculated from quark models. They also point out that Eq. 8 may not be accurate at small x and may not give reliable estimates for sea quark CSV effects. The statistical model of Bhalerao [10] predicts parton CSV since it assumes different statistical parameters for neutrons and protons. The implications of parton charge symmetry, and evidence for this, has been reviewed by Londergan and Thomas [22].

Experimental Status of Parton Charge Symmetry

The most sensitive experimental test of parton charge symmetry to date is the "charge ratio". There is a simple relation between the F_2 structure functions for charged lepton DIS and neutrino charged current reactions on an isoscalar target N_0:

$$R_c(x) = \frac{F_2^{\gamma N_0}(x)}{\frac{5}{18}\overline{F}_2^{WN_0}(x) - \frac{x(s(x)+\bar{s}(x))}{6}} \approx 1 + \frac{3(\delta u(x) + \delta \bar{u}(x) - \delta d(x) - \delta \bar{d}(x))}{10\overline{Q}(x)}$$

$$\overline{Q}(x) = \sum_{j=u,d,s} q_j(x) + \bar{q}_j(x) - \frac{3(s(x)+\bar{s}(x))}{5} \qquad (9)$$

The quantity $\overline{F}_2^{WN_0}(x)$ in Eq. 9 is the average of neutrino and antineutrino charge-changing reactions on isoscalar targets. The relation $R_c(x) = 1$ should hold for all x and Q^2, with no QCD corrections. The only things which break this relation are parton CSV terms (if in Eq. 9 we use neutrino cross-sections on isoscalar targets, there is an additional term proportional to $s(x) - \bar{s}(x)$).

Previous experiments have shown that $R_c(x) \approx 1$, which reinforced the conviction that charge symmetry should be valid for parton distributions, even though the experimental errors in R_c placed upper limits on parton CSV of roughly 15%. However, new experiments measuring both muon-induced DIS and neutrino charge-changing reactions now provide much more stringent upper limits on parton CSV, at least in certain kinematic regions. In Fig. 3a we plot $R_c(x)$, the ratio of NMC μ-D structure functions [1] to CCFR ν-Fe measurements [24].

In order to place strong limits on CSV amplitudes, the experimental results must be corrected for numerous effects. These include: the absolute normalization of cross sections, radiative corrections, isoscalar corrections (the neutrino data is taken on an iron target, which is not isoscalar), heavy quark threshold effects (the neutrino data is obtained at energies where charm quark threshold effects are not negligible), and heavy target corrections. The last corrections include shadowing of the virtual W's at small x, EMC-like effects at moderate x, and Fermi-motion corrections at large x. The solid circles in Fig. 3a show the ratio $R_c(x)$, when the heavy target corrections are calculated specifically for neutrinos. For intermediate values $x > 0.1$ the agreement between structure functions is very good, and we can set upper limits of a few percent on parton CSV. However, in the region $x < 0.1$, R_c

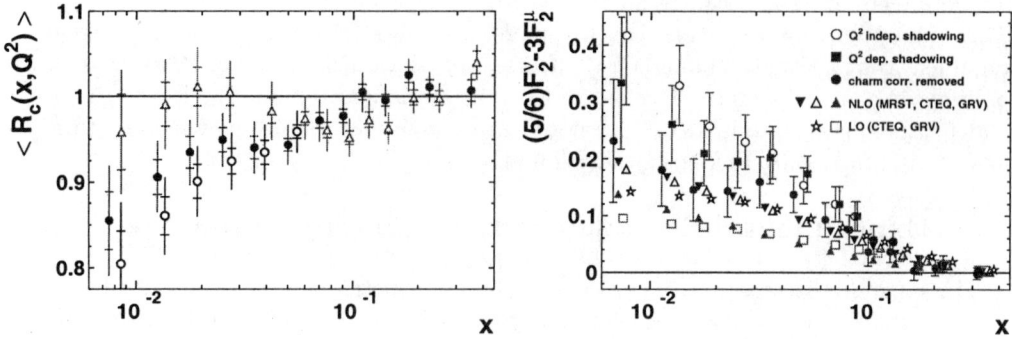

FIGURE 3. (a) Charge ratio $R_c(x)$ of Eq. 9, obtained using NMC μ-D structure functions and CCFR ν-Fe. Solid circles: heavy target shadowing corrections calculated specifically for neutrinos; open circles: ν heavy target corrections taken from shadowing in charged lepton DIS. (b) Comparison of $(5/6)F_2^\nu - 3F_2^\mu$. Solid circles: experimental data without "slow rescaling" correction. Triangles: NLO calculations with phenomenological parton distributions.

deviates from unity by as much as 10%. This is discussed in the paper by Boros et al. at this conference. Since the CCFR data is obtained predominantly from neutrino reactions, it would appear that the low-x discrepancy might be accommodated by allowing $s(x) \neq \bar{s}(x)$. However, if the NMC and CCFR data was combined with opposite-sign dimuon production data from ν reactions (which is used to extract $s(x)$), the resulting discrepancy could not be removed unless $\bar{s}(x) < 0$, which is not physically reasonable [25]. Thus the NMC and CCFR data appeared to be incompatible unless very large CSV sea quark effects were included.

The sea quark CSV amplitude $\delta\bar{d}(x) - \delta\bar{u}(x)$ extracted from this analysis was surprisingly large: it was roughly 25% of the sea quark amplitudes at small x. Such an effect is about two orders of magnitude larger than obtained by theoretical estimates of charge symmetry violation [21]. If CSV amplitudes of this magnitude were present, then they would have significant effects on phenomenological parton distributions. However, for several reasons it is highly unlikely that charge symmetry violating amplitudes are nearly this large. Bodek and collaborators [26] examined possible modifications in low-x parton distributions which would accommodate such large CSV effects. They showed that these modified parton distributions could not account for the $W^\pm$ asymmetry measured in $p - \bar{p}$ collisions at the Tevatron (the W asymmetries are very sensitive to low-x parton distributions).

Large CSV effects would also modify the pp and pD Drell-Yan cross sections measured by the E866 group, and discussed in a previous section. Including the effects of CSV amplitudes on the ratio of DY cross sections changes Eq. 2 to

$$R_{DY} \equiv \sigma_{DY}^{pD}/(2\sigma_{DY}^{pp}) \approx \left[1 + (\bar{d}^p(x_t) - \delta\bar{d}(x_t))/\bar{u}^p(x_t)\right]/2 \ . \tag{10}$$

Since the CSV term $\delta\bar{d}(x)$ extracted from the charge ratio is large and positive, a much larger flavor symmetry violation $\bar{d}/\bar{u}$ would be required in order to reproduce the E866 data.

Other recent analyses suggest that any charge symmetry violating effects must be substantially smaller than suggested by analyzing the NMC and CCFR structure functions at small x. Boros et al. [27] carried out an NLO calculation of heavy quark threshold effects, and found that these effects did not agree with the corrections used by the CCFR group. The CCFR group have re-analyzed their neutrino data and re-calculated a number of corrections, including charm quark threshold effects.

Fig. 3b summarizes the current situation regarding CSV effects. From Eq. 9, the quantity $(5/6)F_2^\nu(x) - 3F_2^\mu(x)$ contains strange quark plus CSV contributions. The triangles in Fig. 3b are the result of an NLO calculation of this quantity, using the latest phenomenological parton distributions, with no CSV terms. The solid circles are the F_2 structure functions from the NMC and CCFR muon and neutrino experiments. Heavy quark threshold effects have been removed from the CCFR results, as it is not appropriate to include these when comparing with NLO analyses. The differences between the data and CTEQ or MRST parton distributions [28,29] now generally lie within the experimental error bars. Thus, any remaining CSV effects at small x are much smaller than were previously suggested [25]. See the talk by Boros et al. in these proceedings for a more detailed discussion of this point. Londergan and Thomas [22] review a number of possible experimental tests of parton CSV. Since CSV effects now appear to be smaller than a few percent of the parton amplitudes, extracting CSV effects will necessarily involve nearly complete cancellations between large observables.

A significant amount of the research summarized here was done in collaboration with C. Boros and A.W. Thomas, of CSSM, Adelaide. The author would also like to thank C. Benesh, J-C Peng, and F. Steffens for useful discussions. He also thanks the Special Research Centre for the Subatomic Structure of Matter for its hospitality. This research was supported in part by the NSF, and by the Australian Research Council.

REFERENCES

1. P. Amaudruz et al. (NMC Collaboration,) *Phys. Rev. Lett.* **66**, 2712 (1991).
2. E.A. Hawker et al. (E866 Collaboration), *Phys. Rev. Lett.* **80**, 3715 (1998).
3. S. Kumano, *Phys. Rev.* **D43**, 59 (1991); **D43**, 3067 (1991); S. Kumano and J.T. Londergan, *Phys. Rev.* **D44**, 717 (1991).
4. E. Eichten, I. Hinchliffe and C. Quigg, *Phys. Rev.* **D45**, 2269 (1992).
5. K. Ackerstaff et al. (HERMES Collaboration), *Phys. Rev. Lett.* **81**, 5519 (1998).
6. J-C. Peng and G.T. Garvey, eprint hep-ph/9912370 (1999).
7. J.D. Sullivan, *Phys. Rev.* **D5**, 1732 (1972).
8. W. Melnitchouk, J. Speth and A.W. Thomas, *Phys. Rev.* **D59**, 014033 (1999).

9. D.I. Diakonov et al., *Phys. Rev.* **D56**, 4069 (1997); B. Dressler et al., eprint hep-ph/9910464.
10. R.S. Bhalerao, N.G. Kelkar and B. Ram, *Phys. Lett.* **B476**, 285 (2000); R.S. Bhalerao, eprint hep-ph/0003075.
11. F.M. Steffens, private communication.
12. A.I. Signal and A.W. Thomas, *Phys. Rev.* **D40**,2832 (1989); A.W. Schreiber et al., *Phys. Rev.* **D42**, 2226 (1990); *Phys. Rev.* **D44**, 2653 (1991).
13. M. Alberg et al., *Phys. Lett.* **B389**, 367 (1996); *Nucl. Phys.* **A644**, 93 (1998).
14. C. Boros, J.T. Londergan and A.W. Thomas, *Phys. Rev.* **D61**, 014007 (2000).
15. K. Ackerstaff et al. (OPAL Collaboration), *Eur. Phys. J.* **C2**, 49 (1998); **C8**, 241 (1999); K. Abe et al. (SLD Collaboration) *Phys. Rev.* **D59**, 052001 (1999); D. Buskulic et al. (ALEPH Collaboration) *Z. Phys.* **C64** 361 (1994); *Phys. Lett.* **B374**, 319 (1996).
16. D. de Florian, M. Stratmann and W. Vogelsang, *Phys. Rev.* **D57**, 5811 (1998).
17. A. Airepetian et al. (HERMES Collaboration), hep-ex/9911017 (1999).
18. G.A. Miller, B.M.K. Nefkens and I. Slaus, *Phys. Rep* **194**, 1 (1990).
19. B.-Q. Ma, *Phys. Lett.* **B274**, 111 (1992).
20. E. Sather, *Phys. Lett.* **B274**, 433 (1992); E. Rodionov, A.W. Thomas and J.T. Londergan, Int.J.Mod.Phys.Lett. **A9**, 1799 (1994); C.J. Benesh and T. Goldman, *Phys. Rev.* **C55**, 441 (1997).
21. C.J. Benesh and J.T. Londergan, *Phys. Rev.* **C58**, 1218 (1998).
22. J.T. Londergan and A.W. Thomas, *Prog. in Part. and Nucl. Phys.* **41**, 49 (1998).
23. F-G Cao and A.I. Signal, hep-ph/0001146 (2000).
24. W.G. Seligman et al. (CCFR Collaboration), *Phys. Rev. Lett.* **79**, 1213 (1997).
25. C. Boros, J.T. Londergan and A.W. Thomas, *Phys. Rev. Lett.* **81**, 4075 (1998); *Phys. Rev.* **D59**, 074021 (1999).
26. A. Bodek et al., *Phys. Rev. Lett.* **83**, 2892 (1999).
27. C. Boros, F.M. Steffens, J.T. Londergan and A.W. Thomas, *Phys. Lett.* **B468**, 161 (1999).
28. H.L. Lai et al., *Eur. Phys. J.* **C12**, 375 (2000).
29. A.D. Martin et al., hep-ph/9907231 (1999).

Studies of Nucleon Spin Structure at HERMES

Bryan Tipton

Kellogg Radiation Laboratory 106-38
1200 E. California Blvd.
Pasadena, CA 91125
E-mail: tipton@krl.caltech.edu

On behalf of the HERMES collaboration

Abstract. The HERMES experiment studies the spin structure of the nucleon using polarized semi-inclusive deep-inelastic positron scattering and polarized quasi-real photoproduction. Previous analyses of nucleon spin structure combine inclusive deep-inelastic scattering data and measured hyperon decay constants assuming $SU(3)_{\text{flavor}}$ symmetry; asymmetries in the processes studied at HERMES may constrain nucleon spin structure more directly. Results are discussed for quark polarizations extracted from semi-inclusive deep-inelastic scattering asymmetries, for gluon polarization extracted from the asymmetry of high-p_T hadron pairs, and for the first observation of single-spin azimuthal asymmetries in pion production.

INTRODUCTION

The HERMES experiment studies the spin structure of the nucleon using polarized scattering of electrons and positrons on nucleons:

$$\vec{l} + \vec{N} \to l' + X. \tag{1}$$

Previous deep-inelastic scattering (DIS) experiments studied this reaction inclusively, detecting only the final lepton. In these experiments, one is sensitive to the polarized structure functions, $g_1(x)$ and $g_2(x)$. The function $g_1(x)$ possesses a simple interpretation in the Quark-Parton Model(QPM) as the sum over all flavors of polarized parton distribution functions(PDFs), Δq:

$$g_1^p(x) = \frac{1}{2}\sum_q e_q^2(q^+(x) - q^-(x)), = \frac{1}{2}\sum_q e_q^2 \Delta q(x). \tag{2}$$

The function, $q^{+(-)}$, represents the distribution of quarks of flavor q with spin parallel (anti-parallel) to the nucleon spin, and e_q is the quark charge in units of elementary charge $|e|$. Here, x is the Bjorken scaling variable $x = Q^2/2M_p\nu$, measured by scattered lepton kinematics; Q^2 and ν are the negative squared four-momentum and energy of the exchanged virtual photon.

One may view the spin of the nucleon in terms of the contributions of its parton constituent:

$$<s_z^N> = \frac{1}{2} = \frac{1}{2}\Delta\Sigma + L_z^q + \Delta G + L_z^G, \tag{3}$$

where $\Delta\Sigma$, ΔG, L_z^q and L_z^G are contributions from quark and gluon intrinsic spin and quark and gluon orbital angular momentum. The quark spin contribution, $\Delta\Sigma$, can be further divided among the quark flavors, and between valence and sea contributions:

$$\Delta\Sigma = \Delta u_v + \Delta d_v + \Delta u_s + \Delta\bar{u} + \Delta d_s + \Delta\bar{d} + \Delta s + \Delta\bar{s}. \tag{4}$$

By measuring $g_1(x)$ from various targets and integrating over x, it has been found that $\Delta\Sigma$ accounts for only about a third of the nucleon spin, and that the strange sea seems negatively polarized. However, the inclusive DIS data on proton and neutron targets do not constrain $\Delta\Sigma$ alone. In reaching this conclusion, one typically combines polarized DIS results with information from hyperon beta decay constants. SU(3)$_{flavor}$ symmetry in the strong interaction is a key assumption: one assumes hyperon data may be related to nucleon structure through a simple rotation in flavor space.

The HERMES experiment attempts to make a detailed study of nucleon spin structure. The approach taken by HERMES is to measure the spin dependence of all high energy lepton scattering final states with a large acceptance detector. A combined analysis of multiple cross-sections, each having different sensitivity to nucleon spin structure, allows a more complete study of the proton spin in one experiment alone. Symmetries of the strong interaction are then no longer assumed, but rather are tested in the course of the investigation. Three results are discussed here. By measuring semi-inclusive hadron production yields and fitting their double-spin asymmetry in the QPM, the flavor dependence of the quark polarizations may be extracted. Photoproduction of high-p_T hadron pairs yields a negative asymmetry, which simulations show is strongly correlated to the polarization of gluons. Finally, single-spin azimuthal asymmetries in pion production hint at a large transversity distribution in the nucleon.

I THE HERMES EXPERIMENT

The HERMES experiment was designed to optimize semi-inclusive DIS studies [1]. The experiment is located at the DESY laboratory in Hamburg, Germany, where a 27.5 GeV positron (or electron) beam circulates in the HERA beam. The positron beam is naturally polarized transverse to the beam direction due to an asymmetry in the emission of synchrotron radiation, and longitudinal polarization at the HERMES interaction point is provided by a pair of matched spin rotators. Beam polarizations averaged 55% during the run, with a relative systematic uncertainty in the measurement of 4.0% (3.4%) for the ^{3}He (H) data.

The HERMES experiment consists of an internal gas target assembly in place along the East Hall interaction point (IP) of the HERA positron storage ring. The HERA beam passes through the center of a 40 cm long storage cell containing polarized ^{3}He, ^{1}H, and ^{2}H. Two different target technologies inject the nuclei into the cell. In 1995, a ^{3}He target was used, where atoms became polarized through metastability-exchange optical pumping, giving an average polarization of 46%. [5] An Atomic Beam Source(ABS) used permanent sextupole magnets and radio-frequency units to select desired hyperfine states of ^{1}H in 1996 and 1997. This source continues to operate with polarized ^{2}H today. Proton polarization averaging 86% was monitored by a Breit-Rabi polarimeter. [6]

Extending 8.5 meters behind the target, a forward acceptance spectrometer captures the high energy particles produced in the positron-gas interaction. Tracking chambers placed before and after a dipole magnet determine momenta and trajectories of those particles. A threshold Čerenkov detector, a Transition-Radiation Detector, a Preshower Counter, and an Electromagnetic Calorimeter allow separation of electrons from hadrons. The Čerenkov further provides clean determination of pions above particle momenta of 5.6 and 3.8 GeV/c in 1995 and 1996-7. Before the 1998 running, this Čerenkov was replaced with a Ring-imaging Čerenkov (RICH), which will provide π, K, and p separation over the entire HERMES kinematic range.

II THE POLARIZATION OF QUARKS

In polarized deep-inelastic scattering, one can construct a cross-section asymmetry of parallel and anti-parallel orientations of the nucleon spin to the lepton beam spin, $A_{\|}$:

$$A_{\|} = \frac{\sigma^{\uparrow\Downarrow} - \sigma^{\uparrow\Uparrow}}{\sigma^{\uparrow\Downarrow} + \sigma^{\uparrow\Uparrow}}. \tag{5}$$

Under the assumption that $g_2(x)$ is zero, the measured double-spin asymmetry directly relates to the asymmetry in nucleon virtual photoabsorption, $A_1(x)$, and to the structure function ratio, $\frac{g_1}{F_1}(x)$:

$$A_1^h = \frac{g_1^h}{F_1^h} = \frac{A_{\|}^h}{D(1+\gamma\eta)}. \tag{6}$$

where h includes both inclusive positron data and coincident positive and negative unidentified hadrons, and D, γ, and η are kinematic factors.

In the QPM and under the assumption of factorization, one sees that the photoabsorption asymmetry probes the helicity difference parton distributions $\Delta q(x)$:

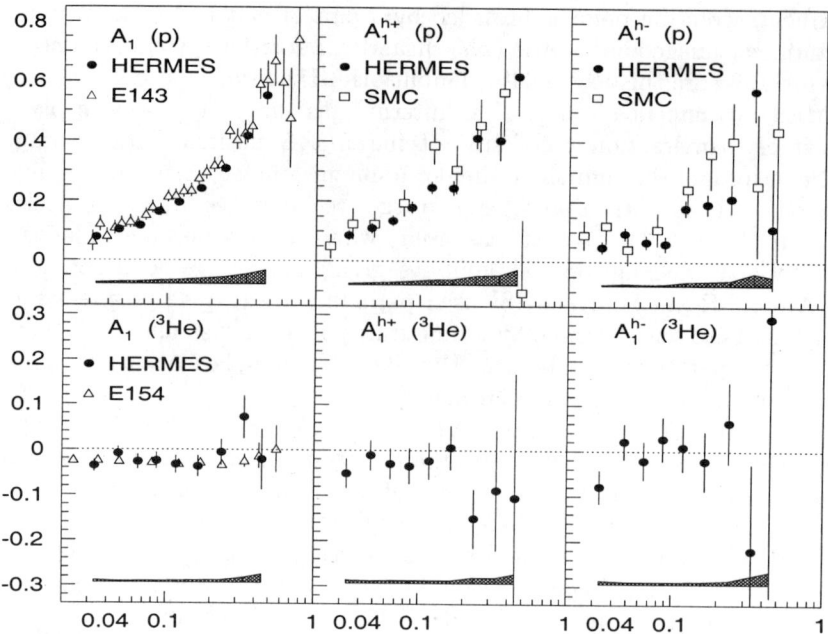

FIGURE 1. Results for the virtual photon asymmetries from HERMES 1995-7 data. The columns show the inclusive, semi-inclusive h^+, and semi-inclusive h^- asymmetries while the rows correspond to the ^{1}H and ^{3}He target. The inclusive asymmetries are compared with the total uncertainties of the SLAC E143 [7] and E154 [8] g_1/F_1 measurements. For the proton semi-inclusive asymmetries, a comparison with the SMC [9] results are shown. All points are plotted at measured x and Q^2.

$$A_1^h(x,z) = \frac{\sum_q e_q^2 \, \Delta q(x) D_q^h(z)}{\sum_q e_q^2 \, q(x) D_q^h(z)} \frac{(1 + R(x,Q^2))}{1 + \gamma^2}. \quad (7)$$

The $D_q^h(z)$ are the (spin-independent) fragmentation functions. In this model, each measured asymmetry has a different correlation to the underlying quark polarization through the charge and fragmentation function factors. By fitting several of these asymmetries on proton and effective neutron targets, one may extract the polarized quark distributions $\Delta q(x)$.

HERMES has measured asymmetries for inclusive and semi-inclusive h^+ and h^- reactions from ^{3}He and ^{1}H targets [2]. These asymmetries are shown in Figure 1. The inclusive DIS sample was determined by cuts on $Q^2 > 1$ GeV^2, $y = \frac{\nu}{E} < 0.85$, and the final photon-nucleon center-of-mass energy, $W^2 > 4$ GeV^2. The sizes of the inclusive event samples satisfying these requirements were $2.2 \cdot 10^6$ and $2.3 \cdot 10^6$ events for ^{3}He and ^{1}H respectively. The analyzed semi-inclusive events form a subsample having a detected charged hadron and $W^2 > 10$ GeV^2. Cuts on the hadron energy fraction, $z > 0.2$, and the Feynman variable, $x_f > 0.1$, also select nucleon fragments from the current region. Beam and target polarization measurement uncertainties, uncertainties in $R(x, Q^2)$, and experimental yield fluctuation uncertainties in 1995 are the dominant contributions to the systematic uncertainty of the measured asymmetries.

One may simply rearrange of Equation 7 to separate the quark polarizations from unpolarized physics factors summarized by the purity, $P_q^h(x)$:

$$A_1^h(x) = \sum_q P_q^h(x) \cdot \left(\frac{\Delta q(x)}{q(x)}\right) \frac{1+R}{1+\gamma^2}. \tag{8}$$

The resulting system of equations may be written in matrix form as,

$$\vec{A}(x) = \mathbf{P} \cdot \vec{Q}(x). \tag{9}$$

The quark polarizations are obtained by least squares minimization of this equation.

The quark purities relevant for the HERMES experiment have been estimated with a Monte Carlo simulation of unpolarized DIS, based on LUND string fragmentation, [11] CTEQ low Q² unpolarized parton distributions, [12] and a model of the HERMES detector. ³He asymmetries convert to effective neutron asymmetries through nuclear corrections. [13]

In solving Equation 9, one would like to separate the spin contributions of three quark flavors u, d, and s, and their anti-quarks $\bar{u}$, $\bar{d}$, and $\bar{s}$. In practice, limited statistics and sensitivity to the sea quark flavors require selection of a model of average sea polarization. A simple assumption of flavor independent sea polarization is used here:

$$\frac{\Delta q_s}{q_s} \equiv \frac{\Delta d_s}{d_s} = \frac{\Delta u_s}{u_s} = \frac{\Delta \bar{d}}{\bar{d}} = \frac{\Delta \bar{u}}{\bar{u}} = \frac{\Delta \bar{s}}{\bar{s}} = \frac{\Delta s}{s}. \tag{10}$$

Note that with further data-taking, this relation may be relaxed in the future and allowing a study of the symmetries implied.

Using the six experimental asymmetries shown in Figure 1, three quark polarizations have been extracted: $(\Delta u + \Delta \bar{u})/(u + \bar{u})(x)$, the total up flavor polarization; $(\Delta d + \Delta \bar{d})/(d + \bar{d})(x)$, the total down flavor polarization; and $(\Delta q_s/q_s)(x)$, the average polarization of the quark sea. After multiplying by x and the unpolarized PDF parameterization, Figure 2 shows the results from this extraction in nine x_{Bj} bins. For $x > 0.3$, it is assumed that sea polarization does not contribute significantly to the measured asymmetries; the sea polarization, $\frac{\Delta q_s}{q_s}(x)$, is set to zero with a small systematic uncertainty. Systematic uncertainties in the extraction originate in the asymmetries' uncertainty and in a study of the sensitivity of the fit to variations in the unpolarized Monte Carlo model.

In Figure 2, the parameterizations of De Florian et. al. ($0.1 < \Delta G < 0.8$, LO) [15] Gehrmann and Stirling (Gluon A, LO), [17] and Glück et. al. (Standard, LO) [16] are provided for comparison. A correction of 1+R to the De Florian and

TABLE 1. The first moments of the extracted polarized parton distributions. The results are given at $Q^2=2.5$ GeV^2 for the measured region $0.023 < x < 0.6$, for the low-x extrapolation and for the total integral. Note that the values for $\Delta s + \Delta \bar{s}$ relies on the assumption in Eq. (10).

	measured region	low-x	total integral	SU(3) prediction [18]	Q^2
$\Delta u + \Delta \bar{u}$	$0.51 \pm 0.02 \pm 0.03$	0.04	$0.57 \pm 0.02 \pm 0.03$	0.66 ± 0.03	2.5
$\Delta d + \Delta \bar{d}$	$-0.22 \pm 0.06 \pm 0.05$	-0.03	$-0.25 \pm 0.06 \pm 0.05$	-0.35 ± 0.03	2.5
$\Delta s + \Delta \bar{s}$	$-0.01 \pm 0.03 \pm 0.04$	0.00	$-0.01 \pm 0.03 \pm 0.04$	-0.08 ± 0.02	2.5

Glück parameterizations is necessary to achieve consistency in model assumptions. The HERMES data agrees with these parameterizations within uncertainties.

Moments of the quark distributions in the measured x range are determined from the area under the measured points. For comparison to previous measurements, the distributions are extrapolated to low x by constant fits to the data; the low x extrapolation is quoted but no value is quoted for the uncertainty due to theoretical ambiguities. The resulting measured region integrals are listed in Table 1. The integrals are compared to predictions from Reference [18], which are corrected to fourth order in QCD; these predictions have been extracted from inclusive data

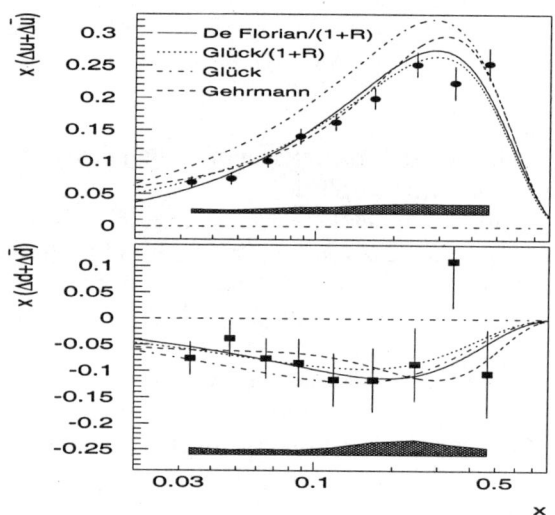

FIGURE 2. Polarized parton distributions for up and down quarks, $x(\Delta u + \Delta \bar{u})$ and $x(\Delta d + \Delta \bar{d})$, compared to parameterizations of world data. The data points have been evolved to $Q^2=2.5$ GeV^2 for the comparison. The parameterizations are explained in the text.

assuming SU(3)$_{\text{flavor}}$ symmetry. The HERMES values are slightly smaller in magnitude than these predictions, and in particular the sea is unpolarized in the fit. Within uncertainties though, the results are not yet in disagreement.

III THE ROLE OF THE GLUON SPIN

As the contributions of the quarks' intrinsic spins does not sum to one half, the remaining angular momentum inside the proton must lie elsewhere. Significant theoretical activity has focused on a possible large gluon contribution to the proton spin. Attempts have been made to constrain this contribution by fitting the Q^2 dependence of the inclusive polarized structure functions with a DGLAP formalism. These efforts yield an indication of positively polarized glue [9], but the statistics of the polarized measurements are as yet too poor to produce a definitive constraint.

New generations of polarized experiments study the gluon contribution by examining the polarization dependence of processes directly sensitive to the gluon distribution in the nucleon. One process is the photoproduction of two hadrons at large transverse momentum, p_T. If the cut on the transverse momentum is sufficiently large, one may reject tree-level quark scattering events with small p_T and isolate events from the hard photon-gluon fusion process [10].

HERMES has measured the double-spin asymmetry in the virtual photoproduction of high-p_T hadrons on a proton target [3]. Events were selected to have one positively and one negatively charged hadron with total momenta greater than 4.5 GeV/c in the HERMES spectrometer. The scattered positron was not required to be detected, and simulations show that very low Q^2 is probed. Further cuts required the reconstructed invariant mass of the hadrons to be above the ρ resonance. The asymmetry is presented as a function of hadron p_T in Figure 3. In each panel in this figure, a hard cut of 1.5 GeV/c is placed on the p_T of one charged hadron, while the p_T of the other is varied. One sees a substantial negative asymmetry when the hard cut is placed on the negative hadron. A much smaller effect is seen in the positive hadron case. If one treats the charged hadron dependence uniformly and averages the plots in Figure 3, one gets the asymmetry in Figure 4.

To relate these asymmetries to gluon polarization, a simulation of the underlying physics processes is required. The PYTHIA Monte Carlo is used to estimate the contribution of various processes to the high-p_T events at HERMES. The processes included vector meson production, the QCD Compton effect, and the photon-gluon fusion. These studies indicate that the cross-section in the region of the negative asymmetry is dominated by photon-gluon fusion. As the photon-gluon fusion asymmetry is directly proportional to $\Delta G/G$, this quantity may be extracted from the data, with a small positive background asymmetry from the QCD Compton effect subtracted. It is important to note that while one expects a negative asymmetry from the photon-gluon fusion process, other processes on a proton target yield either a positive or zero asymmetry. The expected asymmetry at HERMES for three $\Delta G/G$ assumptions are also shown in Figure 4.

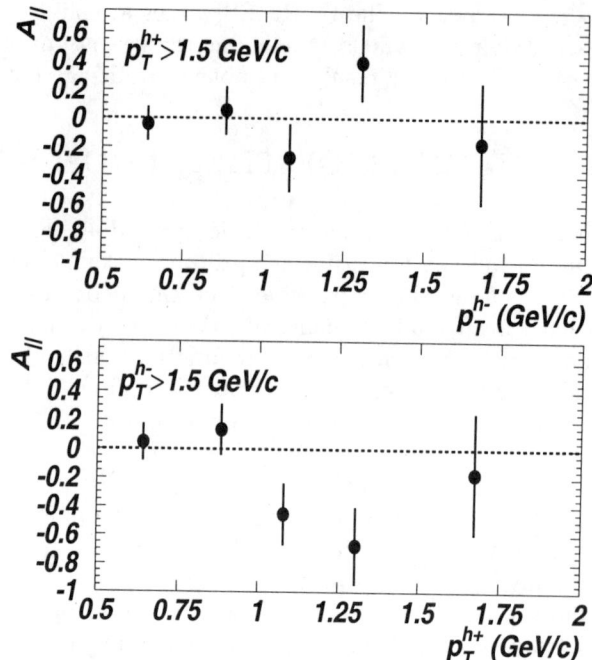

FIGURE 3. The virtual photoproduction asymmetry of high-p_T hadron pairs versus the positive hadron p_T (top) and the negative hadron p_T (bottom).

The points in the region $p_T^{h1} > 1.5\ GeV/c$ and $p_T^{h2} > 1.0\ GeV/c$ give an average asymmetry of $A_\parallel = -0.28 \pm 0.12 \pm 0.02$. Using the above Monte Carlo analysis, this asymmetry yields value of $\Delta G/G$ in LO QCD as $0.41 \pm 0.18(\text{stat.}) \pm 0.03(\text{syst.})$. Only experimental systematics are quoted.

IV SINGLE-SPIN AZIMUTHAL ASYMMETRIES

While the above results probe the structure of the parton spin in along the axis of the proton momentum, the transverse parton spin remains largely unexplored. The structure function, $h_1(x)$, and distribution functions, $\delta q(x)$, associated with transversity are as fundamental in QCD as their longitudinal partners, $g_1(x)$ and $\Delta q(x)$. Study of these distributions involves an examination of *rotational* symmetry in the nucleon. In the non-relativistic limit, one expects that no direction in space is preferred in a nucleon, and therefore $\Delta q(x)$ and $\delta q(x)$ are equal. With relativity, the two distributions may diverge from each other.

Unfortunately, $\delta q(x)$ are chiral odd distributions which are then unobservable in deep-inelastic scattering. One can circumvent this limitation by seeking observables in which the chiral odd distributions are associated with chiral odd fragmentation functions. Final state interactions in fragmentation may allow sensitivity to the spin

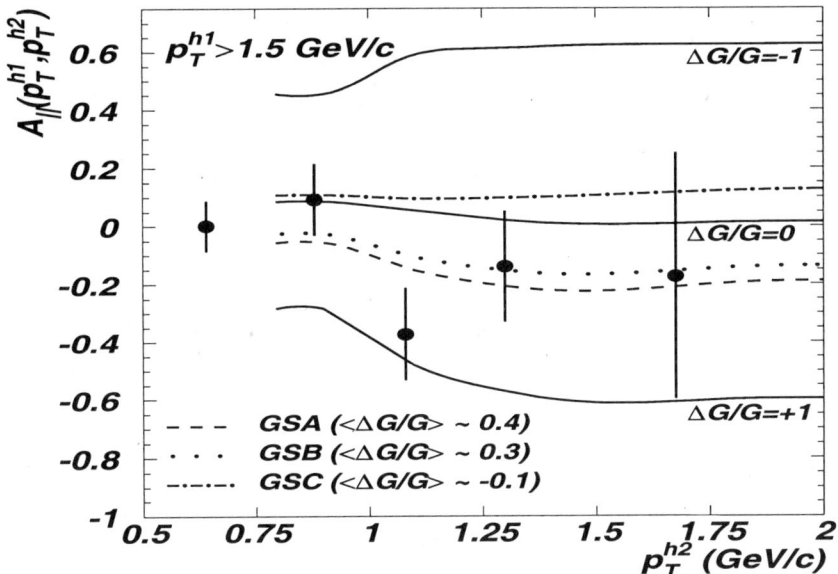

FIGURE 4. The virtual photoproduction asymmetry of high-p_T hadrons compared to Monte Carlo predictions for various $\Delta G/G$ assumptions. The dotted lines represent the predictions of reference [17].

transversity. Single-spin azimuthal asymmetries in semi-inclusive pion production are such observables.

HERMES has measured semi-inclusive pion single-spin asymmetries using a polarized proton target [4]. By weighting the cross-section by a sinusoid function, $W(\phi)$, the analyzing powers $A_{UL}^{W(\phi)}$ may be determined. In Figure 5, these analyzing powers are shown versus x for π^+ and $W(\phi)=\sin\phi$ and $\sin 2\phi$. For π^+ one sees a strong rising x dependence in the $\sin\phi$ moment, similar in shape to the rise in the u-quark longitudinal polarization in Figure 2. The corresponding π^- moment (not shown) is consistent zero. This may be an indication of flavor dependence of the moment, as the π^+ production is dominated by u-quark scattering while the π^- production has more d-quark contributions.

Note that though the proton target was longitudinally polarized, transverse spin effects can still manifest in the data as the virtual photon direction and the proton spin are not collinear for small Q. One expects to see a larger effect from a transversely polarized target. The $\sin 2\phi$ analyzing power, as well as the beam spin correlated analyzing powers $A_{LU}^{W(\phi)}$, have also been investigated, and are found consistent with zero.

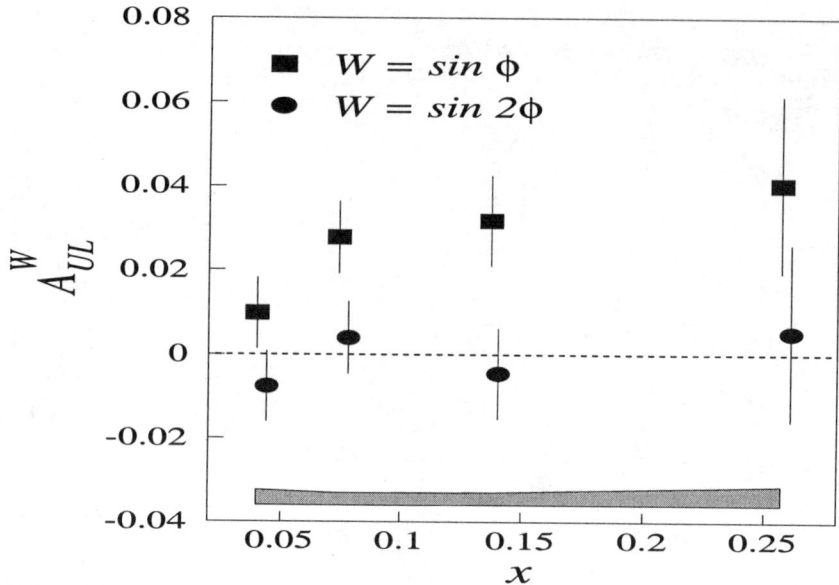

FIGURE 5. Moments of single-spin pion azimuthal asymmetries on a polarized proton target.

V CONCLUSIONS AND OUTLOOK

The first three years of HERMES data taking have yielded high statistics polarized lepton scattering data at 27.5 GeV on ^{3}He and ^{1}H targets. The polarized measurements yield inclusive and semi-inclusive charged hadron asymmetries, which have been presented as a function of x. In the framework of the Quark-Parton Model, these asymmetries are simultaneously fitted to yield helicity distributions for up, down and sea quarks. The current data set begins to test SU(3)$_{\text{flavor}}$ symmetry in the strong interaction, but no significant breaking is seen yet. Gluon polarizations are studied by looking at the asymmetry in high-p_T hadron pairs; an indication of a positive glue is seen. Azimuthal asymmetries may explore the properties of the nucleon spin under spatial rotation. The first single-spin azimuthal pion production asymmetry in semi-inclusive DIS is seen.

HERMES will soon have a new data set with high statistics on a deuterium target, which will significantly improve the precision on the d-quark helicity distribution. Further sensitivity to sea polarizations is gained by identification of different coincident hadron species. Before the deuterium run began, the Threshold Čerenkov was replaced with a Ring Imaging Čerenkov(RICH), which may identify π, K and p particles over nearly all momenta accepted by the spectrometer. After 2000, HERMES will use a transversely polarized ^{1}H and ^{2}H target in order to study spin transversity in more detail.

ACKNOWLEDGMENTS

The author would like to acknowledge the support of the U.S. National Science Foundation.

REFERENCES

1. Ackerstaff, K., et al., Nucl. Instrum. Methods A **417**, 230 (1998).
2. Ackerstaff, K., et al., Phys. Lett. B **464**, 123 (1999).
3. Airapetian, A., et al., Phys. Rev. Lett. **84**, 2584 (2000).
4. Airapetian, A., et al., Phys. Rev. Lett. **84**, 4047 (2000).
5. De Schepper, D., et al., Nucl. Instrum. Methods A **419**, 16 (1998).
6. Braun, B., in Proc. of the Workshop "Polarized gas targets and polarized beams", ed. R.J. Holt and M.A. Miller, AIP Conf. Proc. **421** 69 (1997).
7. Abe, K., et al., Phys. Rev. D **58**, 112003 (1998).
8. Abe, K., et al., Phys. Rev. Lett. **79**, 26 (1997).
9. Adeva, B., et al., Phys. Rev. D **58**, 112001 (1998).
10. Bravar, A., von Harrach, D., and Kotzinian, A., Phys. Lett. B **421**, 349 (1998).
11. Andersson, B., et al., Phys. Rept. **97**, 33 (1983).
12. Lai, H.L., et al., Phys. Rev. D **126**, 1280 (1997).
13. Friar, J.L., et al., Phys. Rev. C **42**, 2310 (1990).
14. Glück, M., et al., Z. Phys. C **67**, 433 (1995).
15. De Florian, D., et al., Phys. Rev. D **57**, 5803 (1998).
16. Glück, M., et al., Phys. Rev. D **53**, 4775 (1996).
17. Gehrmann, T. and Stirling, W. J., Phys. Rev. D **53**, 6100 (1996).
18. Ellis, J. and Karliner, M., Invited Lectures at the International School of Nucleon Spin Structure, Erice, August 1995, hep-ph/9601280.

The Nucleon's Strange Form Factors

Mark L. Pitt

Virginia Tech, Dept. of Physics, Blacksburg, VA 24061, USA

Abstract. Knowledge of the nucleon's strange form factors will provide valuable insight into low energy hadron structure. Measurement of the vector strange form factor of the nucleon is accomplished through parity-violating electron scattering. This paper reviews the current status of this class of experiments.

INTRODUCTION

Understanding the properties of the proton and neutron in terms of their constituents remains an important goal of nuclear physics. Quantum chromodynamics is believed to be the correct theory for describing the interactions among the quarks and gluons that make up the nucleon, but it is not yet possible to solve the theory in the non-perturbative regime relevant to low-energy hadron structure. To gain insight into this structure, it is important to identify all the important degrees of freedom at low energy. One relatively unexplored degree of freedom is the contribution of the sea quarks to low energy nucleon properties. A promising way to do this is by measuring nucleon matrix elements involving strange quarks, which are purely sea degrees of freedom. Analyses of the pion-nucleon sigma term [1] and deep inelastic scattering of polarized leptons on polarized nucleon targets [2] have shown evidence for nonzero values of the strange scaler and axial-vector matrix elements, but the methods of extracting these matrix elements have been subjects of debate in the literature.

On the other hand, it is generally agreed that values of vector strange matrix elements can be extracted reliably from experiment. In particular, it has been shown [3–5] that measurements of the asymmetry in parity-violating electron scattering can be combined with already existing measurements of electromagnetic nucleon form factors to determine the strange form factors associated with the vector strange matrix element - $\langle N|\bar{s}\gamma_\mu s|N\rangle$. There is now a program of parity-violating electron scattering experiments in progress and in preparation at MIT-Bates, Jefferson Lab, and the Mainz microtron. This review briefly covers the formalism underlying these measurements, some theoretical predictions, and details of the experimental programs.

FORMALISM

The vector form factors of the proton can be determined by making measurements with probes that couple to vector currents: the photon and the Z-boson. The strange form factors can then be extracted by determining enough electroweak form factors to make a flavor decomposition.

In general, the matrix element of a vector current, J_μ^V, for a spin-1/2 object like the nucleon can be expressed as

$$\langle N|J_\mu^V|N\rangle = \bar{u}_N \left(F_1 \gamma_\mu + F_2 \frac{i\sigma_{\mu\nu} q^\nu}{2M_N} \right) u_N, \qquad (1)$$

where F_1 and F_2 are the standard Dirac and Pauli vector form factors. Here, we will use the Sachs form factors (electric and magnetic): $G_E = F_1 - \tau F_2$ and $G_M = F_1 + F_2$, where $\tau = Q^2/(2M_N)^2$ with M_N being the nucleon mass. In the Standard Model, the hadronic electromagnetic and neutral weak vector currents can be written as linear combinations of the quark currents with known coupling constants. For example, the hadronic electromagnetic vector current is written as

$$J_\mu^{EM} = \frac{2}{3}\bar{u}\gamma_\mu u - \frac{1}{3}\bar{d}\gamma_\mu d - \frac{1}{3}\bar{s}\gamma_\mu s, \qquad (2)$$

where the currents are weighted by quark charges. A similar expression can be written for the neutral weak current with appropriate neutral weak couplings.

To get the desired flavor decomposition of the nucleon form factors, we first need to define the form factors associated with the individual quark currents. For example, for the strange quark current, we have

$$\langle N|\bar{s}\gamma_\mu s|N\rangle = \bar{u}_N \left(F_1^s \gamma_\mu + F_2^s \frac{i\sigma_{\mu\nu} q^\nu}{2M_N} \right) u_N, \qquad (3)$$

with similar expressions for the u and d quark currents. We will use the Sachs form factors, G_E^s and G_M^s, the strange electric and magnetic form factors of the nucleon. Using the flavor structure of the vector current from Equation 2 and its neutral current analogue, we have the following decomposition of measurable nucleon form factors in terms of the desired quark-nucleon form factors:

$$\begin{aligned} G_{E,M}^{\gamma,p} &= \frac{2}{3} G_{E,M}^{u,p} - \frac{1}{3} G_{E,M}^{d,p} - \frac{1}{3} G_{E,M}^{s,p} \\ G_{E,M}^{\gamma,n} &= \frac{2}{3} G_{E,M}^{u,n} - \frac{1}{3} G_{E,M}^{d,n} - \frac{1}{3} G_{E,M}^{s,n} \\ G_{E,M}^{Z,p} &= (1 - \frac{8}{3}\sin^2\theta_W) G_{E,M}^{u,p} - (1 - \frac{4}{3}\sin^2\theta_W) G_{E,M}^{d,p} - (1 - \frac{4}{3}\sin^2\theta_W) G_{E,M}^{s,p} \end{aligned} \qquad (4)$$

These are the conventional electric and magnetic form factors, $G_{E,M}^{\gamma,p}$ and $G_{E,M}^{\gamma,n}$, of the proton and neutron, and the neutral weak electric and magnetic form factors of the proton, $G_{E,M}^{Z,p}$. To obtain the strange quark form factors, isospin symmetry must be assumed to reduce the number of unknown quantities. The impact of this assumption on the extraction of strange form factors has been studied in several papers [6–10]. In terms of the quark form factors isospin symmetry is expressed as: $G_{E,M}^{u,p} = G_{E,M}^{d,n}$, $G_{E,M}^{d,p} = G_{E,M}^{u,n}$, and $G_{E,M}^{s,p} = G_{E,M}^{s,n}$. Thus, the neutral weak form factors of the proton can be expressed as:

$$G_{E,M}^{Z,p} = \left(1 - 4\sin^2\theta_W\right) G_{E,M}^{\gamma,p} - G_{E,M}^{\gamma,n} - G_{E,M}^{s} \quad (5)$$

Measurements of the neutral weak form factors, $G_{E,M}^{Z,p}$, combined with the previously measured electromagnetic form factors of the proton and neutron can be used to extract the desired strange form factors, $G_{E,M}^{s}$.

The only practical way to measure the neutral weak form factors, $G_{E,M}^{Z,p}$, is through parity-violating electron scattering with longitudinally polarized electrons. The parity-violating asymmetry in the elastic cross-section between right- and left-handed polarized electrons is proportional to the interference between the γ and Z exchange Feynman diagrams. At tree level in the standard model the asymmetry ($\equiv (\sigma_R - \sigma_L)/(\sigma_R + \sigma_L)$) can be written as [11]:

$$A_p = \left[\frac{-G_F Q^2}{\sigma_p \pi \alpha 4\sqrt{2}}\right] \left[\varepsilon G_E^{\gamma,p} G_E^{Z,p} + \tau G_M^{\gamma,p} G_M^{Z,p} - (1 - 4\sin^2\theta_W)\varepsilon' G_M^{\gamma,p} G_A^{e,p}\right], \quad (6)$$

where ε, τ, and ε' are kinematic factors, σ_p is the proton cross-section, and $G_A^{e,p}$ is the neutral weak axial electron-nucleon form factor. At forward electron scattering angles, the asymmetry is dominated by the first two terms, while backward angle measurements mainly access the last two terms. By combining measurements at different kinematics, G_E^s and G_M^s can be separately extracted. The axial e-N form factor, $G_A^{e,p}$, can also be extracted from these measurements. This quantity is of interest because it contains contributions from the nucleon's anapole moment.

THEORETICAL PREDICTIONS

Over the past decade there have been a large number of theoretical efforts to estimate the vector strange form factors. Typically, the quantities that are calculated are the two "static" properties: the strange magnetic moment ($\mu_s \equiv G_M^s(Q^2 = 0)$) and the strangeness radius ($r_s^2 \equiv -6[dG_E^s/dQ^2]_{Q^2=0}$). Figure 1 contains a compilation of theoretical estimates of μ_s and r_s^2 in roughly chronological order from 1989 to the present. Generally, the predicted values of r_s^2 are small, with disagreement on the sign for those models that predict large values. For μ_s, the majority of the estimates predict a negative value, with the results tending to cluster near -0.3 nuclear magnetons. A few calculations do predict large positive results, however.

Values of the strange form factors at finite Q^2 are also important because they contain information on the spatial distribution of strange charge and magnetization. Several of the authors listed in Table 1 also provide estimates of the Q^2 dependence of the strange form factors, typically up to $Q^2 \sim 1$ GeV^2 [12,20–23,25,26,34,35].

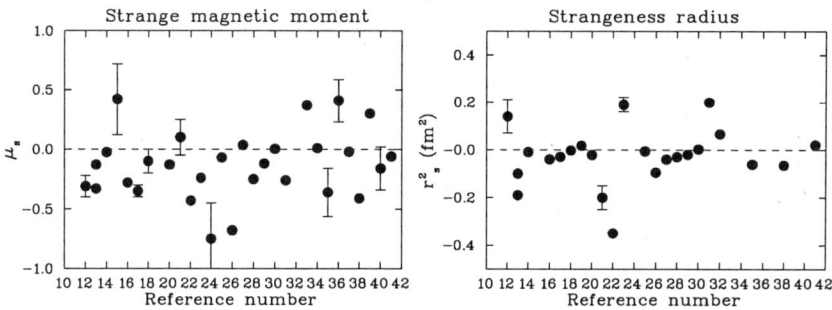

FIGURE 1. Theoretical estimates of μ_s and r_s^2 versus reference number (roughly chronological order). Error bars are included where the author specifies a range of values.

A variety of theoretical methods have been used to make these estimates. The two methods that are easiest to conceptualize are the "poles" and "kaon loops" methods. The pole type calculations are a vector meson dominance picture where the incident virtual photon fluctuates into a ϕ meson, which is predominantly an $\bar{s}s$ pair. The kaon loop calculations estimate the contributions coming from the fluctuation of the nucleon into a K-meson and a hyperon. This can lead to a separation of the s and $\bar{s}$ resulting in finite values of the strange form factors. Early kaon loop calculations only included ground state hyperons and mesons, but recent work [27,42] has shown that excited state hyperons and mesons can be significant.

Heavy baryon chiral perturbation theory has also been recently applied [43] to compute the Q^2 dependence of G_M^s and G_M^s. All counterterms are fixed from data except for two unknown couplings which can be constrained from the SAMPLE and HAPPEX results discussed below. Finally, the first computation of the strange

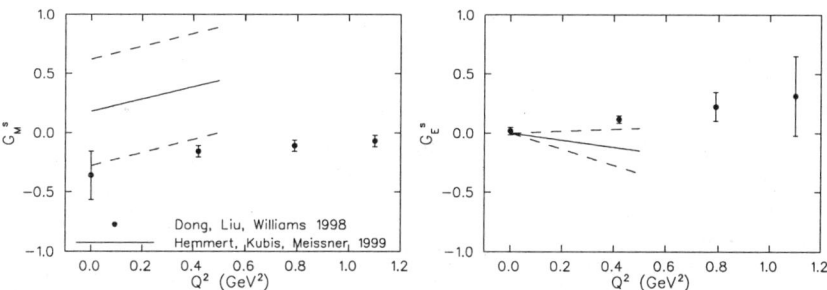

FIGURE 2. Theoretical predictions for the strange form factors from quenched lattice QCD [35] and chiral perturbation theory [43]. The dashed lines represent the range in the χPT predictions resulting from the errors on the two experimental input values (SAMPLE and HAPPEX).

form factors using lattice QCD in a quenched approximation has recently been completed [35]. Results of these calculations are shown in Figure 2.

EXPERIMENTS

The SAMPLE Experiment

The SAMPLE collaboration [44] has carried out measurements at the MIT-Bates Linear Accelerator Center in the period 1995-1999. It measures the asymmetry in backward angle elastic (quasielastic) scattering of polarized electrons from the proton (deuteron) to extract the strange magnetic form factor, G_M^s, at $Q^2 = 0.1$ (GeV/c)^2. The measurement is at low enough Q^2 that it is essentially a measurement of μ_s, ie. the contribution of the strange quark sea to a fundamental nucleon property - the magnetic moment.

The SAMPLE target and detector are shown in Figure 3. The experiment is

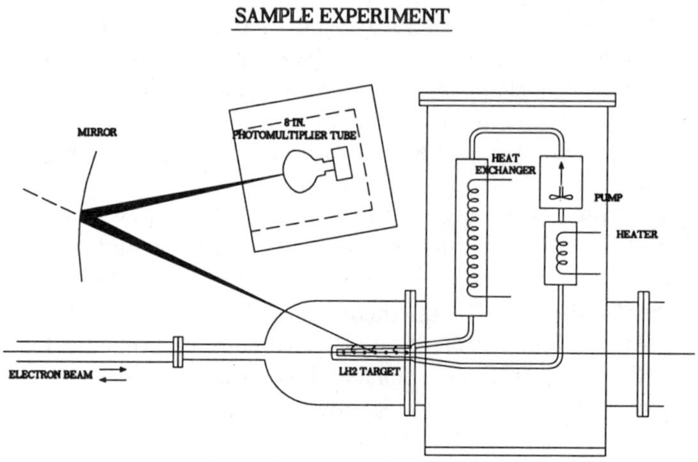

FIGURE 3. Schematic diagram of the SAMPLE experimental setup.

performed using a 200 MeV, pulsed (25 μsec pulses at 600 Hz), 35% polarized electron beam incident on a 40 cm long circulating liquid hydrogen or deuterium target [45]. The scattered electrons are detected in a large solid angle ($\sim$ 1.4 sr) air Čerenkov detector at backward angles (130° < θ < 170°). The detector consists of ten ellipsoidal mirrors which focus the Čerenkov light onto ten 8 inch photomultiplier tubes, each of which is protected by a cast lead shield from direct target and room background. The helicity of the electron beam is flipped pseudo-randomly at 600 Hz, and asymmetries are formed by combining integrated detector signals for opposite helicity states. The fraction of the integrated signal that is due to elastic scattering is measured in special runs where the beam current is reduced enough so that individual scattered particles can be counted.

Due to the small experimental asymmetries, all the experiments described here must pay careful attention to systematic effects that manifest themselves as false asymmetries. These can be controlled by keeping the beam properties (intensity, position, angle, and energy) as identical as possible under helicity reversal. Helicity-correlations in these properties must be continuously monitored, and if necessary feedback systems must be employed to keep them small. In the SAMPLE experiment, active feedback systems are used to reduce the helicity correlations in both the beam intensity and position [46]. The reponse of the detector to beam properties is determined from the random motion of the beam during data-taking. These measured responses are then combined with any residual helicity correlation of the corresponding beam parameter to correct the measured asymmetry.

The SAMPLE experiment took data with a proton target in 1995-96 [47] and 1998 [48]. The 1998 result for the parity-violating asymmetry in the elastic scattering of polarized electrons from the proton was $-4.92 \pm 0.61(\text{stat}) \pm 0.73(\text{syst})$ ppm. The proton asymmetry is sensitive to a linear combination of G_M^s and the isovector G_A^e. The resulting constraints on this linear combination are presented in Figure 4. The isovector axial e-N from factor G_A^e has been shown by Musolf and Holstein [49] to receive substantial higher order corrections. Their calculation of these corrections and its estimated error are also shown in Figure 4.

To reliably extract the strange magnetic form factor, it is desirable to directly measure G_A^e rather than rely on a theoretical estimate. This can be done by mea-

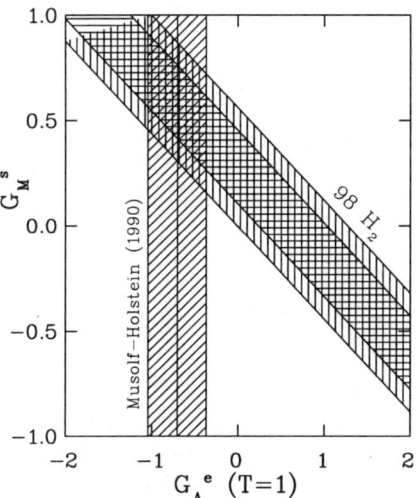

FIGURE 4. Constraints from the 1998 SAMPLE proton data set. The region allowed by the proton asymmetry is shown with the inner hatched region representing the statistical error, while the outer band includes the statistical and systematic errors added in quadrature. The vertical band corresponds to the theoretical estimate of Musolf and Holstein [49] for the isovector axial e-N form factor G_A^e and the hatched region represents their estimate of the theoretical uncertainty associated with their calculation.

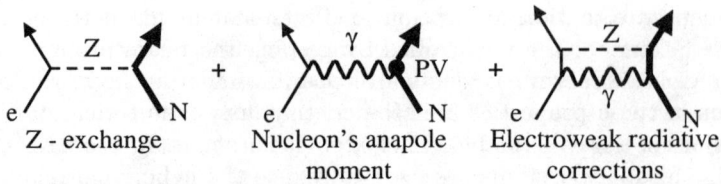

FIGURE 5. Schematic representation of the types of contributions to the electron-nucleon axial form factor G_A^e. In addition to the tree-level Z-exchange diagram, there are non-negligible higher order corrections unique to electron scattering. They include the nucleon's anapole moment and other higher order electroweak radiative corrections.

suring the parity-violating asymmetry in the quasielastic scattering of polarized electrons on deuterium. This measurement provides an additional linear combination of G_A^e and G_M^s (mostly sensitive to G_A^e) so that the two can be determined separately. The SAMPLE collaboration took data with a deuterium target during 1999. Enough data was obtained to get a comparable statistical error to the hydrogen data set. Final results from this data will be released in summer 2000.

The value of the axial electron-nucleon form factor G_A^e is of considerable interest. This is due to the fact that the higher order corrections (see figure 5) to G_A^e are expected to be significant. Among these corrections is the nucleon's anapole moment, which is the parity-violating coupling of the photon to the nucleon. The anapole moment is generated at the fundamental level from the weak interaction between quarks in the nucleon. These corrections were shown to be significant by Musolf and Holstein [49] in 1990, and a recent reevaluation [50] of these corrections using chiral perturbation theory yields similar numerical results. The SAMPLE collaboration was recently approved [51] to obtain an improved determination of G_A^e from a measurement on deuterium at different kinematics ($E = 120$ MeV, $Q^2 \sim 0.04$ GeV^2).

The HAPPEX Experiment

The HAPPEX [52] (Hall A Proton Parity EXperiment) collaboration has taken data in 1998 and 1999 at the Thomas Jefferson National Accelerator Facility (TJ-NAF). This experiment measures the forward angle asymmetry in elastic electron-proton scattering. In contrast to SAMPLE, this experiment is primarily sensitive to the first two terms in Equation 6. Thus, it measures a particular linear combination of the strange electric and magnetic form factors.

The experiment was carried out at 3.335 GeV with the beam incident on a 15 cm long liquid hydrogen target. In 1998 the beam current and polarization were ~ 100 μA and $\sim 37\%$ (bulk GaAs photocathode), while in 1999 these were ~ 35 μA and $\sim 70\%$ (strained GaAs photocathode). Elastically scattered electrons at $\theta = 12.5°$ were focussed by the two identical Hall A high resolution magnetic spectrometers onto a total-absorption lead-lucite sandwich detector in the focal plane. The helicity of the beam was changed pseudo-randomly every 33.3 msec. The parity violating asymmetry was formed by combining the integrated detector signals from two consecutive 33.3 msec periods of opposite helicity.

Correlations of the electron beam helicity with other beam properties were found to be very small at TJNAF during the 1998 run with the bulk GaAs photocathode. During the 1999 run a strained GaAs photocathode was used to provide higher beam polarization. The helicity-correlated electron beam intensity and position differences were generally larger than with bulk GaAs, but the HAPPEX group developed techniques to keep them at an acceptable level. This was the first parity-violating electron scattering experiment to employ a strained GaAs crystal.

The HAPPEX kinematics correspond to a measurement of the strange form factor combination $G_E^s + 0.39G_M^s$ at $Q^2 = 0.47 \text{GeV}^2$. The expected standard model asymmetry with no strange quark contribution is -15.8 ppm. The parity-violating asymmetry measured in the 1998 run [53] was $A = -14.5 \pm 2.0(\text{stat}) \pm 1.1(\text{syst})$ ppm. This result implies that $G_E^s + 0.39G_M^s = 0.023 \pm 0.034(\text{stat}) \pm 0.022(\text{syst}) \pm 0.026(\delta G_E^n)$ at $Q^2 = 0.47$ GeV^2. The last uncertainty comes from assuming a

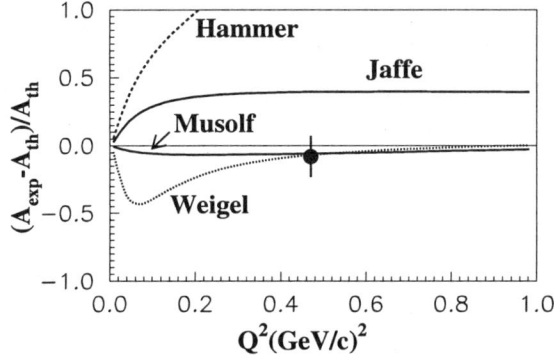

FIGURE 6. Fractional deviation of the asymmetry from that expected with no strange quarks. The data point is the HAPPEX point, while the continuous curves are theoretical predictions of refs. 23,12,17,21.

$\pm 50\%$ error on the neutron electric form factor. A comparison of the HAPPEX data with some theoretical predictions is shown in Figure 6. Also, the quenched lattice QCD prediction [35] discussed above predicts $G_E^s + 0.39G_M^s = 0.07 \pm .04$.

The HAPPEX collaboration completed a run at the same kinematics in 1999 with a similar amount of beam on target to that of 1998, but with roughly twice the polarization. The resulting asymmetry will have a statistical error that is improved by a factor of 2. Finally, the HAPPEX collaboration has been approved [54] to run with a different kinematics ($E = 3.2$ GeV, $\theta_{lab} = 6°$ corresponding to $Q^2 = 0.11$ GeV^2) which will result in a determination of the linear combination $G_E^s + 0.12G_M^s$. When combined with the SAMPLE value of G_M^s, a low Q^2 value of G_E^s is obtained, which can be used to extract a value of the strangeness radius, r_s^2.

The G^0 Experiment

This G^0 experiment [55] is being prepared for Hall C at TJNAF. This experiment will use a dedicated superconducting, toroidal magnetic spectrometer to measure the asymmetry in polarized elastic e-p scattering over a wide range of $Q^2 = 0.2 - 1.0$ GeV^2. The experiment will run in a forward angle mode, where scattered protons will be detected ($\theta_p = 62 - 78°$, corresponding to $\theta_e = 5 - 15°$), and then reversed to run in a backward angle mode where scattered electrons will be detected ($\theta_e = 110°$). This will provide enough measurements to separately extract G^s_M and G^s_E. Backward angle measurements on deuterium are also planned to determine the e-N axial form factor G^e_A

The Mainz A4 Parity Experiment

An experiment [56] to study forward angle parity-violating elastic e-p scattering is in preparation at the Mainz microtron (MAMI). The experiment will employ a forward angle calorimeter consisting of an array of 1022 PbF$_2$ crystals with sufficient energy resolution to adequately separate the elastic and inelastic scattering. The experiment will run with a beam energy of 855 MeV and detect electrons at a mean scattering angle of 35 degrees. This will allow a determination of the linear combination $G^s_E + 0.22 G^s_M$ at $Q^2 = 0.23$ GeV^2.

ACKNOWLEDGEMENTS

I thank Frank Maas and Paul Souder for providing information about their experiments. This work was supported in part by NSF grant PHY-9733772.

REFERENCES

1. J. Gasser, H. Leutwyler, and M.E. Sainio, *Phys. Lett.* B **253**, 252 (1991).
2. K. Abe *et al.*, *Phys. Rev. Lett.* **75**, 25 (1995)., and references therein
3. D. Kaplan and A. Manohar, *Nucl. Phys.* A **310**, 527 (1988).
4. R.D. McKeown, *Phys. Lett.* B **219**, 140 (1989).
5. D.H. Beck, *Phys. Rev.* D **39**, 3248 (1989).
6. V. Dmitrašinović and S.J. Pollock, *Phys. Rev.* C **52**, 1061 (1995).
7. S. Capstick and D. Robson, preprint nucl-th/9708054
8. G.A. Miller, *Phys. Rev.* C **57**, 1492 (1998).
9. B.-Q. Ma, *Phys. Lett.* B **408**, 387 (1997).
10. R. Lewis and N. Mobed, *Phys. Rev.* D **59**, 073002 (1999).
11. M.J. Musolf *et al.*, *Phys. Rep.* **239**, 1 (1994).
12. R. L. Jaffe, *Phys. Lett.* B **229**, 275 (1989).
13. N. W. Park, J. Schechter, and H. Weigel, *Phys. Rev.* D **43**, 869 (1991).
14. W. Koepf, E. M. Henley, and S. J. Pollock, *Phys. Lett.* B **288**, 11 (1992).

15. S. Hong and B. Park, *Nucl. Phys.* A **561**, 525 (1993).
16. T.D. Cohen, H. Forkel, and M. Nelson, *Phys. Lett.* B **316**, 1 (1993).
17. M. J. Musolf and M. Burkardt, *Z. Phys.* C **61**, 433 (1994).
18. S. C. Phatak and Sarira Sahu, *Phys. Lett.* B **321**, 11 (1994).
19. H. Forkel, *et al.*, *Phys. Rev.* C **50**, 3108 (1994).
20. H. Ito, *Phys. Rev.* C **52**, R1750 (1995).
21. H. Weigel, *et al.*, *Phys. Lett.* B **353**, 20 (1995).
22. Chr. V. Christov *et al.*, Prog. Part. Nucl. Phys. **37**, 91 (1996).
23. H.-W. Hammer, *et al.*, *Phys. Lett.* B **367**, 323 (1996).
24. D. Leinweber, *Phys. Rev.* D **53**, 5115 (1996).
25. W. Melnitchouk and M. Malheiro, *Phys. Rev.* C **55**, 431 (1997).
26. H. Kim, *et al.*, *Nucl. Phys.* A **616**, 606 (1997).
27. P. Geiger and N. Isgur, *Phys. Rev.* D **55**, 299 (1997).
28. M.J. Ramsey-Musolf and H. Ito, *Phys. Rev.* C **55**, 3066 (1997).
29. M. J. Musolf, *et al.*, *Phys. Rev.* D **55**, 2741 (1997).
30. U.-G. Meißner *et al.*, PLB **408**, 381 (1997).
31. H. Forkel, *Phys. Rev.* C **56**, 510 (1997).
32. M.J. Musolf and H.-W. Hammer, *Phys. Rev. Lett.* **80**, 2539 (1998).
33. S-T. Hong, B-Y. Park, and D-P. Min, *Phys. Lett.* B **414**, 229 (1997).
34. M. Malheiro and W. Melnitchouk, *Phys. Rev.* C **56**, 2373 (1997).
35. S.J. Dong, K.F. Liu, and A.G. Williams, *Phys. Rev.* D **58**, 074504 (1998).
36. H.-C. Kim, *et al.*, *Phys. Rev.* D **58**, 114027 (1998).
37. D.O. Riska, Few-Body Systems Suppl. **99**, 1 (1998).
38. H.-W. Hammer and M.J. Ramsey-Musolf, *Phys. Rev.* C **60**, 045205 (1999).
39. S-T. Hong and D-P.Min, preprint nucl-th/9909004
40. D.B. Leinweber and A.W. Thomas, preprint hep-lat/9912052
41. L. Hannelius and D.O. Riska, preprint hep-ph/0001325
42. L.L. Barz, *et al.*, *Nucl. Phys.* A **640**, 259 (1998).
43. T. R. Hemmert, B. Kubis, U.-G. Meißner, *Phys. Rev.* C **60**, 045501 (1999).
44. MIT-Bates Expt. 89-06 (D.H. Beck and R.D. McKeown, spokespersons), MIT-Bates Expt. 94-11 (E.J. Beise and M.L. Pitt, spokespersonns)
45. E.J. Beise, *et al.*Nucl. Instrum. Methods Phys. Res. A **378**, 383 (1996).
46. T. Averett, *et al.*, Nucl. Instrum. Methods Phys. Res. A **438**, 246 (1999).
47. B.A. Mueller, *et al.*, *Phys. Rev. Lett.* **78**, 3824 (1997).
48. D.T. Spayde, *et al.*, *Phys. Rev. Lett.* **84**, 1106 (2000).
49. M.J. Musolf and B.R. Holstein, *Phys. Lett.* B **242**, 461 (1990).
50. S-L. Zhu, *et al.*, hep-ph/0002252.
51. MIT-Bates Expt. 00-04 (T. Ito, spokesperson).
52. TJNAF experiment E91-010 (J.M. Finn and P.A. Souder, spokespersons)
53. K.A. Aniol, *et al.*, *Phys. Rev. Lett.* **82**, 1096 (1999).
54. TJNAF experiment E99-115 (K.S. Kumar and D. Lhullier, spokespersons)
55. TJNAF experiment E91-017 (D. Beck, spokesperson)
56. MAMI proposal A4/1-93 (D. von Harrach, spokesperson)

Charge Symmetry Violation in np → dπ⁰

A.K. Opper*

*Ohio University, Department of Physics and Astronomy, Athens OH 45701 USA[1]

Abstract.
   While charge symmetry breaking (CSB) in the strong interaction is ultimately due to the mass difference between u and d quarks, meson-exchange models describe the data most consistently. A high precision measurement of CSB has been undertaken at TRIUMF to complement those data and lead to a richer understanding of the dynamical quark effects in the NN system. The observable of interest is the forward–backward asymmetry (A_{fb}) in np → dπ⁰, which must be zero in the center-of-mass if charge symmetry is conserved and is predicted to be approximately -35×10^{-4} with the dominant contributions being an order of magnitude larger than those of the elastic scattering CSB measurements carried out at TRIUMF and IUCF.
   The measurement used a 279.5 MeV neutron beam, a liquid hydrogen target, and the SASP spectrometer positioned at 0°. With these kinematics and the large acceptance of SASP the full deuteron distribution was detected in one setting of the spectrometer thereby eliminating many systematic uncertainties. A measurement of the pp → dπ⁺ distribution accompanied the primary measurement as a robust test of the analysis and simulation codes. The experimental challenges and status are presented.

INTRODUCTION

Discovering conservation principles and the associated symmetries is a powerful tool in understanding the forces of nature. However, a more subtle and intriguing situation arises when a fundamental symmetry is broken. Within QCD, charge symmetry breaking (CSB) and charge independence breaking (CIB) arise from differences between the current masses of the u and d quarks, and from electromagnetic interactions between the quarks. CSB has been clearly observed in the nucleon–nucleon system, most recently in the difference between the neutron and proton analyzing powers in np elastic scattering [3–5].

Another indication of CSB in the strong interaction would be an asymmetry about 90° in the center-of-mass angular distribution of deuterons (or pions) in the reaction np → dπ⁰.

---

[1] Research supported by the Natural Sciences and Engineering Research Council of Canda and the National Science Foundation of the USA.

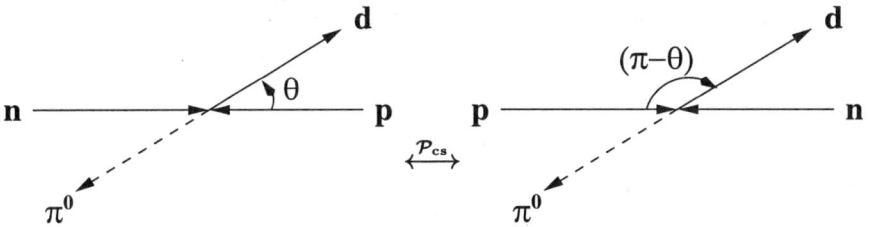

FIGURE 1. Charge symmetric sketches of np → dπ⁰ in the center-of-mass frame.

Fig. 1 shows in an intuitive manner that charge symmetry requires the deuteron distribution in np → dπ⁰ be symmetric about 90° in the center of mass. Both sketches show the np → dπ⁰ reaction in the center of mass and are related by the charge symmetry operator P_{CS}. In going from one sketch to the other, P_{CS} interchanges neutrons and protons. Given that these two figures are invariant under charge symmetry, the deuteron yield at an angle θ in the drawing on the left must be equal to the yield at an angle $(\pi - \theta)$ in the charge symmetric drawing on the right. Said another way, a non-zero forward–backward asymmetry can only occur if charge symmetry is violated. This forward–backward asymmetry in the deuteron distribution is defined as

$$A_{fb}(\theta) \equiv \frac{\sigma(\theta) - \sigma(\pi - \theta)}{\frac{1}{2}[\sigma(\theta) + \sigma(\pi - \theta)]}. \tag{1}$$

Since the deuteron distribution from pp → dπ⁺ must be symmetric in the center of mass due to the indistinguishability of the two protons, this experiment has undertaken a measurement of the pp → dπ⁺ distribution [6] as a test case to verify the analysis and simulation codes.

Niskanen's model of pion production includes CSB effects and he has extended his calculations to energies near threshold [7]. These calculations include contributions from the np mass difference (δ), $\eta - \pi^0$ mixing, $\eta' - \pi^0$ mixing, and $\rho^0 - \omega$ mixing. This calculation also includes the mixing contributions from both the production vertex and the transition potential and 16 time-ordered diagrams. Fig. 2 shows values for A_{fb} extracted from these calculations. Although $\rho^0 - \omega$ mixing is the dominant contribution to the analyzing power difference in elastic np scattering measured at 183 MeV, the contribution from this to A_{fb} near 280 MeV is on the same order as the δ contributions and is not plotted in this figure. Electromagnetic effects are comparably small and are also not plotted. The symbol plotted on this figure indicates the *predicted* value of A_{fb} at the energy of this experiment with *anticipated* error bars. As the figure shows, when only δ contributions are included, A_{fb} is positive and ranges from $(1 \text{ to } 5) \times 10^{-4}$ for the different neutron energies. However, adding the contributions from the $\eta - \pi^0$ and $\eta' - \pi^0$ mixing, neither of which enters in the elastic np scattering analyzing power results, changes the sign, shape, and magnitude of A_{fb} for all neutron energies. It is interesting to note in fig. 2 that the predicted value for A_{fb} increases by only a factor of two as the neutron energy increases from 279.5 MeV to 350 MeV. This is due to the decrease in the vertex contributions above 300 MeV. In fact, both the η and δ contributions change sign around 400 MeV.

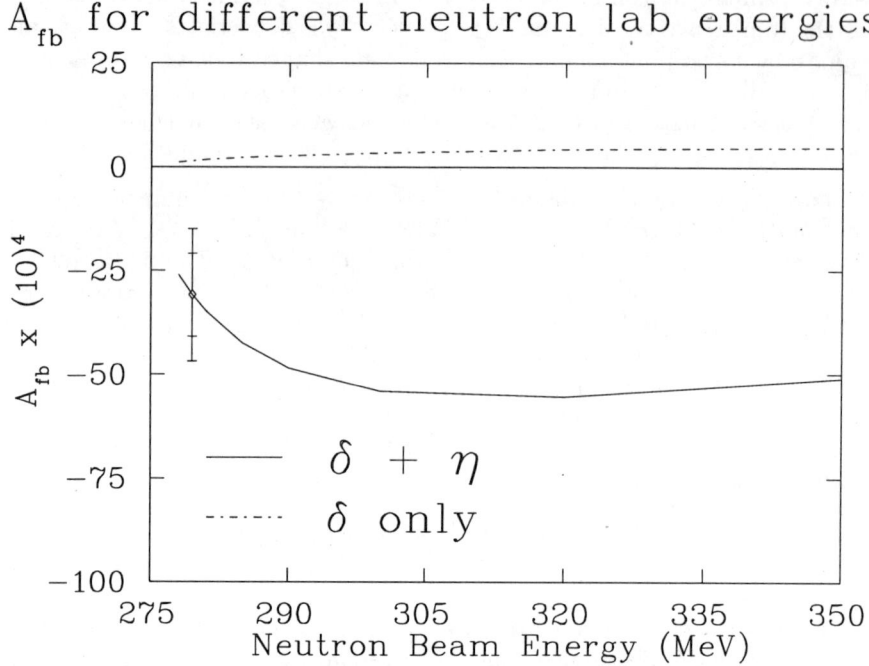

FIGURE 2. Contributions near threshold to A_{fb} from δ, $\eta - \pi^0$ mixing, and $\eta' - \pi^0$ mixing. The open diamond indicates the *predicted* value of A_{fb} at the energy of this experiment with *anticipated* error bars.

The potential for the exchange of an $\eta-\pi$ ($\eta'-\pi$), isospin mixed meson is linearly dependent on the ηNN coupling constant and the $\eta - \pi$ ($\eta' - \pi$) mixing amplitude. While the momentum dependence of the mixing matrices is under debate, the on-shell values are well known from analysis of η and η' decay [8] but the value of the η-nucleon coupling constant ranges from $(g^2_{\eta NN}/4\pi) = 0.2$ to 6.2 [7,9]. With this CSB measurement dominated by η effects, it may be an effective tool to constrain the ηNN coupling.

Further, the meson-baryon coupling is related to the internal structure, and thus the strangeness content, of the nucleon. The effect of an $s\bar{s}$ content in the proton on the ηNN coupling constant would lead to a larger value of A_{fb} than predicted by Niskanen. Recall that in the limit of complete chiral symmetry, $(m_d - m_u) \to$ zero. Van Kolck has shown that the charge-symmetric scattering length is proportional to $(m_d - m_u)$ [10]. Since $NN \to d\pi$ also involves virtual $\pi^0 N$ scattering, the measurement of A_{fb} in np $\to d\pi^0$ may provide information on differences between $\pi - n$ coupling and $\pi - p$ coupling.

EXPERIMENTAL APPARATUS

The TRIUMF CHARGEX facility [11] was used to produce the neutron beam. A 283-MeV primary proton beam striking a ^{7}Li target produced neutrons of 279.5 MeV by the (p,n) reaction. The unscattered beam was deflected 20° towards an external dump by a sweeping magnet. Neutrons were shaped to a 4 cm by 6 cm beam in a tungsten-alloy collimator, in which were set two veto scintillators to eliminate protons scattered from the Li target and deuterons generated by (n,d) reactions on the collimator surfaces. The neutron beam impinged on a liquid hydrogen target of nominal thickness 2 cm. Deuterons from reactions in the target passed through a thin trigger scintillator and three pairs of multi-wire proportional counter planes before entering the Second Arm Spectrometer (SASP) [12]. Vertical drift chambers and three layers of plastic scintillators were located at and beyond the SASP focal plane. Energy loss and time-of-flight allowed a very clean separation of deuterons from protons.

The pp $\to$ dπ^+ reaction must have A_{fb}=0 due to the identity of beam and target particles, and will be used to check for many of the potential systematic errors in the np $\to$ dπ^0 experiment. Interleaved measurements of the pp $\to$ dπ^+ reaction were carried out with the following changes from neutron beam mode: the ^{7}Li target was retracted and the sweeping magnet turned off; a 1-mm thick Ta foil was inserted in the beamline approximately 13 m upstream of the liquid hydrogen target; the primary beam intensity was lowered below 1 nA and its energy changed to account for the difference in reaction thresholds for the neutral and charged pion; SASP fields were increased to put the deuteron locus at the same position in the focal plane; veto counters were used in coincidence mode.

EXTRACTING A_{fb} FROM LAB VARIABLES

Near threshold the deuterons from np $\to$ dπ^0 form a small forward cone centered about the beam direction and a kinematic locus with a narrow range of momentum and scattering angle in the lab. This locus is easily seen in fig. 3. The background surrounding the locus is from (n, d) reactions on nuclei such as carbon in materials near the target. The deuterons with higher (lower) momentum are produced in the forward (backward) direction in the center-of-mass frame so that comparing the top half with the bottom half of the locus gives an indication of A_{fb}.

In the center of mass frame the np $\to$ dπ^0 cross section can be described by three parameters, A_0, A_1, and A_2:

$$d\sigma/d\Omega = A_0 + A_1 P_1(\theta) + A_2 P_2(\theta), \qquad (2)$$

where θ is the deuteron centre-of-mass angle and P_i are Legendre polynomials [13]. In this expression, the second term describes any charge symmetry breaking with the coefficient A_1 giving the asymmetry.

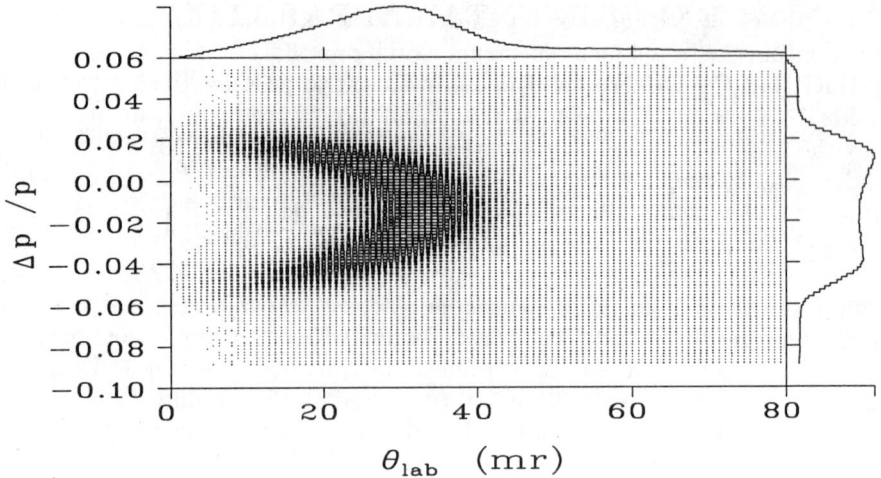

FIGURE 3. Data for np → dπ⁰ through Summer 1998 showing the kinematic locus of momentum vs scattering angle in the lab.

All the experimental features (target, detectors, magnetic fields, ...) are described in a GEANT simulation with the np → $d\pi^0$ center-of-mass cross section given in equation 1. After the deuteron and π are produced, their positions and momenta are converted to the lab frame, they are tracked through the detector system and a kinematic locus is produced. This simulated locus is compared to data and the coefficients of the Legendre polynomials extracted through a χ^2 minimization procedure. This gives the CSB observable since the angle integrated A_{fb} is essentially equal to A_1/A_0.

STATUS

As of August 1999, almost seven million np → $d\pi^0$ events and 10 million pp → $d\pi^+$ events have been put to tape. Additionally, calibration measurements of detector efficiencies and various sources of background were carried out. Analysis of these auxillary measurements was finished at the end of 1999 and the results (e.g. efficiency profiles for each run period) will be incorporated into the simulation.

The sum of 8/97 through 5/99 target full locus data over the nominal "full acceptance" target variable region of 16 cm² [(-2,2) cm in the vertical position, XI, by (-2,2) cm in the horizontal position, YI] was fit using a sum of a single GEANT background-free locus and an 8-term background functional form. The fit yielded a χ^2/d.f. of 4.4 (see fig. 4). This description of the background gave a χ^2/d.f. of 1.5 when compared with the target empty data. The increase in χ^2 from the target empty fit is due to much greater statistical sensitivity to the locus signal; thus at this stage of the simulation the locus is relatively poorly described in comparison to the smooth background. Features of the full-acceptance fit include the following:

- The "shelf" of background at the low- and high-momentum ends of the momentum-projected locus spectrum is well reproduced. This gives some confidence that the background shape is sufficiently well described.

- The positions of the locus "lobes" of the momentum-projected spectrum for reaction angles less than 10 mr are well reproduced by the simulation. The centroid of the distribution in reaction angle (i.e. the "turnaround angle" of the locus) is also well reproduced. The yield near the turnaround is overpredicted by the model.

- The overall width of the momentum-projected locus spectrum is fairly well reproduced.

- The simulation underpredicts the strength in the high-momentum end of the locus.

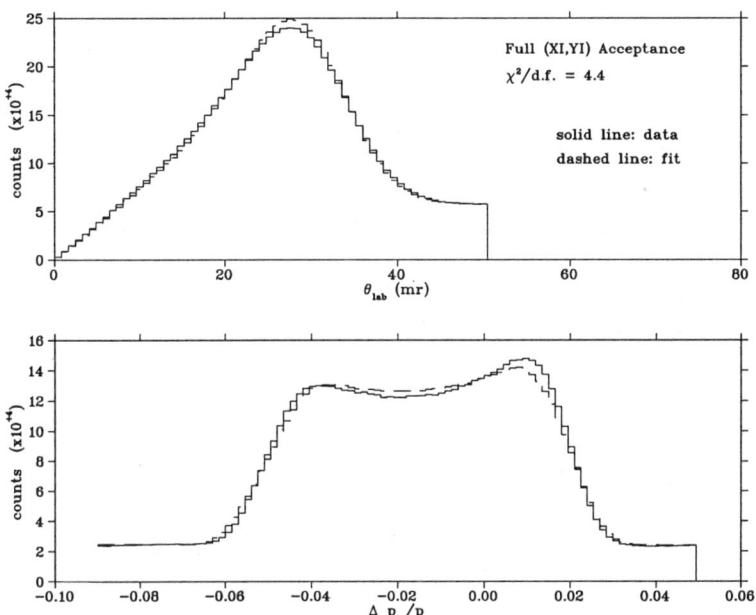

FIGURE 4. GEANT fit to sum of 8/97 through 5/99 np $\rightarrow$ dπ^0 data, full (XI,YI) acceptance. Upper panel is the locus projection on the reaction angle axis for relative momenta in the range -9% to 5%. Lower panel is the projection on the momentum axis for reaction angles in the range 0 to 50 mr.

The fact that the simulation does not reproduce either the gross yields or the instrumental asymmetry observed in the data is presumably evidence that the SASP angular acceptance, and the momentum variation of this acceptance, is not yet correctly described. Investigations of the SASP acceptance model via simulation of np elastic data are in progress and demanding consistancy in the agreement of the simulation of these data and the deuteron production data will aid in extracting a result for A_{fb}.

In addition to development of the np elastic simulation, the position and alignment of all experimental equipment in the simulation has been verified and made consistent with blueprints and in situ measurments. Further improvements to be added in the next few months are: inclusion of deuteron reaction losses, additional charged particles in the pp $\to$ dπ^+ simulation, and an improved description of the magnetic fields. By the summer of 2000 the GEANT code should be fine tuned to allow generation of the final sets of simulation data and a result available by the end of 2000. Given the large number of events on tape, the statistical uncertainty in A_{fb} is 9.4×10^{-4}; as many of the systematic effects are themselves determined by statistics we anticipate a systematic uncertainty of 12×10^{-4}.

REFERENCES

1. G.A. Miller and W.T.H. Van Oers, in *Fundamental Symmetries,* edited by W.C. Haxton and E.M. Henley (World Scientific, Singapore, 1995), p 127.
2. G.A. Miller, B.M.K. Nefkens, and I. Slaus, Phys. Rep. **194**, 1 (1990)
3. R. Abegg, *et al.*, Phys. Rev. D **39**, 2464 (1989).
4. S.E. Vigdor, *et al.*, Phys. Rev. C **46**, 410 (1992).
5. J. Zhao *et al.*, Phys. Rev. C 57 (1998) 2126.
6. E. Korkmaz, *et al.*, Nucl. Phys. **A535**, 637 (1991).
7. J.A. Niskanen, *nucl-th/9809009.*
8. S.A Coon, BHJ McKellar, and MD Scadron, Phys. Rev. D **34**, 2784 (1986).
9. Benmerrouche, Mukhopadhyay, and Zhang, Phys. Rev. D **51**, 3237 (1995).
10. U. Van Kolck, PhD thesis (1993): Univ of Washington preprint DOE/ER/40427-13-N94: and *private communication.*
11. R. Helmer *et al.*, Can J. Phys. **65**, 588 (1987).
12. P. Walden *et al.*, to be published in Nucl. Instr. and Meth.
13. D.A. Hutcheon, *et al.*, Nucl. Phys. **A535**, 618 (1991).

Isospin Symmetry Breaking in Nuclei
— ONS Anomaly —

Koichi Saito

Tohoku College of Pharmacy, Sendai 981-8558, Japan

Abstract. We study the binding energy differences of the valence proton and neutron of the mirror nuclei, ^{15}O - ^{15}N, ^{17}F - ^{17}O, ^{39}Ca - ^{39}K and ^{41}Sc - ^{41}Ca, using the quark-meson coupling model. The calculation involves nuclear structure and shell effects explicitly. It is shown that binding energy differences of a few hundred keV arise from the strong interaction, even after subtracting all electromagnetic corrections. The origin of these differences may be ascribed to the charge symmetry breaking effects set in the strong interaction through the u and d current quark mass difference. In this report, we first review the quark-meson coupling model. In particular, we discuss about the nucleon mass in nuclear medium. Then, we present details of the charge symmetry breaking in finite nuclei, especially the Okamoto-Nolen-Schiffer anomaly.

INTRODUCTION

The discrepancy between the calculated binding energy differences of mirror nuclei and those measured is a long-standing problem in nuclear physics. It is known as the Okamoto-Nolen-Schiffer (ONS) anomaly [1,2]. Although it was first thought that electromagnetic effects could almost account for the observed binding energy differences, it is now believed that the ONS anomaly has its origin in charge symmetry breaking (CSB) in the strong interaction [3]. In addition to calculations based on charge symmetry violating meson exchange potentials [3–5], a number of quark-based calculations have been performed [6,7] in an attempt to resolve this anomaly. In such calculations, CSB enters through the up (u) and down (d) current quark mass difference in QCD. Despite these efforts, the difficulty of producing a realistic description of nuclear structure on the basis of explicit quark degrees of freedom has hindered the direct calculation of the binding energy differences.

In this study we report the results for the binding energy differences of the valence (excess) proton and neutron of the mirror nuclei, ^{15}O - ^{15}N, ^{17}F - ^{17}O, ^{39}Ca - ^{39}K and ^{41}Sc - ^{41}Ca, calculated using a quark-based model involving explicit nuclear structure and shell effects, namely the quark-meson coupling (QMC) model [8]. This model has been successfully applied not only to traditional nuclear problems [8] but also to other new areas as well [9]. Although some exploratory QMC results on the ONS anomaly have already been reported [7], an early version of the model was used there, and it was applied to finite nuclei only through local density approximation, rather than a consistent shell model calculation.

THE QUARK-MESON COUPLING MODEL

In this section, we introduce the QMC model, and then report the medium modification of the nucleon properties in finite nuclei [8].

Effect of Nucleon Structure

Let us suppose that a free nucleon (at the origin) consists of three light (u and d) quarks under a (Lorentz scalar) confinement potential, V_c. Then, the Dirac equation for the quark field ψ_q is given by

$$[i\gamma \cdot \partial - m_q - V_c(r)]\psi_q(r) = 0, \tag{1}$$

where m_q is the bare quark mass.

Next we consider how Equation (1) is modified when the nucleon is bound in static, uniformly distributed (iso-symmetric) nuclear matter. In the QMC model [8] it is assumed that each quark feels scalar, V_s^q, and vector, V_v^q, potentials, which are generated by the surrounding nucleons, as well as the confinement potential. This assumption seems appropriate when the nuclear density ρ_B is near around normal nuclear matter density ($\rho_0 = 0.15$ fm^{-3}). If we use the mean-field approximation (MFA) for the meson fields, Equation (1) may be rewritten as

$$[i\gamma \cdot \partial - (m_q - V_s^q) - V_c(r) - \gamma_0 V_v^q]\psi_q(r) = 0. \tag{2}$$

The potentials generated by the medium are constants because the matter distributes uniformly. As the nucleon is static, the time-derivative operator in the Dirac equation can be replaced by the quark energy, $-i\epsilon_q$. By analogy with the procedure applied to the nucleon in QHD [10], if we introduce the effective quark mass by $m_q^* = m_q - V_s^q$, the Dirac equation (2) can be rewritten in the same form as that in free space with the mass m_q^* and the energy $\epsilon_q - V_v^q$, instead of m_q and ϵ_q. In other words, the vector interaction has *no effect on the nucleon structure* except for an overall phase in the quark wave function, which gives a shift in the nucleon energy. This fact *does not* depend on how to choose the confinement potential, V_c. Then, the nucleon energy at rest in the medium is given by $E_N = M_N^*(V_s^q) + 3V_v^q$, where the effective nucleon mass M_N^* depends on *only the scalar potential*.

We can extend this idea to finite nuclei [8]. Let us suppose that the scalar and vector potentials in Equation (2) are mediated by the σ and ω mesons, and introduce their mean-field values, which now depend on position $\vec{r}$, by $V_s^q(\vec{r}) = g_\sigma^q \sigma(\vec{r})$ and $V_v^q(\vec{r}) = g_\omega^q \omega(\vec{r})$, respectively, where g_σ^q (g_ω^q) is the coupling constant of the quark-σ (ω) meson. Furthermore, we shall add the isovector vector meson, ρ, and the Coulomb field, A, to describe finite nuclei realistically. Then, the effective Lagrangian density for finite nuclei would be given by [8]

$$\mathcal{L}_{QMC} = \bar{\psi}[i\gamma \cdot \partial - M_N^\star - g_\omega \omega \gamma_0 - g_\rho \frac{\tau_3^N}{2} b \gamma_0 - \frac{e}{2}(1+\tau_3^N)A\gamma_0]\psi \qquad (3)$$
$$- \frac{1}{2}[(\nabla \sigma)^2 + m_\sigma^2 \sigma^2] + \frac{1}{2}[(\nabla \omega)^2 + m_\omega^2 \omega^2] + \frac{1}{2}[(\nabla b)^2 + m_\rho^2 b^2] + \frac{1}{2}(\nabla A)^2,$$

where ψ and b are respectively the nucleon and the ρ fields. m_σ, m_ω and m_ρ are respectively the masses of the σ, ω and ρ mesons. g_ω and g_ρ are respectively the ω-N and ρ-N coupling constants, which are given by $g_\omega = 3g_\omega^q$ and $g_\rho = g_\rho^q$ (where g_ρ^q is the quark-ρ coupling constant).

If we define the field-dependent σ-N coupling constant, $g_\sigma(\sigma)$, by [8]

$$M_N^\star(\sigma(\vec{r})) \equiv M_N - g_\sigma(\sigma(\vec{r}))\sigma(\vec{r}), \qquad (4)$$

where M_N is the free nucleon mass, it is easy to compare with QHD [10]. The difference between QMC and QHD lies only in the coupling constant g_σ, which depends on the scalar field in QMC while it is constant in QHD. However, this difference leads to a lot of favorable results [8].

Now we need a model for the structure of the nucleon involved to perform actual calculations. We here use the MIT bag model. In the present model, the bag constant, B, and the z parameter for the nucleon are fixed to reproduce the free nucleon mass ($M_N = 939$ MeV) and the free bag radius $R_N = 0.8$ fm. In the following we choose $m_q = 5$ MeV and set $m_\sigma = 550$ MeV, $m_\omega = 783$ MeV and $m_\rho = 770$ MeV. (Variations of the quark mass and R_N only lead to numerically small changes in the calculated results [8].) We then find that $B^{1/4} = 170.0$ MeV and $z = 3.295$.

For infinite nuclear matter, from the Lagrangian density (3), we can easily find the total energy per nucleon, E_{tot}/A, and the mean-field values of ω and ρ (which are respectively given by baryon number conservation and the difference in proton and neutron densities). The scalar mean-field is given by a self-consistency condition, $\partial E_{tot}/\partial \sigma = 0$ [8]. The coupling constants, g_σ^2 and g_ω^2, are fixed to fit the average binding energy (-15.7 MeV) at ρ_0 for nuclear matter. Furthermore, the ρ-N coupling constant is used to reproduce the bulk symmetry energy, 35 MeV. We then find [8]: $g_\sigma^2/4\pi = 5.40$, $g_\omega^2/4\pi = 5.31$, $g_\rho^2/4\pi = 6.93$, and the nuclear incompressibility, $K \simeq 280$ MeV. Note that the model gives the variation of the nucleon bag radius, $\delta R_N^\star/R_N = -0.02$, the lowest quark eigenvalue, $\delta x_q^\star/x_q = -0.16$ and the root-mean-square radius of the quark wave function, $\delta r_q^\star/r_q = +0.02$, at saturation density.

Using these parameters, we can solve a finite nuclear system. As an example, we show charge density distribution of ^{40}Ca in Figure 1. The QMC model can reproduce the properties of not only nuclear matter but also finite nuclei (for more details, see [8]).

Nucleon Mass in Nuclear Matter

Here we consider the nucleon mass in matter furthermore. The nucleon mass is a function of the scalar field. Because the scalar field is small at low density the mass may be expanded in terms of σ as

$$M_N^* = M_N + \left(\frac{\partial M_N^*}{\partial \sigma}\right)_{\sigma=0} \sigma + \frac{1}{2}\left(\frac{\partial^2 M_N^*}{\partial \sigma^2}\right)_{\sigma=0} \sigma^2 + \mathcal{O}(\sigma^3). \qquad (5)$$

Since the interaction Hamiltonian between the nucleon and the σ field at the quark level is given by $H_{int} = -3g_\sigma^q \int d\vec{r}\, \overline{\psi}_q \sigma \psi_q$, the derivative of M_N^* with respect to σ is $-3g_\sigma^q \int d\vec{r}\, \overline{\psi}_q \psi_q \equiv -3g_\sigma^q S_N(\sigma)$, where we have defined the quark scalar charge in the nucleon, $S_N(\sigma)$, which is itself a function of σ. Because of a negative value of the derivative, the nucleon mass decreases in matter at low density.

Furthermore, we define the scalar-charge ratio, $S_N(\sigma)/S_N(0)$, to be $C_N(\sigma)$ and the σ-N coupling constant in free space to be g_σ (i.e., $g_\sigma = g_\sigma(\sigma = 0) = 3g_\sigma^q S_N(0)$). Using these quantities, we find

$$M_N^* = M_N - g_\sigma \sigma - \frac{1}{2}g_\sigma C_N'(0)\sigma^2 + \mathcal{O}(\sigma^3). \qquad (6)$$

In general, C_N is a decreasing function because the quark in matter becomes more relativistic than in free space. Thus, $C_N'(0)$ takes a negative value. If the nucleon were structureless C_N would not depend on σ. Therefore, only the first two terms in the RHS of Equation (6) remain, which is exactly the same as the effective nucleon mass in QHD [10].

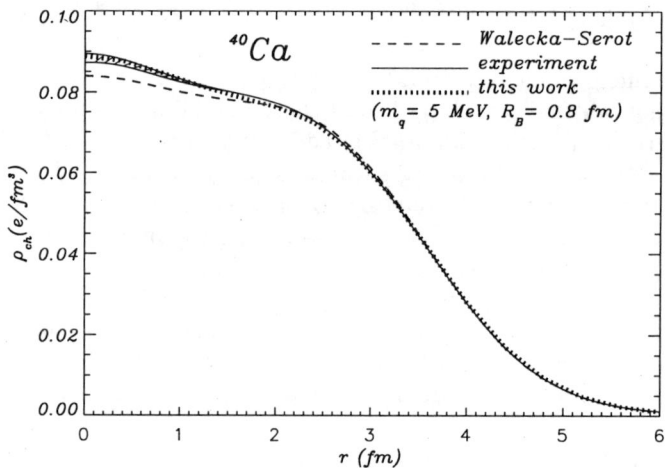

FIGURE 1. Charge density distribution for ^{40}Ca [8] compared with the experimental data and that of QHD.

TABLE 1. Inputs, parameters and some of the quantities calculated in the present study. The quantities with a star, *, are those quantities calculated at ρ_0. We take $m_u = 5$ MeV.

	M_j (MeV)	R_j (fm)	$B^{1/4}$ (MeV)	z	M_j^* (MeV)	R_j^* (fm)
p (CSB)	937.6423 (input)	0.8 (input)	169.81	3.305	751.928	0.7950
n (CSB)	939.6956 (input)	0.8000	169.81	3.305	753.597	0.7951
N (SU(2))	939.0 (input)	0.8 (input)	169.97	3.295	754.542	0.7864

CHARGE SYMMETRY BREAKING IN QMC

Now we introduce the charge symmetry breaking in the QMC model [7,11]. The charge symmetry is explicitly broken at the quark level through their masses. We use different values for the u and d current quark masses, and the effective proton, M_p^*, and neutron, M_n^*, masses. At position $\vec{r}$ in a nucleus (the coordinate origin is taken at the center of the nucleus), the Dirac equations for the quarks in the proton or neutron bag are given by

$$\left[i\gamma \cdot \partial_x - \left(\begin{pmatrix} m_u \\ m_d \end{pmatrix} - V_\sigma^q(\vec{r})\right) - \gamma^0 \left(V_\omega^q(\vec{r}) \pm \frac{1}{2} V_\rho^q(\vec{r})\right)\right] \begin{pmatrix} \psi_u(x) \\ \psi_d(x) \end{pmatrix} = 0, \quad (7)$$

where $|\vec{x} - \vec{r}| \leq R_j^*$ (j specifies proton or neutron). Note that we have assumed that the scalar potential is common to both the u and d quarks. The nucleon and meson fields are calculated self-consistently by solving a set of coupled non-linear differential equations, derived from the effective Lagrangian density (3) with the proper modifications caused by the different proton and neutron (or u and d quark) masses in MFA. Thus, the present calculation is free from the sort of double counting questioned by Auerbach [12], and includes the shell effects, which were discussed by Cohen et al. [13].

Before discussing the results obtained, we again need to specify the parameters and inputs used in the calculation [11]. They are summarized in Table 1. The bag constant, B, and the z parameter are determined by the bare proton mass, after allowing for the electromagnetic self-energy correction +0.63 MeV, with the bag radius, $R_p = 0.8$ fm, in free space. For the neutron, the procedure is the same as that for the proton, allowing for the electromagnetic self-energy correction, -0.13 MeV, but using the values of B and z determined above and calculating the d current quark mass and the bag radius for the neutron. Thus, the u current quark mass ($m_u = 5$ MeV) is the basic input parameter used to fix the model parameters so as to reproduce the bare proton and neutron masses in free space. We found $m_d = 9.2424$ MeV in the present calculation.

The coupling constants, g_σ^q and g_ω^q, are determined so as to fit the saturation properties of symmetric nuclear matter [11]. In Table 1, SU(2) stands for the parameters and inputs obtained and used for the calculation when SU(2) symmetry is assumed, namely $m_u = m_d = 5$ MeV. We then found: $(g_\sigma^q, g_\omega^q) = (5.698, 2.744)$ for CSB, and $(5.685, 2.721)$ for SU(2). For the quark-ρ meson coupling constant, to make a realistic estimate, we here use the phenomenological value, $g_\rho^q = 4.595$, the value at zero three-momentum transfer corresponding to Hartree approximation, from Table 4.1 of Ref. [14]. (Note that because the QMC model does not contain the ρ-nucleon tensor coupling [8], this gives an unrealistically large value for the coupling constant [11].)

Proton and Neutron Masses in Nuclear Matter

As in Equation (5), the proton and neutron masses are again given by functions of σ in matter, and may be expanded in terms of σ at low ρ_B

$$M_p^* = M_p + (3g_\sigma^q)\frac{1}{3}[2S_{u/p}(0) + S_{d/p}(0)]\sigma + \mathcal{O}(\sigma^2), \qquad (8)$$

$$M_n^* = M_n + (3g_\sigma^q)\frac{1}{3}[S_{u/n}(0) + 2S_{d/n}(0)]\sigma + \mathcal{O}(\sigma^2). \qquad (9)$$

Because $m_u \neq m_d$, the u-quark scalar charge is no longer the same as the d-quark scalar charge. We have therefore introduced four kinds of quark scalar charges in the expansion: $S_{i/j}(\sigma) = \int_{V_j} d\vec{r}\, \overline{\psi}_{i/j}\psi_{i/j}$, where i denotes u or d quark, V_j is the volume of j ($=$ p or n) and $\psi_{i/j}$ is the i quark wave function in j. Since the proton consists of two u quarks and one d quark, the derivative of M_p^* with respect to σ is given by $2S_{u/p} + S_{d/p}$. Similarly, the derivative for the neutron is given by $S_{u/n} + 2S_{d/n}$.

Taking the difference between the in-medium neutron and proton masses, we find

$$\Delta_{np}^* \equiv M_n^* - M_p^* = \Delta_{np} - (3g_\sigma^q)[S_n(0) - S_p(0)]\sigma + \mathcal{O}(\sigma^2), \qquad (10)$$

where $\Delta_{np} = M_n - M_p$, $S_n(0) = \frac{1}{3}[S_{u/n}(0) + 2S_{d/n}(0)]$ and $S_p(0) = \frac{1}{3}[2S_{u/p}(0) + S_{d/p}(0)]$. Here we may expect that $S_{u/j} < S_{d/j}$ because the u quark is *more relativistic* than the d quark in nuclear matter ($m_u < m_d$) — note that the quark scalar charge is given in terms of $\overline{\psi}_q\psi_q$ in matter. Thus, we find that $S_n(0) > S_p(0)$ and then $\Delta_{np}^* < \Delta_{np}$ in nuclear medium.

In Figure 2 we show the neutron-proton effective mass difference calculated in symmetric nuclear matter, including the electromagnetic self-energy corrections. One notices that the mass difference becomes smaller as the density increases. This behavior works in the right direction to resolve the ONS anomaly.

The ONS Anomaly in Mirror Nuclei

Now we are in a position to show our results of the ONS anomaly in mirror nuclei [11]. We first present the calculated single-particle energies for ^{17}F and ^{17}O in Table 2. These mirror nuclei have a common core nucleus, ^{16}O, and have an extra valence proton for ^{17}F and neutron for ^{17}O. In order to focus on the strong interaction effect for the valence proton and neutron, the Dirac equations are solved without the Coulomb and ρ-meson potentials, or the electromagnetic self-energy corrections, and keeping only the charge symmetric σ and ω mean field potentials. Consistently, the valence nucleon contributions are not included in the Coulomb and ρ-mean field source densities in the core nucleus. However, for the nucleons in the core nucleus, electromagnetic self-energy corrections and the Coulomb potential as well as the ρ mean field potential are included in addition to the σ and ω mean field potentials in solving the Dirac equations. Results are shown for two cases in Table 2: calculation with charge symmetry breaking (denoted by CSB) and calculation performed assuming SU(2) symmetry (denoted by SU(2)).

The SU(2) results for ^{17}F and ^{17}O agree perfectly with each other as they should. Single-particle energies in the cores of ^{17}F and ^{17}O are slightly different for CSB. This difference is induced by the different (effective) masses for the valence proton and neutron, arising from the charge and density dependence of their coupling to the self-consistent scalar mean field. This also causes a second order effect on the Coulomb and ρ-meson potentials through the self-consistency procedure.

It is interesting to compare the binding energy differences between the valence proton in ^{17}F and neutron in ^{17}O. In CSB, the result gives, $E(p)(1d_{5/2}) - E(n)(1d_{5/2}) \simeq 0.18$ MeV, while the SU(2) case is zero as it should be. This amount already shows a magnitude similar to that of the observed binding energy differ-

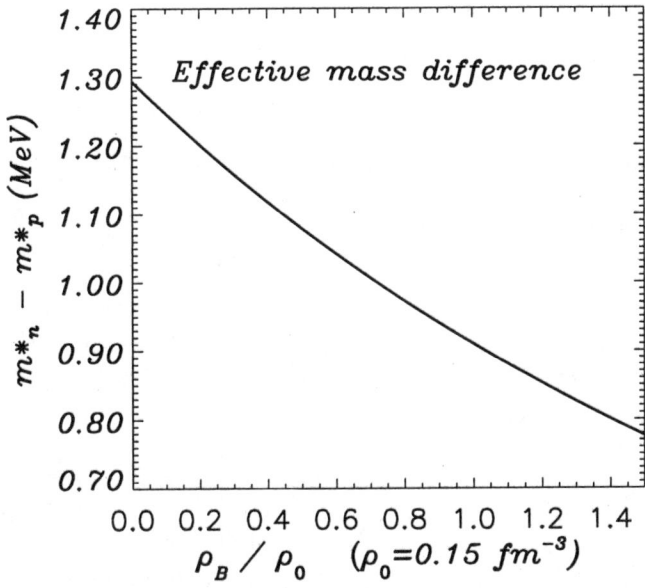

FIGURE 2. Neutron-proton effective mass difference in symmetric nuclear matter with the electromagnetic self-energy corrections.

TABLE 2. Calculated single-particle energies (in MeV) for ^{17}F and ^{17}O.

	CSB		SU(2)	
	^{17}F	^{17}O	^{17}F	^{17}O
p 1s$_{1/2}$	-28.800	-28.805	-28.663	-28.663
1p$_{3/2}$	-14.154	-14.158	-14.032	-14.032
1p$_{1/2}$	-12.495	-12.499	-12.383	-12.383
n 1s$_{1/2}$	-33.367	-33.372	-32.967	-32.967
1p$_{3/2}$	-18.259	-18.263	-17.918	-17.918
1p$_{1/2}$	-16.587	-16.590	-16.258	-16.258
valence	p	n	p	n
1d$_{5/2}$	-3.918	-4.099	-3.848	-3.848

ences.

In Table 3, we summarize the calculated single-particle energies for the valence proton and neutron of several mirror nuclei (in CSB) [11]. Comparing the ρ-potential contributions for the hole states with core plus valence states, one notices the shell effects due to the ρ-potentials. These results reflect the difference in the shell structure, namely the hole states tend to have larger ρ-potential contributions than the core plus valence nucleon states.

The binding energy differences obtained indicate that the prime CSB effects originate in the u-d current quark mass difference. The calculated binding energy differences give of the order of about a few hundred keV. This is precisely the order

TABLE 3. Calculated single-particle energies of mirror nuclei. For each nucleus, the top row shows the single-particle energy of the valence proton or neutron (the orbit is also indicated). δE_ρ stands for the contribution from the ρ-meson central and spin-orbit potentials of the core nucleus. The discrepancies between the experimental values and the theoretical expectations in the absence of charge symmetry violating strong interactions are taken from Table II of Ref. [5], by averaging over the theoretical values.

	^{15}O(p)	^{15}N(n)	^{17}F(p)	^{17}O(n)
1p$_{1/2}$ or 1d$_{5/2}$(MeV)	-14.397	-14.631	-3.918	-4.099
δE_ρ(MeV)	-0.055	0.056	-0.005	0.005
Total(MeV)	-14.452	-14.575	-3.923	-4.094
$\delta E = E(p) - E(n)$	$\delta E =$	123(keV)	$\delta E =$	171(keV)
	observed =	230(keV)	observed =	220(keV)
	^{39}Ca(p)	^{39}K(n)	^{41}Sc(p)	^{41}Ca(n)
1d$_{3/2}$ or 1f$_{7/2}$(MeV)	-16.407	-16.689	-6.970	-7.210
δE_ρ(MeV)	-0.087	0.088	-0.006	0.006
Total(MeV)	-16.494	-16.601	-6.976	-7.204
$\delta E = E(p) - E(n)$	$\delta E =$	108(keV)	$\delta E =$	228(keV)
	observed =	340(keV)	observed =	460(keV)

of magnitude which is observed as the ONS anomaly [3,5].

SUMMARY

Using the QMC model, we have discussed CSB in nuclear medium and calculated the ONS anomaly in mirror nuclei, including the quark degrees of freedom explicitly. We stress that the present contribution to the ONS anomaly is based on a very simple but novel idea, namely the slight difference between the quark scalar densities of the u and d quarks in a bound nucleon, which stems from the u and d quark mass difference [7,11]. This implies that the in-medium proton-σ and neutron-σ coupling constants differ from their values in free space and that the neutron-proton effective mass difference is reduced in matter.

Our results were obtained within an explicit shell model calculation, based on quark degrees of freedom. They show that once CSB is set through the u and d current quark mass difference so as to reproduce the proton and neutron masses in free space, it can produces binding energy differences for the valence (excess) proton and neutron of mirror nuclei of a few hundred keV. The origin of this effect is so simple that it is natural to conclude that a sizable fraction of CSB in mirror nuclei arises from the density dependence of the u and d quark scalar densities in a bound nucleon.

It is a fascinating challenge for the future to compare this result with the traditional mechanism involving $\rho - \omega$ mixing [4]. This will involve the issue of the possible momentum dependence of the $\rho - \omega$ mixing amplitude [3,15]. In addition, one would need to examine whether there is any deeper connection between these apparently quite different sources of charge symmetry violation.

ACKNOWLEDGMENTS

The author would like to thank K. Tsushima, A.W. Thomas and A.G. Williams for valuable discussions. This work was supported by the Australian Research Council and the Japan Society for the Promotion of Science. In the present paper, Fugure 1 was reprinted from Nucl. Phys. A 609 (1996), Saito et al., "Self-consistent description of finite nuclei based on a relativistic quark model", p.352 (Figure 4), and Figure 2 and Table 1-3 were reprinted from Phys. Lett. B 465 (1999), Tsushima et al., "Charge symmetry breaking in mirror nuclei from quarks", p.38 (Figure 1 and Table 1), p.39 (Table 2) and p.41 (Table3), with permission from Elsevier Science.

REFERENCES

1. Okamoto, K., *Phys. Lett.* **11**, 150-153 (1964).
2. Nolen, Jr. A.J., and Schiffer, J.P., *Ann. Rev. Nucl. Sci.* **19**, 471-526 (1969).

3. See for example, Miller, G.A., and Van Oers, W.T.H., *"Symmetries and Fundamental Interactions in Nuclei"*, edited by W.C. Haxton and E.M. Henley, World Scientific, 1995, pp. 127-167.
4. Blunden, P.G., and Iqbal, M.J., *Phys. Lett.* **B 198**, 14-18 (1987); Suzuki, T., Sagawa H., and Arima, A., *Nucl. Phys.* **A 536**, 141-158 (1992); Krein, G., Menezes, D.P., and Nielsen, M., *Phys. Lett.* **B 294**, 7-13 (1992).
5. Shahnas, M.H., *Phys. Rev.* **C 50**, 2346-2350 (1994).
6. Henley, E.M., and Krein, G., *Phys. Rev. Lett.* **62**, 2586-2588 (1989); Hatsuda, T., Hogaasen, H., and Prakash, M., *Phys. Rev. Lett.* **66**, 2851-2854 (1991); Adami, C., and Brown, G.E., *Z. Phys.* **A 340**, 93-100 (1991); Fiolhais, M., et al, *Phys. Lett.* **B 269**, 43-48 (1991); Schäfer, T., Koch, V., and Brown, G.E., *Nucl. Phys.* **A 562**, 644-658 (1993).
7. Saito, K., and Thomas, A.W., *Phys. Lett.* **B 335**, 17-23 (1994).
8. Guichon, P.A.M., Saito, K., Rodionov, E., and Thomas, A.W., *Nucl. Phys.* **A 601**, 349-379 (1996); Saito, K., Tsushima, K., and Thomas, A.W., *Nucl. Phys.* **A 609**, 339-363 (1996); Saito, K., Tsushima, K., and Thomas, A.W., *Phys. Rev.* **C 55**, 2637-2648 (1997).
9. For example, Tsushima, K., Lu, D.H., Thomas, A.W., Saito, K., and Landau, L.H., *Phys. Rev.* **C 59**, 2824-2828 (1999); Steffens, F.M., Tsushima, K., Thomas, A.W., and Saito, K., *Phys. Lett.* **B 447**, 233-239 (1999).
10. Serot, B.D., and Walecka, J.D., *Advan. Nucl. Phys.* **16**, 1-327 (1986).
11. Tsushima, K., Saito, K., and Thomas, A.W., *Phys. Lett.* **B 465**, 36-42 (1999).
12. Auerbach, N., *Phys. Lett.* **B 282**, 263-266 (1992).
13. Cohen, T.D., Furnstahl, R.J., and Banerjee, M.K., *Phys. Rev.* **C 43**, 357-360 (1991).
14. Machleidt, R., *Adv. Nucl. Phys.* **19**, 189-376 (1989).
15. Goldman, T., Henderson, J.A., and Thomas, A.W., *Few Body Syst.* **12**, 123-132 (1992).

Charge Symmetry Violation in Parton Distributions

C. Boros[1], F.M. Steffens[2], J.T. Londergan[3] and A.W. Thomas[1]

[1] *Special Research Center for the Subatomic Structure of Matter and Department of Physics and Mathematical Physics The University of Adelaide, SA 5005, Australia*
[2] *Instituto de Fisica, USP, C. P. 66 318, 05315-970,SP, Brasil*
[3] *Department of Physics and Nuclear Theory Center Indiana University, Bloomington, IN 47408, USA*

Abstract. Charge symmetry is the invariance of the strong Hamiltonian under rotations of 180 degrees about the 2-axis in isospace. It implies the equality of the d-distribution in the proton, d^p, and the u-distribution in the neutron, u^n etc. In this talk, we discuss possible charge symmetry violation in parton distributions and review a recent experimental test of charge symmetry.

In nuclear physics, charge symmetry which interchanges protons and neutrons is respected to a high degree of precision. Most low-energy tests of charge symmetry find that it is good to at least 1% in reaction amplitudes [1]. Therefore, charge symmetry is usually assumed to be valid in discussions of strong interactions. Currently all phenomenological analyses describe deep inelastic scattering [DIS] data using charge symmetric parton distributions. Until recently this assumption seemed to be justified, since high energy experimental data were consistent with parton charge symmetry [1]. However, most recently, a careful comparison of the F_2 structure functions measured in ν reactions by the CCFR Collaboration [2] and μ deep inelastic scattering by the NMC group [3], has revealed [4] a discrepancy that suggests a violation of charge symmetry in the parton distributions of the nucleon, at a level significantly larger than expected.

In this talk, we mainly focus on these experimental findings and discuss the assumptions and corrections made in the analysis of the experimental data. We show that in order to make precise tests of charge symmetry with the neutrino data, two conditions must be satisfied. First, the nuclear shadowing calculations must be made explicitly for neutrinos, not simply taken from muon data on nuclei. Second, the contribution of strange and charm quarks should be calculated explicitly using next-to-leading order [NLO] QCD, and the "slow rescaling" charm threshold correction should not be applied to the neutrino data. When these criteria are satisfied, the comparison is consistent with charge symmetry within the experimental errors and the present uncertainty in the strange quark distribution of the nucleon.

Comparison of structure functions measured in neutrino and charged lepton deep inelastic scattering can be used to test basic symmetry properties of parton distribution functions such as the validity of charge symmetry [1]. Such comparisons are based on the interpretation of these structure functions in terms of parton distribution functions. In the quark-parton model the structure functions measured in neutrino, anti neutrino and charged lepton DIS on an iso-scalar target, N_0, are given by

$$F_2^{\nu N_0}(x) = x[Q(x) + 2s(x) - \delta u(x) - \delta \bar{d}(x)]$$
$$F_2^{\bar{\nu} N_0}(x) = x[Q(x) + 2\bar{s}(x) - \delta d(x) - \delta \bar{u}(x)]$$
$$F_2^{\ell N_0}(x) = \frac{5}{18}x[Q(x) + \frac{2}{5}(s(x) + \bar{s}(x)) - \frac{4(\delta d(x) + \delta \bar{d}(x))}{5} - \frac{(\delta u(x) + \delta \bar{u}(x))}{5}] \quad (1)$$

Here, we introduce the notation $Q(x) = u(x) + \bar{u}(x) + d(x) + \bar{d}(x)$ and express everything in terms of parton distributions functions in the proton and charge symmetry violating distributions which are defined as the differences between the up (down) quark distribution in the proton and the down (up) quark distribution in the neutron

$$\delta u(x) = u^p(x) - d^n(x); \quad \delta d(x) = d^p(x) - u^n(x). \quad (2)$$

If charge symmetry is valid these terms are zero.

Recent experimental measurements made by the NMC and CCFR Collaborations allow a precise comparison between $F_2^\nu(x, Q^2)$ and $F_2^\mu(x, Q^2)$. A direct comparision of the neutrino and muon structure functions shows that while, in the region of intermediate values of Bjorken x ($0.1 \leq x \leq 0.4$), the two structure functions are in very good agreement, in the small x-region ($x < 0.1$), they differ by as much as 10-15% when nuclear corrections are taken into account. This can be seen in Fig.1 where we plot the calculated charge ratio

$$R_c(x) \equiv \frac{F_2^{\mu N_0}(x)}{\frac{5}{18}F_2^{\nu N_0}(x) - x[s(x) + \bar{s}(x)]/6}. \quad (3)$$

Deviation $R_c(x) \neq 1$, at any value of x, indicates that one of the assumptions made in Eq. 3, are not valid.

This discrepancy could be interpreted as evidence for charge symmetry violation. However, several corrections have to be applied to the data before any definitive conclusions may be drawn. The CCFR Collaboration made a careful study of overall normalization, charm threshold and iso-scalar correction effects [2]. Here, we discuss nuclear corrections for neutrinos, $s(x) \neq \bar{s}(x)$ effects and re-examine next-to-leading order (NLO) QCD corrections to the structure functions.

Nuclear corrections for neutrinos are generally calculated using correction factors from charged lepton reactions at the same kinematic values. A priori, there is

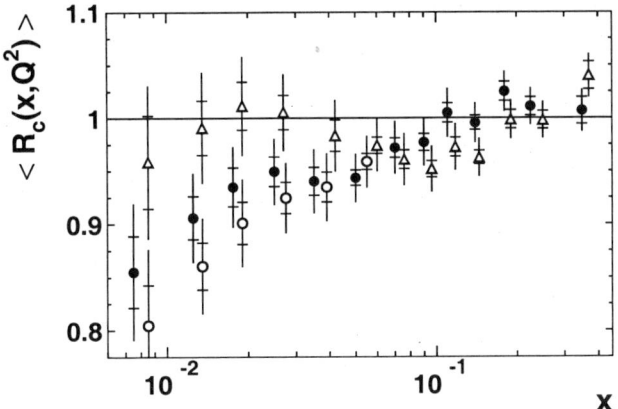

FIGURE 1. The "charge ratio" R_c of Eq. 3 vs. x calculated using CCFR [2] data for neutrino and NMC [3] data for muon structure functions. Open triangles: no heavy target corrections; open circles: ν data corrected for heavy target effects using corrections from charged lepton scattering; solid circles: ν shadowing corrections calculated in the "two phase" model.

no reason that neutrino and charged lepton heavy target corrections should be identical, especially if such corrections depend strongly on the properties of the exchanged object (photon, W) used to probe the structure of the target. Since this is the case for nuclear shadowing corrections in the small x_B region for small to moderately large Q^2-values we re-examined shadowing corrections to neutrino DIS focusing on the differences between neutrino and charge lepton scattering and on effects due to the Q^2-dependence of shadowing. We used a two phase model which has been successfully applied to the description of shadowing in charged lepton DIS [5,6]. In generalizing this approach to weak currents, subtle differences between shadowing in neutrino and charged lepton DIS arise because of the partial conservation of axial currents (PCAC) and the coupling of the weak current to both vector and axial vector mesons. PCAC requires that that the divergence of the axial current does not vanish but is proportional to the pion field for $Q^2 = 0$. This is Adler's theorem, which relates the neutrino cross section to the pion cross section on the same target for $Q^2 = 0$. Thus, for low $Q^2 \approx m_\pi^2$ shadowing in neutrino scattering is determined by the absorption of pions on the target.

For larger Q^2-values the contributions of vector and axial vector mesons become important. Differences between the coupling of the weak current to the vector and axial vector mesons and that of the electro-magnetic current to vector mesons lead to differences in ν and μ shadowing. For large Q^2-values, shadowing due to Pomeron exchange (which is of leading twist) becomes dominant, leading to identical (relative) shadowing in neutrino and charged lepton DIS.

At small x, careful consideration of neutrino shadowing corrections decreases,

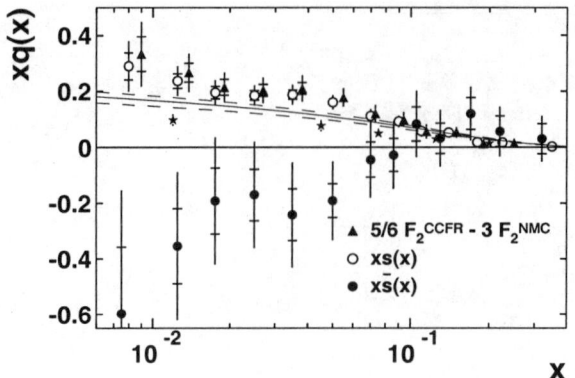

FIGURE 2. The strange quark distribution $x\,s(x)$ (open circles) and antistrange distribution $x\,\bar{s}(x)$ (solid circles) extracted from the CCFR and NMC structure functions. The difference between the CCFR neutrino and NMC muon structure functions $\frac{5}{6}F_2^{CCFR} - 3F_2^{NMC}$ (see Eq. 5) is shown as solid triangles. The strange quark distribution extracted by CCFR in a LO-analysis is shown as solid stars, while that from a NLO-analysis [8] is represented by the solid line, with a band indicating $\pm 1\sigma$ uncertainty in the distribution.

but does not resolve, the low-x discrepancy between the CCFR and NMC data as can be seen in Fig. 1. More details can be found in Ref. [7].

Since the CCFR-Collaboration uses both neutrino and anti neutrino events in the structure function analysis the extracted structure function F_2^{CCFR} is a flux weighted average between ν and $\bar{\nu}$ structure functions [2]

$$F_2^{CCFR}(x, Q^2) = \alpha F_2^{\nu}(x, Q^2) + (1 - \alpha) F_2^{\bar{\nu}}(x, Q^2). \tag{4}$$

Here, we define the relative neutrino flux as $\alpha = \Phi_\nu / (\Phi_\nu + \Phi_{\bar{\nu}})$, where Φ_ν and $\Phi_{\bar{\nu}}$ are the ν and $\bar{\nu}$ fluxes, respectively. The value of α depends on the energy of the incident neutrinos and anti neutrinos in the CCFR-experiment. In the relevant kinematic region, at small x which corresponds to large incident energies, it is ≈ 0.83 so to a good approximation $F_2^{CCFR}(x, Q^2)$ can be regarded as a neutrino structure function.

We examine the role played by the strange quark distributions in connection with the CCFR-NMC discrepancy. Assuming charge symmetry, $s(x)$ and $\bar{s}(x)$ are given by a linear combination of neutrino and muon structure functions,

$$\frac{5}{6}F_2^{CCFR}(x, Q^2) - 3F_2^{NMC}(x, Q^2) = \frac{1}{2}x\left[s(x) + \bar{s}(x)\right]$$
$$+ \frac{5}{6}(2\alpha - 1)x\left[s(x) - \bar{s}(x)\right]. \tag{5}$$

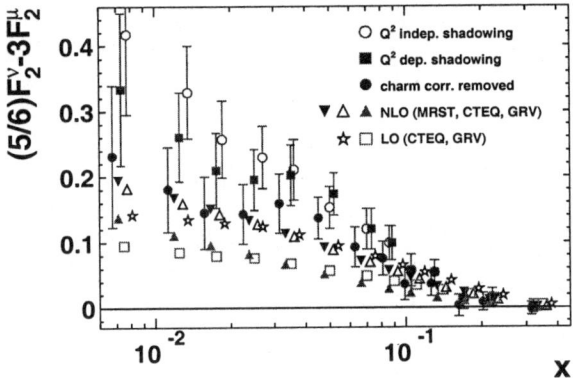

FIGURE 3. Comparison between theory and experiment for the difference $\frac{5}{6}F_2^\nu - 3F_2^\mu$, which is sensitive to deviations from charge symmetry in the parton distributions. The open circles use the original CCFR data, where the nuclear corrections to the ν data are taken from muon measurements [2]. The solid squares involve the same data, but the shadowing corrections have been made explicitly for neutrinos [4,7]. The solid circles are the same as the solid squares except that the "slow rescaling" correction has been removed. The open squares, stars and the triangles are respectively LO (massless) and NLO calculations, including charm mass effects and using different parametrizations for the parton distributions. (Note that the theoretical calculations are all made at the *same* x and Q^2 as the data, but displaced slightly for clarity.)

Opposite sign dimuon production in deep inelastic ν and $\bar{\nu}$ scattering provides a direct determination of both $s(x)$ and $\bar{s}(x)$. The CCFR Collaboration made an independent measurement using charmed hadrons as triggers for events where the W is scattered either on a strange or an antistrange quark [8]. Defining $\alpha' = N_\nu/(N_\nu + N_{\bar{\nu}})$, where $N_\nu = 5,030$, $N_{\bar{\nu}} = 1,060$ ($\alpha' \approx 0.83$) are respectively the ν and $\bar{\nu}$ events from the dimuon production experiment [8], the flux-weighted experimental distribution $xs(x)^{\mu\mu}$ from dimuon production is

$$xs^{\mu\mu}(x) = \frac{1}{2}x\left[s(x) + \bar{s}(x)\right] + \frac{1}{2}(2\alpha' - 1)x\left[s(x) - \bar{s}(x)\right]. \tag{6}$$

Eqs. (5) and (6) form a pair of linear equations which can be solved for $s(x)$ and $\bar{s}(x)$. We can simultaneously test the compatibility of the various experiments. Compatibility of the two experiments requires that physically acceptable solutions for $\frac{1}{2}x[s(x) + \bar{s}(x)]$ and $\frac{1}{2}x[s(x) - \bar{s}(x)]$, satisfying both Eq. 5 and Eq. 6, can be found.

In Fig. 2 we show the results obtained for $xs(x)$ (open circles) and $x\bar{s}(x)$ (solid circles) by solving the linear equations, Eqs. 5 and 6, with $\alpha \approx \alpha' \approx 0.83$. We see that the results are completely unphysical, since the equations require $\bar{s}(x) < 0$.

The experimental results are incompatible, even if $\bar{s}(x)$ is completely unconstrained.

Up to this point, our discussion was based on the LO-QCD parton model. Clearly, NLO corrections can be sizable, especially in the small x region where boson-gluon fusion plays an important role. There are subtle differences between neutrino and muon charm production processes. While the photon-fusion process always creates a quark antiquark pair with the same flavor, $\gamma g \to s\bar{s}(c\bar{c})$ the W-gluon fusion involves quarks antiquark pairs with different flavors $W^+g \to \bar{s}c$. Mass differences of the respective final states lead to different threshold and kinematic conditions for the two processes. These differences have been implemented in leading order in the analysis of the CCFR Collaboration by simple correcting the neutrino structure function using the slow rescaling formalism. The slow rescaling formalism replaces the variable x in the leading order process $W^+s \to c$ by the rescaled variable $\xi = (1+m_c^2/Q^2)x$. In the following, we remove these corrections and use a complete NLO calculation for both the neutrino and the muon structure functions. These calculations can be compared to the uncorrected experimental data. This is shown in Fig. 3 where we plot the measured linear combination $5/6 F_2^\nu - 3 F_2^\mu$ (solid cicles) compared to the calculated values (solid triangles) [9].

In summary, we see that a direct comparison of the CCFR ν data and the NMC μ data with a NLO QCD calculation leads to a much more consistent picture, if the nuclear corrections on Fe are made specifically for neutrinos. In order to make the comparison directly between the NLO calculation and the data, the "slow rescaling" correction had to be removed from the data. This new analysis leads to a much less dramatic discrepancy than earlier work. We observe that the data is still systematically above the NLO calculation based on the GRV98 distributions, while it is in quite good agreement with calculations based on either the MRST or CTEQ distributions. For the present, the possibility of detecting any residual charge symmetry violation depends on resolving this uncertainty in our knowledge of the strange quark distribution.

REFERENCES

1. J. T. Londergan and A. W. Thomas, in *Progress in Particle and Nuclear Physics*, Volume 41, p. 49, ed. A. Faessler (Elsevier Science, Amsterdam, 1998).
2. CCFR-Collaboration, W. G. Seligman *et al.*, Phys. Rev. Lett. **79**, 1213 (1997).
3. NMC-Collaboration, M. Arneodo *et al.*, Nucl. Phys. **B483**, 3 (1997).
4. C. Boros, J. T. Londergan and A. W. Thomas, Phys. Rev. Lett., **81**, 4075, 1998 and Phys. Rev. **D59**, 074021 (1999).
5. J. Kwiecinski and B. Badelek, Phys. Lett. **B208**, 508 (1988).
6. W. Melnitchouk and A. W. Thomas, Phys. Lett. **B317**, 437 (1993).
7. C. Boros, J. T. Londergan and A. W. Thomas, Phys. Rev. **D58**, 114030, 1998.
8. S. A. Rabinowitz *et al.*, CCFR-Collaboration, Phys. Rev. Lett. **70**, 134 (1993); CCFR-Collaboration, A. O. Bazarko *et al.*, Z. Phys. **C65**, 189 (1995).
9. C. Boros, F.M. Steffens, J. T. Londergan and A. W. Thomas, Phys. Lett. **B 468**, 161 (1999)

A precision Measurement of the Michel Parameter $\xi"$ in Polarized Muon Decay

P. Van Hove*, N. Danneberg[‡], J. Deutsch*, J. Egger[†], W. Fetscher[‡],
F. Foroughi[†], J. Govaerts*, M. Hadri[‡], C. Hilbes[‡], K. Kirch[‡],
P. Knowles*[1], J. Lang[‡], M. Markiewicz[‡], R. Medve*, X. Morelle[†],
O. Naviliat[‡2], A. Ninane*, R. Prieels*, T. Schweizer[‡], J. Sromicki[‡]

*Université catholique de Louvain la Neuve, B-1348 Louvain-la-Neuve, Belgium
[†]Paul Scherrer Institut, CH-5232 Villigen, Switzerland
[‡] Eidgenössische Technische Hochschule Zürich, CH-8093 Zürich, Switzerland

Abstract. The present form of the Standard Model (SM) describes weak process in terms of vector and axial-vector type interaction, with fully left handed massless neutrinos. Any experimental deviation from the SM predictions will in special cases reveal the need for extensions to the SM. The longitudinal polarization of the positron emitted from polarized muon present a simple and sensitive case. We present hereafter the status of a longitudinal polarization measurement of these positrons.

INTRODUCTION

The Michel parameters are phenomenological quantities which describe the various observables in muon-decay [1]. They can be related to the leptonic coupling constants, only one of them (g_{LL}^V) being non-zero in the Standard Model (SM). As a consequence, precision measurements of these parameters test physics beyond the SM.

Most of the Michel parameters are known to have values close to the SM ones with precisions better than a few percent [1]. One notable exception is the parameter $\xi"$, or the combination ($\xi"/\xi\xi' - 1$) which is zero in the SM. Its present experimental value is ($\xi"/\xi\xi' - 1 = -0.35 \pm 0.39$) [2].

This combination governs the angular and energy dependence of the positron longitudinal polarization in polarized muon decay:

$$P_L(x,z) = \xi' + \frac{P_\mu z \xi \xi' (2x-1)}{(3-2x) + P_\mu z \xi (2x-1)} \left(\frac{\xi"}{\xi\xi'} - 1\right), \tag{1}$$

[1]) Now at "Institut de Physique, Universit de Fribourg, CH-1700, Fribourg, Switzerland"
[2]) Now at "Laboratoire de Physique Corpusculaire - ISMRa, F-14050 CAEN-CEDEX France"

where $0 \leq x \leq 1$ is the normalized energy and $z = \cos\theta$, θ being the angle between the muon spin and the positron momentum. The parameters ξ, ξ', $\xi"$ are all equal to 1 in the SM. As can be seen in equation (1), values of x close to 1 and z close to -1, strongly enhance the impact of a non vanishing $(\xi"/\xi\xi' - 1)$ on P_L for highly polarized muons.

The experiment, designed to measure P_L at various energies near the end point of the positron spectrum for polarized $(P_\mu^\star)$ and unpolarized (P_μ°) muons, aims to decrease the uncertainty on $\xi"/\xi\xi' - 1$ by more than one order of magnitude.

EXPERIMENTAL SET-UP AND PERFORMANCES

A three week test with one day of real data taking was performed in November 1999 at the $\pi E3$ beam line in PSI. The 95% polarized muons, separated from beam positrons by a Wien filter, are stopped in targets that either maintain (Al) or destroy (S) muon polarization. A spectrometer made from three solenoidal magnets selects the high energy decay positrons, provides a region of uniform magnetic field for tracking and energy measurement, and finally focuses the positrons on the polarimeter (see fig. 1).

Aluminum is known to keep the polarization of the stopped muons. Sulfur has been tested in our experimental condition with a 1kG stray field on the stopping target; it yields a remaining polarization of about 10%.

The three solenoidal magnets select positrons from 45 to 53 MeV/c emitted at angles between 155 and 180 degrees to the muon polarization.

The tracking inside the uniform magnetic field of 2 Tesla is performed by three planes of double sided silicon strip detectors with a 1 mm pitch. Knowing the

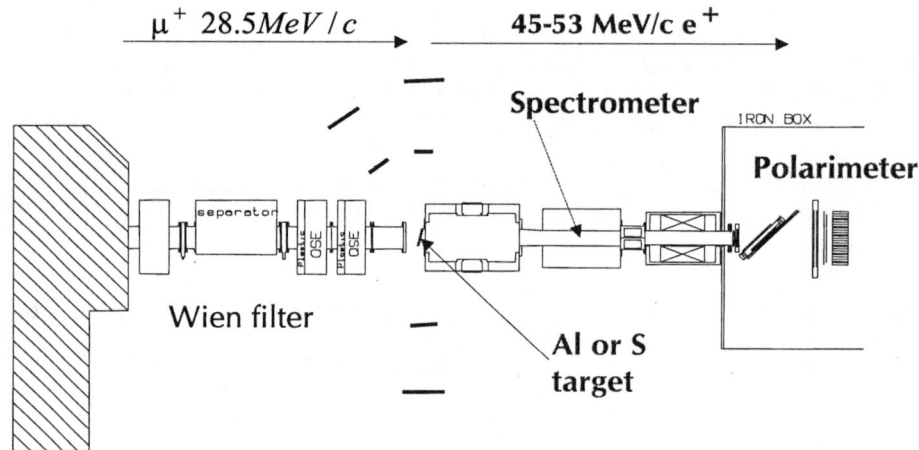

FIGURE 1. Top view of the experimental set-up in $\pi E3$ area with three telescopes viewing the target.

magnetic field, the distances between the planes and three positions (one in each plane), we are able to reconstruct the positron momentum with a 1 MeV/c FWHM resolution; this correspond to the prediction of a Monte-Carlo simulation. The experimental efficiency of the tracking is about 75% meaning that each plane works with a 95% efficiency. An additional consistency check of the momentum reconstruction rejects 10% of the events leading to a global experimental efficiency of the momentum reconstruction of about 67%

The polarimeter uses the spin dependence of Bhabha scattering (Bhabha) and annihilation in flight (AIF) of positrons on polarized electrons. Two iron foils of opposite and saturated magnetization are used with interleaved wire chambers determining the nature of the interaction and the foil in which it took place. A hodoscope in coincidence with a 127 element BGO wall identifies e^+e^- or $\gamma\gamma$ clusters. It should be noted that Bhabha and AIF interactions have opposite analyzing powers. Monte-Carlo simulations predicts analyzing power ($\mathcal{A}$) and luminosity ($\mathcal{L}$) for Bhabha and AIF:

Type	$\mathcal{A}$	$\mathcal{L}$
AIF	3.7 %	7.5 10^{-4}
Bhabha	-1.5 %	16.0 10^{-4}

The experimental yields of this first test are in agreement with Monte-Carlo simulations; the analysis is in progress.

EXPERIMENT AND THEORY

With the performance of the detectors as obtained from the recent test and from Monte-Carlo simulations, we are able to predict how different values of $\xi''/\xi\xi' - 1$ (0 in the SM) would modify the final experimental results. From the experiment, we will obtain:

- The number of transmitted positrons for polarized and unpolarized muons as a function of the reconstructed momentum: $N^\star(Rec\_P)$ and $N^\circ(Rec\_P)$

- The longitudinal polarization of these transmitted positrons: $P_L^\star(Rec\_P)$ and $P_L^\circ(Rec\_P)$

With these, we compute the ratio of longitudinal polarization ($R(Rec\_P)$) and the enhancement factor ($EnF(Rec\_P)$):

$$R(Rec\_P) = \frac{P_L^\star(Rec\_P)}{P_L^\circ(Rec\_P)} \quad (2)$$

$$EnF(Rec\_P) = \frac{N^\star(Rec\_P) - N^\circ(Rec\_P)}{N^\star(Rec\_P)} \quad (3)$$

which are related by theory:

$$R(Rec\_P) = 1 - EnF(Rec\_P) \left(\frac{\xi"}{\xi\xi'} - 1 \right) \qquad (4)$$

The measured values of $EnF(Rec\_P)$ and the expected values of $R(Rec\_P)$ for two different values of $\xi"/\xi\xi' - 1$ (0.005 and 0.03) are shown in fig.2.

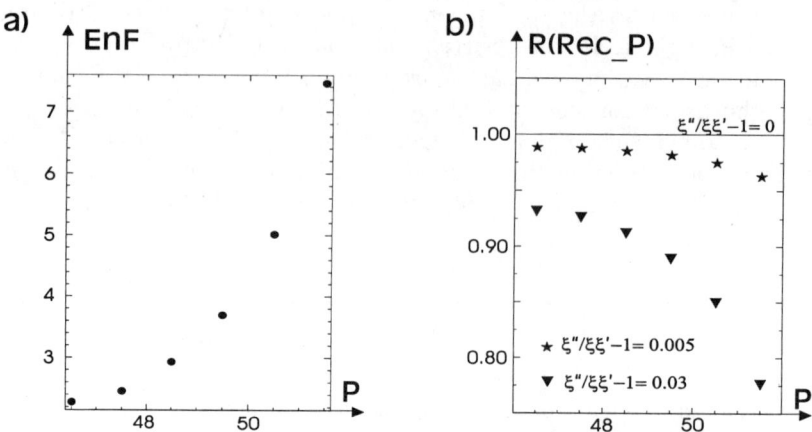

FIGURE 2. a) the enhancement factor $EnF(Rec\_P)$ b) the ratio of longitudinal polarization $R(Rec\_P)$

With the present rough analysis and without any improvement of the detectors, the results of the present set of measurement but extended to 21 days of data taking would yield error bars of about 0.15 on each point of fig.2b. This would then correspond to a standard deviation of 0.015 on $(\xi"/\xi\xi' - 1)$ which is already more than 25 times better than the present published limit. However, trusting our Monte-Carlo simulations (and we do trust it since it is in good agreement with many experimental results), improvement of our analysis and detectors should reduce the error bars to about 0.05 so that the standard deviation on $(\xi"/\xi\xi' - 1)$ would then reach 0.005. A first production period of six weeks is foreseen in august 2000.

REFERENCES

1. W. Fetscher and H.J. Gerber in *Precision Tests of the Standard Model*, edited by P.Langacker, World Scientific, 1995, pp.655-705.
2. H. Burkard et al., *Phys. Lett.*, **150B**, 242(1985).

Nonleptonic decays of supermultiplets

R. Delbourgo and Dongsheng Liu

School of Mathematics and Physics, University of Tasmania[1],
Hobart, Australia 7001

Abstract. By regarding weak interactions as a perturbation of strong interactions, which are themselves well described within a relativistic supermultiplet scheme, we derive expressions for nonleptonic weak decay amplitudes in terms of constituent quark masses and CKM angles, with no other parameters. The method is applied to typical pseudoscalar meson decays and seems to lead naturally to $\Delta I = 1/2$ dominance. For heavy hadrons it also requires $1/\sqrt{M}$ mass damping factors, consistent with heavy quark symmetry.

INTRODUCTION

With so much data available for heavy hadrons, a veritable industry has arisen [1] devoted to calculating nonleptonic weak decays. Yet much confusion remains in this subject, the only striking fact being the dominance of the mysterious $\Delta I = 1/2$ rule in kaon and hyperon decays. In this article we will try to shed some light by trying a different approach, one which has long been applied to electromagnetic interactions; namely, we shall regard electroweak interactions as a perturbation of the strong interactions, instead of the other way around. The difficulty with this approach lies in knowing the initial strong interaction amplitudes. Although they are given reliably by QCD at high energy, at low energy we usually resort to models which respect the symmetries of QCD, rather than QCD itself, because of unknown effects from confinement. One of the simplest models is the linear or nonlinear sigma-type model, embodying chiral symmetry [2] for massless bare quarks. Another is the relativistic spin-flavour supermultiplet model [3], which not only applies to heavy quarks but to light ones remarkably well at tree level (often better than 10%); this will be our platform. A preliminary version of this method was presented in [5] and here we will provide the full results for various contributions to nonleptonic amplitudes, skipping the intermediate technical details, for reasons of brevity.

Semileptonic decays are well-understood via W-boson exchange to leptons and we have nothing to say about them; our focus is on nonleptonic processes where

---

[1] Email: Bob.Delbourgo@utas.edu.au & D.Liu@utas.edu.au

weak and strong interactions are intertwined [4]. Our attitude towards flavour changing weak amplitudes is that, to first order in the Fermi coupling G_F, the process is given by a set of W-exchanges between the quarks in the hadrons, with the hadronic amplitude satisfactorily determined by higher spin-flavour symmetry. The rules for computation are thus fixed and our final answers can only depend on the masses of the *constituent quarks* and the CKM mixing angles. There are no adjustable parameters in our scheme and no gluon corrections with associated 'penguin diagrams': the gluons are already accounted for when dressing bare quark into constituent quark fields.

We therefore begin by assuming that hadrons are reasonably well described by multiquark, constituent composites having supermultiplet wavefunctions $A = \alpha a$, where α is a Dirac spinor label and a stands for flavour:

$$\Phi_A^B(p) = [(1+\gamma.v)(\gamma_5 P_a^b + \gamma^\mu V_{\mu a}^b)]_\alpha^\beta; \quad m = m_a + m_b, \, p = mv \qquad (1)$$

$$\Psi_{(ABC)}(p) = [(1+\gamma.v)\gamma^\mu C]_{\alpha\beta} u_{\mu(abc)}(p) + \{[(1+\gamma.v)\gamma_5 C]_{\alpha\beta}\epsilon_{abd} u^d_{c\gamma}(p) + \text{perms}\}$$
$$m = m_a + m_b + m_c, \, p = mv \qquad (2)$$

of ground state mesons and baryons, with tree-level interactions given by momentum-conserving effective Lagrangians like,

$$g\bar{\Psi}^{(ABC)}\Psi_{(ABD)}\Phi_C^D, \quad f\Phi_A^B\Phi_B^C\Phi_C^A. \qquad (3)$$

Such algebraic contractions of indices can be represented pictorially as a joining of the flavour quantum numbers via duality diagrams, the remaining contractions over spinorial indices serving to provide the Lorentz structure of the amplitude— in keeping with higher symmetry requirements. In (3) the coupling constant g is dimensionless while f has dimensions of mass.

Now consider the addition of electroweak perturbations on such effective interactions, by invoking the standard electroweak Lagrangian,

$$\mathcal{L} = \sum_{\text{flavour}} \bar{\psi}[\gamma.(i\partial + \mathcal{W}) - g_H H]\psi - \vec{\mathcal{W}}^{\mu\nu}\cdot\vec{\mathcal{W}}_{\mu\nu}/4 + \mathcal{L}_{\text{Higgs}}(H) + \ldots, \qquad (4)$$

where the supermultiplet boson field is

$$\mathcal{W} = -eQA + g_W VW(1-i\gamma_5)/\sqrt{8} + \ldots; \quad V = \text{CKM matrix}, \, Q = \text{charge}. \qquad (5)$$

Above, we have not bothered to include the Z-bosons because our main concern is flavour-changing processes. The most significant aspect of (4) is that it should be applied to current quark fields, not the more massive constituent quark fields which are dynamically induced via strong interactions. It becomes questionable then whether (5) is applicable to the effective quark fields[2] present in (3). The principal

---

[2] Indeed previous experience shows that we can expect nontrivial corrections of around 30% due to renormalization effects; for example the chiral interaction should be changed from the left-projection $P_L = (1-i\gamma_5)/2$ to about $(1-\frac{3}{4}i\gamma_5)/2$ so that the axial vector component for the nucleon is reduced from 5/3 (arising the higher symmetry D/F ratio of 3/2) to the experimental value of $g_A/g_V \simeq 5/4$.

goal of this paper is to see if one can get sensible estimates of *all* nonleptonic amplitudes without invoking extra parameters, so we will ignore relatively small renormalization effects on the axial current but not the dependence on the mass of the quark constituents in the hadrons. Refinements come later.

One can recognise three types of virtual W-exchange, where

- a charge-conserving transition occurs on one quark line, either initially or finally (wave-function renormalizations) or via a vertex-like correction; see Figures 1a,b,c.

- the charge exchange takes place between the participating quarks in the hadron, again either as a self-energy or a pair of vertex corrections; see Figures 2a,b,c. (If these participating quarks comprise a meson, it must be uncharged.

- the W acts as a quark annihilation intermediary (see Figure 3) and is just what one encounters in semileptonic processes. It requires a charged meson to be one of the participating particles, of course.

Since there is a well-defined prescription for treating the strong vertex—where the quark tramlines join up—it only remains to estimate the Feynman integrals corresponding to the W-loop exchange. This we shall soon do, using the Feynman-'tHooft gauge field propagator, $\Delta_{\mu\nu}(k) = -\eta_{\mu\nu}/(k^2 - m_W^2)$. The left matrix $\gamma_{L\mu} = \gamma_\mu(1 - i\gamma_5)/2$ which the propagator multiplies has a number of handy properties; aside from the obvious utility of projection, there is the bonus that Fierz identities allow $\gamma_L \otimes \gamma_L$ to be shuffled from one pair of fermion lines into another pair.

We shall focus on pseudoscalar mesons decays, without striving for total accuracy; rather, we are content ourselves if we can capture *all* the nonleptonic amplitudes to within about 30% or better. The scale is set by the coupling factor f of the strong three-meson vertex, found via $\rho \to 2\pi$, where

$$\mathcal{L}_{\rho\pi\pi} = f \operatorname{Tr}[\Phi_P(p_3)\Phi_P(p_2)\Phi_V(p_1)/\sqrt{2} - (1 \leftrightarrow 2)]/4$$
$$= f \operatorname{Tr}[(1 + \gamma \cdot v_3)\gamma_5(1 + \gamma \cdot v_2)\gamma_5(1 + \gamma \cdot v_1)\gamma \cdot \epsilon_1 - (1 \leftrightarrow 2)]/4\sqrt{2}$$
$$= (p_3 - p_2) \cdot \epsilon_1 f(m_1 + m_2 + m_3)/\sqrt{2m_2 m_3} \equiv g_{\rho\pi\pi}(p_3 - p_2) \cdot \epsilon_1.$$

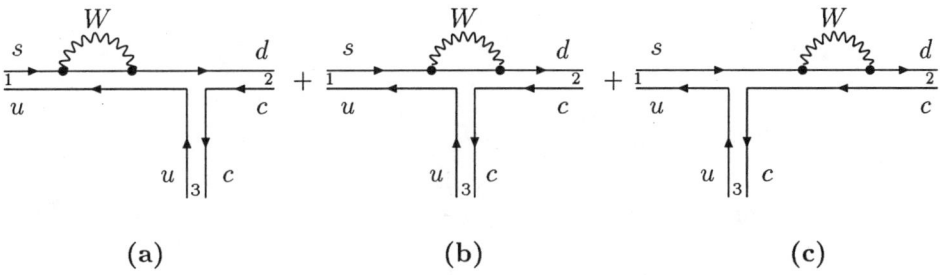

FIGURE 1. Single quark line transition.

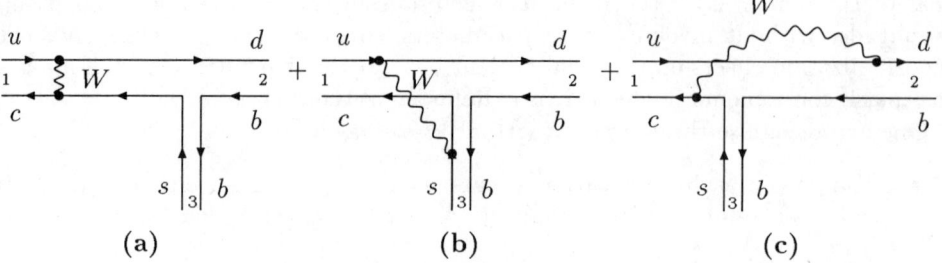

FIGURE 2. W-exchange across quark lines.

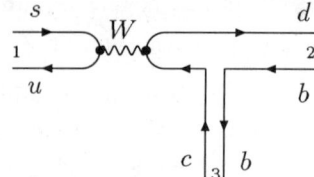

FIGURE 3. Annihilation diagram.

The experimental ρ-decay width indicates $g_{\rho\pi\pi} \simeq 6.03$. Assuming nonstrange quark masses are not far from $m_u \simeq m_d \equiv \hat{m} \sim 0.34$ GeV, $f(6\hat{m})/4\hat{m}^2 = \sqrt{2} g_{\rho\pi\pi}$, or $f = \sqrt{8} \hat{m} g_{\rho\pi\pi}/3 \sim 2$ GeV. It turns out that $f(m_1 + m_2 + m_3)$ is a ubiquitous factor in relativistic supermultiplet theory, so we shall adopt this combination as the 'universal' value which therefore equals about 4.1 GeV^2.

W-EXCHANGE EFFECTS

First we consider changes of flavour (but not of charge) on a single quark [6], which can take place as self-energy-like diagrams at each hadronic leg (Figures 1a, 1c) or as a vertex correction across the legs (Figure 1b). In both cases, one must sum over the internal flavours; a generic case is the $s - d$ transition, where we have to sum over u, c and t. Quite generally, between i and k quarks, the self-energy part is $\Sigma_{ik}(p) \equiv \sum_j V_i^j V_k^{j*} \Sigma_j(p)$ where, neglecting form factors,

$$\Sigma_j(p) = i\frac{g_W^2}{2} \int \frac{\bar{d}^4 k}{k^2 - m_W^2} \gamma_{L\mu} \frac{\eta^{\mu\nu}}{\gamma.(p+k) - m_j} \gamma_{L\nu} \equiv p.\gamma_L \mathcal{F}_j(p^2, m_W, m_j). \quad (6)$$

Now $\mathcal{F}$ contains a logarithmic divergence,

$$\mathcal{F}_j = -ig_W^2 \int_0^1 dx \int \frac{(1-x)\,\bar{d}^4 k}{[k^2 + p^2 x(1-x) - m_j^2 x - m_W^2(1-x)]^2}. \quad (7)$$

but it is innocuous through unitarity of the CKM matrix and the fact that we are dealing with flavour changing transitions ($i \neq k$). Subtracting $\mathcal{F}(0)$, we arrive at

$$\mathcal{F}_j(m_j) - \mathcal{F}_j(0) \simeq -\frac{1}{2}\left(\frac{g_W m_q}{4\pi m_W}\right)^2, \tag{8}$$

except for the top quark, where there is an extra factor of about 1/4. Remembering $G_F/\sqrt{2} = g_W^2/8m_W^2 = 8.25 \times 10^{-6}$ GeV^{-2}, we end up with

$$\Sigma_{ik}(p) \equiv p.\gamma_L \mathcal{F}_{ik} = -\frac{G_F p.\gamma_L}{4\sqrt{2}\pi^2}\sum_j V_i^j V_k^{j*} m_j^2 \rho_j, \tag{9}$$

where the weight factor $\rho_q = 1$ for all but the top quark, when $\rho_t \simeq 1/4$.

Turning to Figure 1b and the (matrix) vertex integral,

$$\Gamma_{ik} = -i\sum_j V_i^j V_k^{j*}\frac{g_W^2}{8}\int\frac{d^4k}{k^2 - m_W^2}\gamma_{\mu L}\frac{1}{\gamma.(p_i+k)-m_j}\frac{1}{\gamma.(p_k+k)-m_j}\gamma_L^\mu, \tag{10}$$

we can simplify the answer to $\Gamma_{ik} \equiv \frac{1}{2}(p_i+p_k).\gamma_L \mathcal{G}_{ik} = \frac{G_F(p_i+p_k).\gamma_L}{4\sqrt{2}\pi^2}\sum_j V_i^j V_k^{j*} m_j \sigma_j$, where the weight factor $\sigma_q = 1$ for all quarks but the top, when $\sigma_t \simeq 1/8$.

Evaluation of the transition factors, $\mathcal{F}_{ik}$ and $\mathcal{G}_{ik}$ is reasonably straightforward, using known values of $|V_i^j|$ taken from the Particle Data Group and the (GeV) values $m_u \simeq m_d \simeq 0.34$, $m_s \simeq 0.48$, $m_c \simeq 1.5$, $m_b \simeq 4.7$, $m_t \simeq 175$. The values are listed in Table 1, and it should be noted that transition elements for down-type quarks depend significantly on the contribution from the intermediate top quark; the effect from t is smallest for the $s-d$ transition, but even so t competes well with the u and c; mostly it *dominates* the other contributions, in spite of the fact that off-diagonal V_t^q terms are quite small.

Summing the Figures 1a,1b and 1c terms between free spinors $\bar{u}(p_k)..u(p_i)$,

$$\mathcal{T}_{ik}^I = -\frac{1}{2(\gamma.p_i - m_k)}\Sigma_{ik}(p_i) + \Gamma_{ik}(p_i,p_k) - \Sigma_{ik}(p_k)\frac{1}{2(\gamma.p_k - m_i)}$$
$$= -\frac{\mathcal{F}_{ik}}{4}\left[1 + i\gamma_5\frac{m_k - m_i}{m_k + m_i}\right] + \frac{\mathcal{G}_{ik}}{4}[(m_k+m_i) + i\gamma_5(m_i - m_k)].$$

Finally we contract $\mathcal{T}^I$ over the hadronic wavefunctions. For three 0^- mesons, with the flavour labels of Figure 1, we get the generic amplitude:

$$M_{sdcu}^I = f\,\text{Tr}[\Phi_P(p_2)\mathcal{T}^I\Phi_P(p_1)\Phi_P(p_3)]/m_1 m_2 m_3$$
$$= i\frac{(m_d - m_s)f}{4m_1 m_2 m_3}\left[\mathcal{G}_{sd} + \frac{\mathcal{F}_{sd}}{m_s + m_d}\right]\text{Tr}[(\gamma.p_2 + m_2)(\gamma.p_1 + m_1)(\gamma.p_3 - m_3)]$$
$$= i\frac{f(m_1 + m_2 + m_3)(m_s - m_d)[m_3^2 - (m_1 - m_2)^2]}{2m_1 m_2 m_3}\left[\mathcal{G}_{sd} + \frac{\mathcal{F}_{sd}}{m_s + m_d}\right]. \tag{11}$$

TABLE 1. Flavour changing transition elements (in GeV units) .

$\mathcal{F}_{sd}$	6.42×10^{-7}	$\mathcal{F}_{uc}$	-4.27×10^{-9}	$\mathcal{G}_{sd}$	-1.06×10^{-7}	$\mathcal{G}_{uc}$	1.14×10^{-8}
$\mathcal{F}_{sb}$	-6.07×10^{-5}	$\mathcal{F}_{ut}$	-1.49×10^{-8}	$\mathcal{G}_{sb}$	3.23×10^{-7}	$\mathcal{G}_{ut}$	0.67×10^{-8}
$\mathcal{F}_{db}$	1.44×10^{-5}	$\mathcal{F}_{ct}$	1.78×10^{-7}	$\mathcal{G}_{db}$	-0.76×10^{-7}	$\mathcal{G}_{ct}$	-6.88×10^{-8}

Turning to the Figure 2 graphs, the first of these corresponds to a $u\bar{c}$ transition into two mesons, dominated by an intermediate $d\bar{s}$ state. Neglecting the small W-momentum relative to its mass, this contribution is given by the generic contraction,

$$M_{udbsc}^{IIA} = -\frac{g_W^2 f}{32 m_W^2} V_u^d V_c^{s*} \text{Tr}[(\Phi_P(p_2)(\gamma \cdot p_d + m_d)\gamma_{L\mu} \Phi_P(p_1) \gamma_L^\mu (\gamma \cdot p_s + m_s) \Phi_P(p_3)]$$

$$= \frac{iG_F f[m_1^2 - (m_2+m_3)^2] m_d m_s V_u^d V_c^{s*}}{2\sqrt{2} m_1 m_2 m_3} \left[1 - \left(\frac{m_u + m_c}{m_s + m_d}\right)^2\right](m_d - m_s). \quad (12)$$

Competing with this answer are the vertex corrections of Figures 2b,2c. The latter involve Feynman integrals which are technically more difficult to calculate. Fortunately they contain cancelling logarithmic divergences, irrespective of CKM unitarity. Figures 2b and 2c yield, in the soft limit

$$M_{udbsc}^{IIBC} \simeq -\frac{iG_F V_u^d V_c^{s*} f}{16\sqrt{2}\pi^2 m_2 m_3} \left[m_1^2 - (m_2+m_3)^2\right]$$

$$\left[(m_u^2 - m_c^2) \ln\left(\frac{m_c^2 m_u^2}{m_W^4}\right) + \frac{1}{2}(m_s^2 - m_d^2)\ln\left(\frac{m_s^2 m_d^2 m_c^2 m_u^2}{m_W^8}\right)\right]. \quad (13)$$

The significant point about this last result is that it is of the same order of magnitude as M^{IIA}; even though there is a suppression factor of $1/4\pi^2$ from the loop integral, it is compensated by a number of logarithms which are typically in the range $\ln(m_W^2/m_s^2) \sim 10$. In fact, cancellations between these sorts of terms are responsible for the small size of the $\Delta I = 3/2$ amplitude in K decays.

Finally, the W-annihilation process only applies to charged mesons, incoming or outgoing. A typical example is drawn in Figure 3 and is easy to calculate, by taking advantage of Fierz reshuffling, $(\bar{\psi}_1 \gamma_L^\mu \psi_2) \cdot (\bar{\psi}_3 \gamma_{L\mu} \psi_4) = -(\bar{\psi}_3 \gamma_L^\mu \psi_2) \cdot (\bar{\psi}_1 \gamma_{L\mu} \psi_4)$. We find that Figure 3 is the same as Figure 2a so far as the Lorentz contraction is concerned and just contains an extra colour factor of 3. Thus we arrive at the generic amplitude,

$$M_{sdbcu}^{III} = \frac{3iG_F f[m_1^2 - (m_2+m_3)^2] m_d m_c V_u^s V_c^{d*}}{2\sqrt{2} m_1 m_2 m_3} \left[1 - \left(\frac{m_s + m_u}{m_c + m_d}\right)^2\right](m_d - m_c). \quad (14)$$

COMPARISON WITH EXPERIMENT

Before testing our theoretical results, it is worth pointing out that experimental couplings deduced from *all measured channels* range between $g_{K^+\pi^+\pi^0} \simeq 2 \times 10^{-8}$

GeV to $g_{B^0 D^- D_s^+} \simeq 10^{-6}$ GeV. (In stating this we have not included couplings for which there exists an upper bound, because the decay channel has not yet been detected.) The range is not so large, bearing in mind the big variation in CKM elements V, which is a bit surprising. On the theoretical side, we notice that the transition elements listed in Table 1, through the combination $\mathcal{G}_{ik} + \mathcal{F}_{ik}/(m_i + m_k)$, can vary by as much as 1000; specifically the $s - b$ element is largely governed by the factor $G_F m_t^2 V_t^s V_t^{b*}$ and dominates other comparable terms.[3] Another source of theoretical variation in all amplitudes M is the factor $(m_i - m_j)$ which can be as small as $m_s - m_u \simeq 0.1$ GeV or as large as $m_b - m_s \simeq 4.7$ GeV, even if there are no external top composites.

For the (light) K-meson, the theoretical results are not too bad. Upon adding the appropriate duality diagram contributions, we find for the $\Delta I = 3/2$ process,

$$M_{K^+ \pi^+ \pi^0} = [M^{II}_{usudu} + M^{III}_{usuud}]/\sqrt{2} \simeq -2.4 \times 10^{-8} \text{ GeV} \quad ; \text{cf } g^{\text{expt}}_{K^+ \pi^+ \pi^0} \simeq 1.8 \times 10^{-8},$$

and for the $\Delta I = 1/2$ dominated process,

$$M_{K^0_s \pi^+ \pi^-} = \sqrt{2}[M^I_{sdud} + M^{II}_{sudud} + M^{III}_{uudsd}] \simeq 3.2 \times 10^{-7} \text{ GeV; cf } g^{\text{expt}}_{K^0_s \pi^+ \pi^-} \simeq 3.9 \times 10^{-7}.$$

However for heavier mesons, theoretical results are *greatly* overestimated. Typically,

$$M_{D^0 K^- \pi^+} = M^{II}_{csudu} + M^{III}_{dsucu} \simeq 6.1 \times 10^{-5} \text{ GeV; cf } |g^{\text{expt}}_{D^0 K^- \pi^+}| \simeq 2.5 \times 10^{-6},$$

$$M_{B^+ \bar{D}^0 \pi^+} = M^{II}_{cbudu} + M^{III}_{dbucu} \simeq 4 \times 10^{-5} \text{ GeV; cf } |g^{\text{expt}}_{B^+ \bar{D}^0 \pi^+}| \simeq 8 \times 10^{-7},$$

$$M_{B^+ K^0 \pi^+} = M^{II}_{sbdu} + M^{III}_{sbduu} \simeq -3.2 \times 10^{-5} \text{ GeV; cf } |g^{\text{expt}}_{B^+ K^0 \pi^+}| \simeq 5 \times 10^{-8},$$

bear witness that our theory is seriously awry for heavy hadrons.

The reason is not hard to find: we have been using supermultiplet wavefunctions which have *not* been scaled down by a factor of $\sqrt{M}$, where M is the hadronic mass, as required [7] by heavy quark theory.[4] The consequence of this observation is that all our estimates must be reanalyzed with the inclusion of such factors. The main effect is that intermediate top will *not* swamp the other quarks as it has done hitherto. We are optimistic that the mass reduction factors—which have little effect on the light hadrons—will serve to bring down theoretical results to acceptable values[5]; for instance, their inclusion in the decay $B^+ \to K^0 \pi^+$ will damp M^I_{sbdu} by about $1/\sqrt{m_B m_t^2} \sim 1/300$ and M^{III}_{sbduu} by about $1/\sqrt{m_B^3} \sim 1/20$, giving the right order of magnitude for the final coupling.

---

[3] As we shall see, there is no evidence, on the experimental side, of such a large contribution. Furthermore the inclusion of form factors in W-interactions, such as $1/(1 - k^2/m_t^2)$, which may diminish the results by a factor of 4 for intermediate top, cannot alleviate this problem.
[4] The evidence for this scaling is clearest from photon-mediated leptonic decays with matrix element, $\langle 0|J^{\text{em}}_\mu|V(p)\rangle = em_V f_V \epsilon_\mu(p)$, where the decay constant f_V scales conspicuously as $\sqrt{m_V}$. In turn this implies that the wavefunction Φ should carry an extra factor of $1/\sqrt{m_V}$.
[5] and in any event, we would claim that the effect of intermediate top on weak transition rates is by no means as dominant as commonly stated in the literature.

These and all the other nonleptonic pseudoscalar decay amplitudes remain to be recalculated properly. Once understood and incorporated into our Lorentz-invariant scheme, it will be a relatively simple matter to extend the scheme to the vector mesons and the baryons.

ACKNOWLEDGMENTS

This work was supported by the Australian Research Council under grant number A69800907.

REFERENCES

1. Ali, A., Kramer, G., and Lu, C-D., *Phys. Rev.* **D58**: 094009 (1998); *ibid* **D59**: 014005 (1999); Lu, C-D., *Nucl. Phys. Proc. Suppl.* **74**, 227 (1999); Neubert, M., and Stech, B., "Nonleptonic weak decays of B-mesons", in *Heavy Flavours II*, World Scientific, Singapore, 1997, pp. 294-344; Stech, B., *Twenty Beautiful Years of Bottom Physics*, Chicago, 1997, hep-ph/9709280.
2. Scadron, M.D., *Nuovo Cim.* **110A**, 865 (1997) and references therein.
3. Salam, A., Delbourgo, R., and Strathdee, J., *Proc. R. Soc. London*, **A284**, 146 (1965); Sakita, B., and Wali, K.C., *Phys. Rev.* **139**, B1355 (1965); Delbourgo, R., and Liu, Dongsheng, *Phys. Rev* **D53**, 6576 (1996).
4. Neubert, M., *Physics Rept.* **245**, 259 (1994); Neubert, M., "B-decays and heavy quark expansion" in *Heavy Flavours II*, World Scientific, Singapore, 1997, pp 239-293.
5. Delbourgo, R. and Liu D., " Nonleptonic decays: amplitude analysis and supermultiplet schemes", in *Nonperturbative Methods in Quantum Field Theory*, ed. by A.W. Schreiber, A.G. Williams and A.W. Thomas, World Scientific, Singapore, 1998, pp 249-260.
6. Delbourgo, R., and Scadron, M.D., *Nuovo Cim. Lett.* **44**, 193 (1985); Fuchs, N.H., and Scadron, M.D., *Nuovo Cim.* **A93**, 205 (1986).
7. Isgur, N., and Wise, M.B., *Phys. Lett.* **B232**, 113 (1989); *ibid* **B237**, 527 (1990); Stech, B., *Z. Phys.* **C75**, 245 (1997).

Distinguished features of muon colliders physics potential

F.F. Tikhonin

Institute for High Energy Physics, Protvino, 142284, Russia

Abstract. Features of the muon collider physics potential are briefly discussed.

INTRODUCTION

The muon collider concept was first mooted many years ago [1–4]. The main impetus for such an idea was the craving to get rid of the powerful synchrotron radiation from which cycling electron-positron colliders suffer so much. Muons being as much as ≈ 200 times heavier than electrons undergo this disease to a much less extent because the intensity of the synchrotron radiation is proportional to $mass^{-4}$. The interest of last several years to the muon colliders has hugely grown due to several reasons. From the point of view of the machine design consideration it was revealed that the ionization cooling concept [3,4] offers the possibility of making a high luminosity accelerator. On the other hand, the physics potential of the muon collider was enriched by the possibility to build a "Higgs boson factory" in analogy with existing "Z^0 boson factories", and, therefore, such a facility might provide a unique laboratory for particle physics research (for a recent review of both the machine design and physics potential see e.g. Ref. [5]).

Apparently one of the main objectives of the present-day elementary particle physics is to find the Higgs boson and investigate its properties with as high precision as possible. It is hoped that this particle will be revealed at the forthcoming LHC machine. However, there exists the so-called intermediate mass region, extended from $m_H > 95$ GeV to $m_H \leq 2m_Z$, which is the most difficult for experimental research. In spite of this difficulty it is hoped that LHC will allow to know the Higgs mass with the precision sufficient to tune to the resonance with the muon collider. Happily, just at this region the SM Higgs boson has a narrow width, more exactly, only if its mass is less then the threshold of $W^{\pm}$ - pair production (see below). Thereby, the $\mu^+\mu^-$ collider would not merely fill the gap but improve the precision of the studies in this region to a great extent.

MOTIVATION

As mentioned at the end of Introduction, the Higgs boson width in the SM is narrow only in the region of c.m. energy up to $\sqrt{s} \sim 2m_W$, so the Higgs factory is feasible only in this region. To see this let us calculate the cross section of the process $\mu^+\mu^- \to W^+W^-$ which proceeds only through the Higgs boson exchange in the s-channel. Using the standard notations, the result for this cross section reads as follows:

$$\sigma^{\mu^+\mu^- \xrightarrow{H} W^+W^-} = \frac{\pi\alpha^2}{\sin^4\theta_W} \frac{m_\mu^2}{16} \frac{s - 4m_\mu^2}{(m_H^2 - s)^2 + \Gamma_H^2 m_H^2} \left(\frac{1}{2}\frac{s^2}{m_W^4} - 2\frac{s}{m_W^2} + 6\right). \quad (1)$$

It can seen from the equation above that the cross section due to the Higgs boson exchange reaches its maximum, $\sigma_{max} \cong 0.067 pb$ at the c.m. energy $\sqrt{s} \sim 2m_Z$ while the "conventional" cross section (due to γ, ν and Z^0 exchange) reaches at this point the value of ≈ 15 pb. Partly in view of this it is expedient to search for other processes where the interaction of the Higgs within the lepton sector would be involved. Two examples of that consideration was presented in the paper [6], where processes $\mu^+\mu^- \to HZ^0$ and $\mu^+\mu^- \to H\gamma$ were proposed as complementary to that of the resonance Higgs scalar production. Note that the latter of the two processes above is negligibly small at the tree level in the case of electron-positron collision, but there is hope to observe it at the muon collider.

ASSOCIATED HZ PRODUCTION IN SM

Let us begin with the Bjorken process having muons as the initial state particles, $\mu^+\mu^- \to ZH^0$.

Usually in the course of cross section calculations one uses the the s-channel diagram alone. Let us do it and take into account masses of initial muons. Then we obtain the following asymptotics of this process at $\sqrt{s} \to \infty$:

$$\sigma^{(s-channel),as}_{\mu^+\mu^- \to ZH^0}|_{m_\mu \neq 0} = \frac{2\pi\alpha^2}{\sin^4(2\theta_W)} \cdot g_A^2 \cdot \frac{m_\mu^2}{m_Z^4}. \quad (2)$$

It can be seen that despite the fact that this diagram is pure s-channel one, the corresponding cross section is not falling with energy but approaches a constant limit, whose value is equal to $\cong 1.2 \cdot 10^{-2} fb$. Concerning the angular dependence of this cross section, it could be seen that this distribution is **flat**, indicating that it comes entirely from the $J = 0$ partial wave. It is obvious that this behaviour contradicts the unitarity condition which requires $\sigma_{J=0} \leq s^{-1}$ at high energy.

Now we calculate the contribution to this cross section given by the cross channels diagrams, t- and u- ones. It turns out that the corresponding contribution is again equal exactly to the value of $\cong 1.2 \cdot 10^{-2} fb$. The corresponding angular distribution is also flat. At last, let us take into account the interference term between the two classes of the diagrams above. We found that it is equal exactly to $\cong -2.4 \cdot 10^{-2} fb$. Adding all three contributions we obtain the result which removes a seeming contradiction. As it must be, the asymptotic form of the cross section for the process under consideration at $\sqrt{s} \to \infty$ acquires the "desired" form, i.e. it falls with energy

$$\sigma^{as}_{\mu^+\mu^- \to H^0 Z} = \frac{1}{3} \cdot \frac{\pi \alpha^2}{sin^4(2\theta_W)} \cdot (g_V^2 + g_A^2) \cdot \frac{1}{s}. \tag{3}$$

Attention should be drawn to the difference between factors containing the coupling constants entering into Eqs. (2) and (3). The obtained cancellation reflects the most fundamental property of the electroweak theory. This is a consequence of the unitarity condition at the tree level which must be fulfilled in any non-Abelian gauge theory with the symmetry broken in a manner like the Higgs mechanism [7].

In order to extract information about the Higgs – lepton sector interplay let us look once more at the individual contributions to the discussed cross section. In this respect it is worthwhile to note that all three contributions reach their constant asymptotic values not simultaneously. Those stemming from the sum of the t-channel and u-channel go to the plateau at the energy around 1 TeV. The negative contribution reaches its minimum value at $\sqrt{s} \cong 2.5 TeV$, while the cross section, corresponding to the s-channel, becomes constant (at finite muon mass) far away from 1-2 TeV region. However, in spite of different characters of the behaviour of these contributions, it seems unlikely that this difference will be observed experimentally owing to the small value of the μ- meson mass. Partly because of that, in the next paragraph we will consider the process where the Higgs-Gauge-Boson vertex is not involved.

ASSOCIATED $H^0\gamma$ PRODUCTION. *ARBITRARY ENERGIES CASE*

We now turn to a process analogous to the one just considered but which is free from the s-channel diagram complication. At first, it is instructive to consider the arbitrary energies case which covers the low energy region where scalar and pseudoscalar couplings of the Higgs with muons yield different cross sections as opposed to the high energy case. In addition, the expressions for cross sections obtained are very simple. We omit here differential cross sections and give only the integrated expressions. For the scalar case, we have obtained the following cross section without neglecting masses of the initial particles

$$\sigma^{Scalar}_{\mu^+\mu^- \to H^0\gamma} = \frac{\pi \alpha^2}{2 \sin^2 \theta_W} \frac{m_\mu^2}{M_W^2} \frac{1}{s^2} \frac{1}{\beta} \frac{1}{s - m_H^2} \times$$
$$\left\{ -2m_H^2 s \beta_H^2 + (s^2 \beta^4 + m_H^4 \beta_H^2)\frac{1}{\beta} \ln \frac{1+\beta}{1-\beta} \right\}, \tag{4}$$

where in addition to the usual $\beta = \sqrt{1 - \frac{4m_\mu^2}{s}}$ we have introduced the notation $\beta_H = \sqrt{1 - \frac{4m_\mu^2}{m_H^2}}$ with $\sqrt{s}$ as the c.m. energy. For the case of the pseudoscalar coupling of the Higgs with the muon, we have

$$\sigma^{Pseudoscalar}_{\mu^+\mu^- \to H^0\gamma} = \frac{\pi\alpha^2}{2\sin^2\theta_W} \frac{m_\mu^2}{M_W^2} \frac{1}{s^2} \frac{1}{\beta} \frac{1}{s - m_H^2} \times$$
$$\left\{ -2m_H^2 s + (s^2 + m_H^4 \beta_H^2) \frac{1}{\beta} \ln \frac{1+\beta}{1-\beta} \right\}. \qquad (5)$$

It is the γ_5 non-invariance, which is responsible for the different behaviours of the two cross sections, see eqs. (4) and (5), at low energies. The formulae obtained are very important, for example, in searching for the light axion or even light Higgs at Υ decays to the γ + nothing, i.e. in the $b\bar{b}$ annihilation, in which case the cross section is greatly enhanced due to the large mass of the b-quark. At higher energies this difference disappears as will be seen in the next section.

ASSOCIATED $H^0\gamma$ PRODUCTION. *HIGH ENERGY CASE.*

The differential cross section of the process $\mu^+\mu^- \to H\gamma$ for the case, when the photon is hitting a non-forward detector, and when we neglect the muon mass (apart from where it affects the muon-Higgs boson coupling constant) reads as follows:

$$\frac{d\sigma^{\gamma H}}{d(\cos\theta)} = \frac{\pi\alpha^2}{2\sin^2\theta_W} \frac{m_\mu^2}{M_W^2} \cdot \frac{s^2 + M_H^4}{s^2(s - M_H^2)} \frac{1}{(1-\cos^2\theta)}, \quad \cos\theta \leq 1. \qquad (6)$$

The corresponding integrated cross section is as follows (we have neglected here the muon mass apart from where it is a part of the muon- Higgs boson coupling constant):

$$\sigma^0_{\gamma H} = \frac{\pi\alpha^2}{2\sin^2\theta_W} \frac{m_\mu^2}{M_W^2} \frac{1}{s^2} \frac{1}{s - m_H^2} \left[-2m_H^2 s + (s^2 + m_H^4)\ln\frac{s}{m_\mu^2} \right], \qquad (7)$$

where superscript 0 means that the cross section is calculated at the tree-diagram level.

With the yearly integrated luminosity of $\mathcal{L} \cong 10^3$ fb^{-1} expected at future $\mu^+\mu^-$ colliders, one could collect 20 to 30 $H^0\gamma$ events (the detector efficiency is supposed equal 1, and the acceptance – 4π). The signal which mainly consists of a photon and $b\bar{b}$ pairs in the low Higgs mass range or WW/ZZ pairs for Higgs masses larger than $\cong 200$ GeV, is extremely clean. The background should be very small since the photon must be very energetic and the $b\bar{b}$ or WW/ZZ pairs should peak at an invariant mass M_H. Therefore, despite the low rates, a clean signal gives a good possibility to detect these events.

Another mechanism for associated γH production is the one of the electroweak loops, considered in Ref. [8] and in recently published [9,10]. This last mechanism is equally applicable to the case of colliding $\mu^+\mu^-$ beams and to the e^+e^- case. However, the both mechanisms give the cross sections with very different c.m. energy behaviours. As is seen from Eq. 2 the tree-level cross section grows when

$\sqrt{s} \to m_H$ due to kinematical factor $\frac{1}{s-m_H^2}$ in front of it. Contrary to this case, the one-loop cross section is negligible at the threshold and rises with energy. Comparative pictures of the two types cross section behaviours are depicted on Fig. 3 of Ref. [10] at some representative Higgs boson mass values. A remarkable feature of these figures is equality of the tree-level and one-loop cross sections at the practically invariable point $m_H \approx \frac{\sqrt{s}}{2}$, after which the tree-level cross section falls rapidly and the process is dominated by the one-loop amplitudes, while up to this point the main contribution to the cross section comes from the tree-level graphs. At the first sight this feature delivers a good possibility of studying the Higgs behaviour in lepton sector. But we must realize that the tree-level cross section has the "bad behaviour" in the vicinity of the point where $\sqrt{s} \succeq m_H$, so we need to take care of this region in order to smooth the front edge of the cross section curve. To this end we need to calculate the radiative corrections (RC) to the process under consideration.

In other words, to make a consistent comparison between the lowest order cross section for the process $\mu^+\mu^- \to H\gamma$ and that due to the one-loop amplitudes, we need to calculate quantum electrodynamics (QED) correction to the tree-level amplitudes. That is the main aim of the next section.

RADIATIVE CORRECTIONS TO THE PROCESS $\mu^+\mu^- \to H\gamma$.

Due of the lack of the space here only a very concise sketch of RC will be given. For the comprehensive information I refer to [12]. For the process under study 1) the first order leading logarithmic correction were obtained, and 2) the contribution of the higher order perturbation theory were taken into account in the leading logarithmic approximation. It is expedient to give here the expression for the case 2). The master formula for the radiatively corrected cross section has the form of the Drell-Yan cross section. So, we suggest to write the result as a convolution of the modified lepton structure functions with the shifted cross-section of the hard subprocess. It reads

$$\frac{d\sigma}{d\cos\theta} = \int_{z_1^{min}}^{1} dz_1 \widetilde{D}(z_1) \int_{z_2^{min}}^{1} dz_2 \widetilde{D}(z_2) \frac{d\widetilde{\sigma}_0(z_1 p_1, z_2 p_2)}{d\cos\theta} \left(1 - \frac{\alpha}{2\pi} K\right) \Theta(\omega - \omega_{th}), \quad (8)$$

where a part of non-leading terms is taken into account by the so-called K-factor (again, see the ref. [12]); ω_{th} is the experimental energy threshold of the photon registration. A *smoothed* representation for the modified D-function was used:

$$\widetilde{D}(z, L) = \frac{1}{2}\beta(1-z)^{\beta/2-1}(1+z^2)(1+\mathcal{O}(\beta^2)), \qquad \beta = \frac{2\alpha}{\pi}(L-1). \quad (9)$$

The energy conservation law gives us the energy of the detected photon

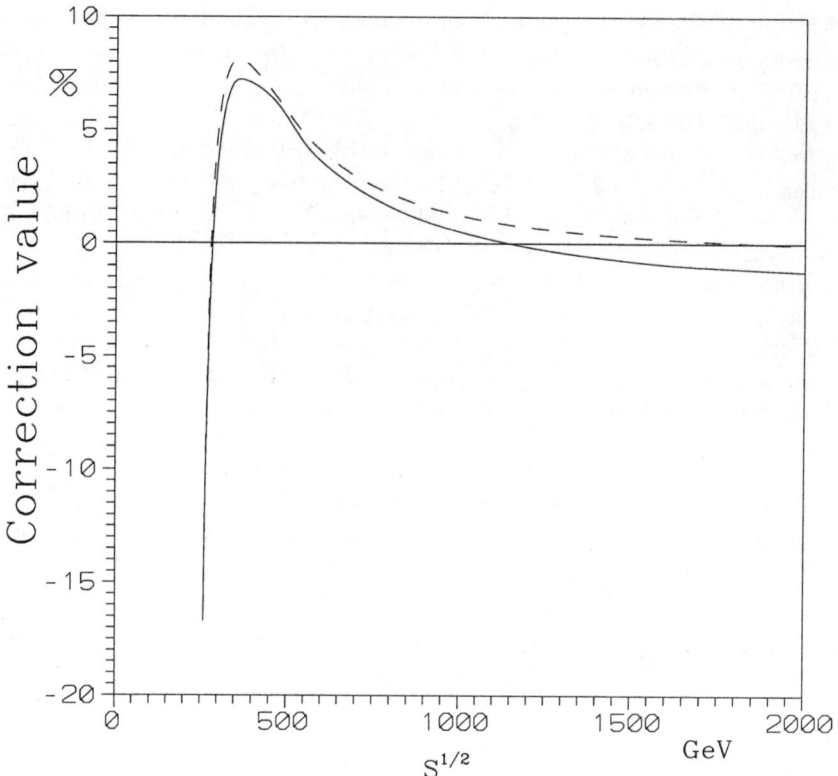

FIGURE 1. Radiative corrections to the process $\mu\bar{\mu} \to H\gamma$.

$$\omega = \frac{sz_1 z_2 - M_H^2}{2\varepsilon(z_1 + z_2 - c(z_1 - z_2))}. \tag{10}$$

The lower limits of integration over $z_{1,2}$ are to be defined also just from the above expression by imposing the condition $\omega > \omega_{th}$:

$$z_1^{min} = \frac{M_H^2 + \sqrt{s}\omega_{th}(1+c)}{s - \sqrt{s}\omega_{th}(1-c)}, \qquad z_2^{min} = \frac{M_H^2 + \sqrt{s}z_1\omega_{th}(1-c)}{sz_1 - \sqrt{s}\omega_{th}(1+c)}. \tag{11}$$

In Fig. 1 we presented the values of radiative corrections as functions of the center-of-mass energy

$$\delta(\sqrt{s}) = \left[\frac{\int_{c_{min}}^{c_{max}} dc(d\sigma/dc)}{\int_{c_{min}}^{c_{max}} dc(d\sigma_0^{\gamma H}/dc)} - 1\right] 100\%. \tag{12}$$

We took $M_H = 250$ GeV, the value of photon energy threshold $\omega_{th}=5$ GeV, and the angular range for photon detection $-0.999 < c < 0.999$. The dashed line represents the first order leading logarithmic correction. The solid line shows the values of the complete RC according to Eq. (8).

We see that the tree-level cross section dominates in the vicinity of $\sqrt{s} \approx m_H$. This region is appropriate for the $\mu\bar{\mu}H$ vertex study. At the same time, the loop-induced cross section dominates in the region of $\sqrt{s}$ which is far from m_H value. This region is appropriate, for example, for the $t\bar{t}H$ vertex study.

CONCLUSIONS

Muon colliders will deliver an excellent opportunity for the particle physics investigations.

ACKNOWLEDGEMENTS

I am very grateful to Prof. A. Thomas for the invitation to participate in the work of Conference and Organizing Committee for the excellent job.

REFERENCES

1. Tikhonin F.F., Preprint JINR P2–4120. Dubna, 1968.
2. Budker G.I., Proceedings of the Intern. Conf. on High Energy Physics, Kiev, 1970, Dubna, 1970, p.1017.
3. Skrinsky A.N. and Parkhomchuk V.V., Sov.J. Nucl. Physics, 12 (1981) 3.
4. Neuffer D., Particle Accelerators 14 (1983) 74.
5. Gunion J.F.,Talk given at 5th Intern. Conf. on Physics Beyond the Standard Model, Balholm, Norway, 29 Apr - 4 May 1997, e-Print Archive: **hep-ph/9707379** .
6. Litvin V.A. and Tikhonin F.F., Preprint IFVE-97-24, Protvino 1997, e-Print Archive: **hep-ph/9704417**.
7. Llewellyn-Smith C.H., Phys.Lett. B46 (1973) 233; Cornwall J.F. et all., Phys. Rev. Lett. 30 (1973) 1268.
8. Barroso A., Pulido J. and Romao J.C., Nucl. Phys. B 267 (1986) 509 .
9. Abbasabadi A., Bowser-Chao D., Dicus D., and Repko W., Phys. Rev. D 52 (1995) 3919.
10. Abbasabadi A., Bowser-Chao D., Dicus D., and Repko W., e-Print Archive: **hep-ph/9706335**.
11. Abbasabadi A., Bowser-Chao D., Dicus D., and Repko W., e-Print Archive: **hep-ph/9708328**.
12. Arbuzov A.B., Kuraev E.A., Tikhonin F.F., and Shaikhatdenov B.G., Phys. Atom. Nucl. 62, (1999) 1393; Yad. Fiz. 62 (1999) 1477.

New Facility for Fundamental Physics with Polarized Cold Neutrons

K. Bodek *¶, P. Böni †, N. Danneberg *, W. Fetscher *, C. Hilbes *, St. Kistryn ¶, J. Lang *, M. Lüthy ‡, M. Markiewicz *, A. Pusenkov ∥ A. Schebetov ∥, A. Serebrov ∥, J. Sromicki *

* Institute of Particle Physics, ETH Hönggerberg, 8093 Zürich, Switzerland
† Laboratory for Neutron Scattering, ETH and PSI, 5232 Villigen, Switzerland
‡ Paul Scherrer Institute, 5232 Villigen, Switzerland
∥ St. Petersburg Nuclear Physics Institute, 188350 Gatchina, Russia
¶ Jagellonian University, 30059 Cracow, Poland

Abstract. A new facility for particle physics with polarized cold neutrons has been taken into operation at the spallation source SINQ at PSI. After extraction of the first beam, its intensity and polarization have been measured as a function of the neutron wavelength. The beam characteristics are among the best in the world for studies of neutron decay. An experimental area was constructed with infrastructure support for convenient experimentation. The physics program will focus on detailed investigations of the free neutron decay process, in particular fundamental symmetries of the weak interaction. The first approved experiment is a novel search for time reversal violation.

INTRODUCTION

The neutron and its decay are still mysterious enough to hide important secrets of particle physics. In the last decade great progress has been achieved in determinations of the neutron lifetime and decay asymmetry parameter A. However, the more complicated observables involving the spin of the neutron (B, D, R, G, in the notation of ref. [1,2]) are not yet known with desired precision; some of them have not yet been measured at all. These experiments bear first order sensitivity to the "Physics Beyond the Standard Model". For instance, the question of time reversal violation, addressed to static properties of the neutron as well as to the decay process, attracts considerable attention [3].

SPALLATION SOURCE SINQ AT PSI

In 1997, the high power, continuous flow spallation source SINQ was taken into operation at the Paul Scherrer Institute, Switzerland. SINQ is fed by a 590 MeV proton beam delivered by an isochronous ring cyclotron, which is simultaneously used for abundant production of pions and muons ("meson factory"). In the recent decade a massive increase of the proton beam current has taken place (250 μA in 1988 to 1700 μA in 1999) from which all three fields of research take profit today. Approximately 70 % of the beam intensity with 0.7 MW power, is directed vertically upward on the spallation target. In the recent shutdown 1999/2000 a steel-clad lead-rod target was inserted, which provides $\sim 5 \cdot 10^{16}$ fast neutrons per second. These neutrons are slowed down to thermal energies in a 3 m diameter heavy water moderator and to lower energies in the 25 K, 20 liter volume liquid deuterium source. Production of cold neutrons is considered a particular strength of the SINQ complex.

BEAMLINE FOR POLARIZED COLD NEUTRONS

Neutron decay experiments are limited by counting statistics, therefore a high intensity beam of very slow neutrons is of utmost importance. In 1999 a new facility for fundamental physics with polarized cold neutrons has been added to the SINQ complex (Fig. 1).

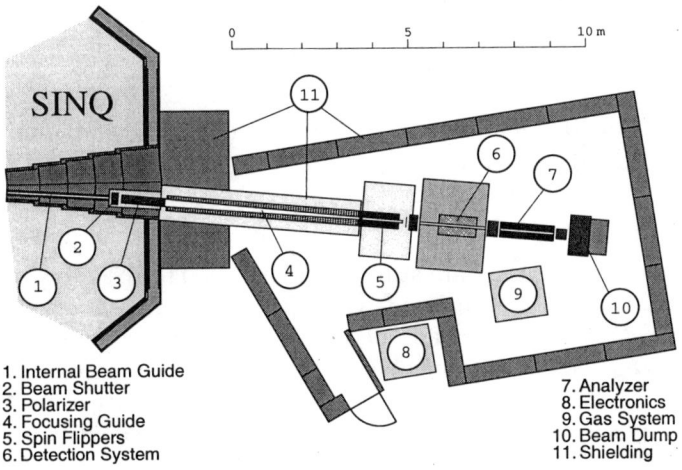

Fig. 1. Layout of the polarized cold neutron facility at PSI. The described beam tract starts from the exit port of the cold source, at the border of the heavy water moderator tank, which is placed in the center of the 12 m diameter biological SINQ shielding on the left.

We have equipped one of the neutron channels, facing the liquid deuterium cold source, with a large aperture and large momentum acceptance beam guide [4] extending from the exit of the cold source tubing in the D_2O moderator tank. The guide is covered by supermirrors consisting of 450 layers of Ni/Ti. Best quality mirrors with a high index of angular reflection ($m \approx 3$) are used also in the vicinity of the moderator tank. We will monitor the time dependence of the neutron flux and polarization to provide data about possible damage of the supermirrors in the hard radiation field. After half a year of operation no evidence of such effects is found.

The "real" flux density of unpolarized neutrons, measured at the border of the SINQ shielding (pos. 2, Fig. 1) is around 10^9 n/(cm²·s·mA); the "thermal equivalent flux" is $3 \cdot 10^9$ n/(cm²·s·mA). The total number of unpolarized neutrons with the characteristic cold spectrum exceeds 10^{11} n/(s·mA).

The main beam shutter (pos. 2, Fig. 1) is integrated into the SINQ shielding block. A large size polarizer and a beam bender, alltogether 1.6 m long, and an external beam tract [4] (pos. 3 and 4, respectively), guide the neutrons to the experimental station. This complex equipment consists of a multislit glass-plate polarizer, covered with magnetic supermirrors, a cold neutron beam stop, a focusing beam guide (pos. 4) tapered from 8×15 cm² down to 4×15 cm², radiofrequency spin flippers (pos. 5), a chopper device for time of flight measurements, and two additional cold neutron beam dumps placed downstream. The beam tract following the polarizer is immersed in a guiding magnetic field for the neutron spin, provided by permanent magnets with appropriate yoke structure. The front of this beam channel, with precisely adjusted polarizer and bender, is integrated into the bunker of the SINQ, to contain the radiation associated with the neutron capture reactions.

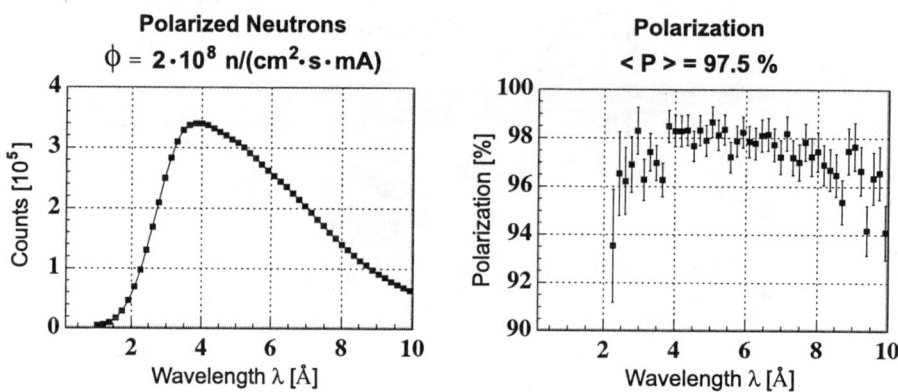

Fig. 2. Left: A characteristic cold neutron spectrum measured using the time of flight technique and a thin ³He detector. Right: Wavelength dependence of the neutron polarization in the central part of the beam.

The external beamline has its own, independent vacuum system. A sandwich radiation shield (brass/polyethylene/iron/lead) slows down and absorbs fast neutrons penetrating the walls of the guide as well as secondary gamma rays. The residual radiation is contained in a 40 cm thick biological shield mounted around the neutron guide. The main functions of this compact (total length of $\sim$ 10 m) neutron optic system are: increase of the neutron flux density, reduction of transport losses, and provision of enough room for different experiments.

The commisioning data measured at the location of experiments show a "real" flux of polarized cold neutrons of $2 \cdot 10^8$ n/(cm$^2 \cdot$s$\cdot$mA) with a polarization in the central part of the beam cross section exceeding 97% (Fig. 2).

Detailed investigations of the beam properties, i.e. the flux and polarization distributions and divergence, are in progress. Fig. 3 presents the first results from the measurements of the horizontal beam profiles and the computer representation of the beam intensity for the unpolarized beam (pos. 2, Fig. 1) and polarized, compressed beam at the experimental station (pos. 6, Fig. 1). An obvious improvement in the beam quality downstream is visible in the reduced cross section and better beam collimation.

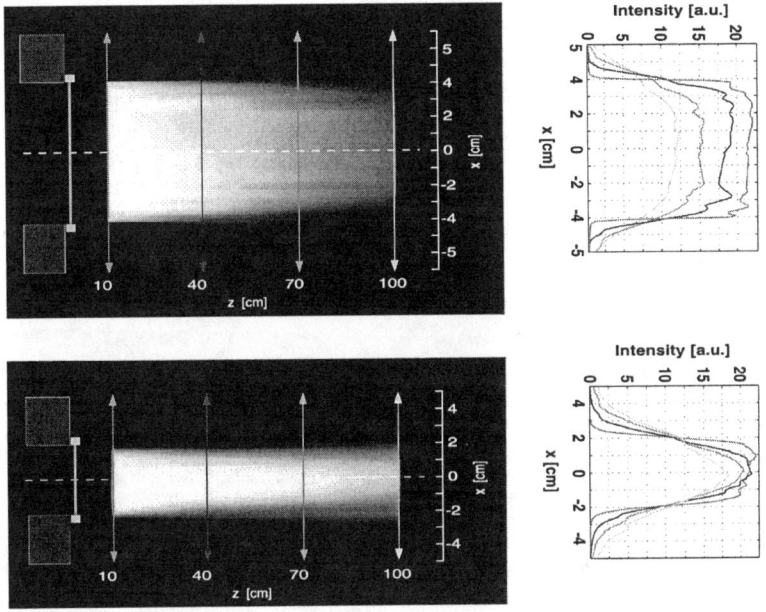

Fig. 3. Measured beam profiles (right) and a representation of the beam intensity obtained by interpolation (left). Upper part: measurement in pos. 2, Fig 1. Lower part: measurement in pos. 6, Fig. 1. Note different scales for "z" and "x" direction, which indicate the distance from the corresponding beam exit and the horizontal position across the beam, respectively.

NEUTRON DECAY EXPERIMENTS

The physics program of this new facility will focus on free neutron decay studies. The first approved experiment is a novel search for time reversal violation (TRV) [5]. The experiment determines the so-called R-parameter, which is a measure of the transverse polarization of electrons emitted at right angle with respect to the neutron polarization axis. The challenge of this experiment is the measurement of the polarization of low energy electrons from neutron decay in flight. Mott scattering from lead foils will be used as a spin analysis process for electrons. The decay electrons are tracked in a detector insensitive to gamma rays in order to identify desired events in a difficult background environment.

Recently, after successfull commissioning of the experimental area, including the biological shielding and the security system, a short series of experiments has been performed. The first neutron decay events were detected using a prototype tracking detector consisting of 12 active planes of wires with an effective volume of $12 \times 12 \times 8$ cm^3. Fig. 4 shows two projections of an observed event with scattering of the electron from the analyzer foil.

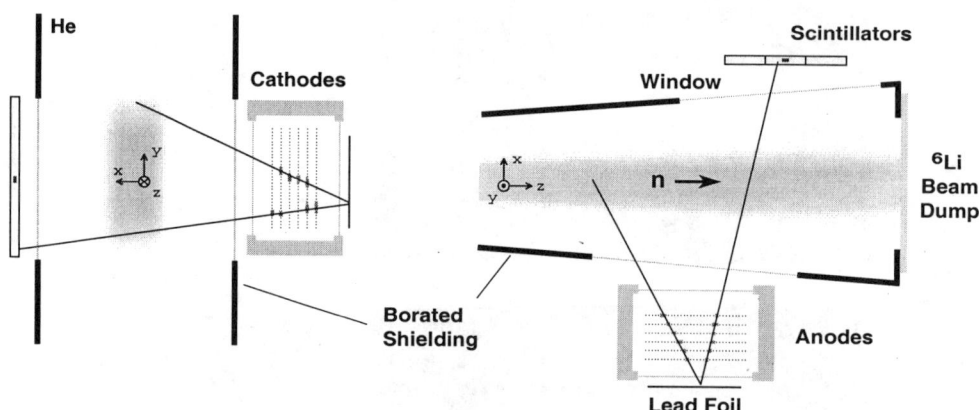

Fig. 4. One of the first candidate events for neutron decay and scattering of the emitted electron from the lead foil. The event was detected by a test wire chamber installed at the side of the neutron beam. A scintillator located on the other side of the neutron beam provided the trigger condition. Left: beam axis view. Right: top view.

Further development of the experiment with larger size tracking detectors placed on both sides of the neutron beam is foreseen for autumn 2000.

CONCLUSIONS

Present neutron decay experiments require the highest spatial density of neutrons and a very high polarization. Neutron beams extracted from liquid hydrogen/deuterium sources at high flux research reactors transport more than 10^3 neutrons/cm^3 into the vicinity of experimental stations. For the first time, such an intense beam is produced using spallation neutrons. Specific advantages of the SINQ spallation source at PSI and modern technology were used to enhance the beam quality and intensity. They include: a large phase space accepted from the cold source due to the highest quality supermirror guides, polarizer and bender, a relatively short beam tract with a small number of wall reflections, reduced losses in the beam transport and beam compression in the horizontal direction. The beamline was optimized for a maximal intensity of *polarized* neutrons, which are of particular interest for studies of fundamental symmetries of the weak interaction. In spite of the much weaker thermal flux in the moderator, the characteristics of the polarized cold beam at SINQ are comparable to those at the ILL reactor.

ACKNOWLEDGEMENTS

The construction of this complex facility was a challenging task and investment. It was possible within a short time due to the extraordinary engagement of the PSI and PNPI Institutes and their technical staff. We are grateful to PSI management for continuous support and for providing financial resources.

REFERENCES

1. J.D. Jackson, S.B. Treiman, H.W. Wyld, Phys. Rev. **106**(1957)517.
2. J.D. Jackson, S.B. Treiman, H.W. Wyld, Nucl. Phys. **4**(1957)206.
3. P. Harris, M. Pendelbury *Electric Dipole Moment*; J. Gordon *"D"–Correlation*; J. Sromicki *"R"–Correlation*, International Workshop on Particle Physics with Slow Neutrons, ILL, Oct. 1998, Nucl. Instr. Meth. **A440**(2000).
4. A. Schebetov, A. Serebrov, J. Sromicki, P. Böni, V Pusenkov et al., *Polarizers with Condensers for Sector 50 at SINQ*, PNPI Reports, 1997, 1998.
5. I.C. Barnett, K. Bodek, P. Böni, D. Conti et. al., *Search for Time Reversal Violation in the Decay of Free, Polarized Neutrons*, PSI Proposal, June 1997.

A Proposed Measurement of the β Asymmetry in Neutron Decay with the Los Alamos Ultra-Cold Neutron Source

B. Tipton*, A. Alduschenkov, K. Asahi, T. Bowles, B. Filippone, M. Fowler, P. Geltenbort, F. Hartmann, R. Hill, A. Hime, M. Hino, S. Hoedl, G. Hogan, T. Ito, C. Jones, T. Kawai, A. Kharitonov, K. Kirch, T. Kitagaki, S. Lamoreaux, M. Lassakov, C-Y. Liu, M. Makela, J. Martin, R. McKeown, C. Morris, A. Pichlmaier, M. Pitt, Yu. Rudnev, A. Saunders, S. Seestrom, A. Serebrov, D. Smith, K. Soyama, M. Utsuro, A. Vasilev, B. Vogelaar, P. Walstrom, J. Wilhelmy, A.R. Young, J. Yuan

*Kellogg Radiation Laboratory 106-38
1200 E. California Blvd.
Pasadena, CA 91125
E-mail: tipton@krl.caltech.edu.edu*

Abstract. This article reviews the status of an experiment to study the neutron spin-electron angular correlation with the Los Alamos Ultra-Cold Neutron (UCN) source. The experiment will generate UCNs from a novel solid deuterium, spallation source, and polarize them in a solenoid magnetic field. The experiment spectrometer will consist of a neutron decay region in a soleniod magnetic field combined with several different detector possibilities. An electron beam and a magnetic spectrometer will provide a precise, absolute calibration for these detectors. An A-correlation measurement with a relative precision of 0.2% is expected by the end of 2002.

INTRODUCTION

Fundamental studies of neutron beta decay provide new constraints on electroweak interactions at low energies. The leading order expression for neutron beta decay is

$$\frac{d\sigma}{d\Omega_e d\Omega_\nu} = F(p_e)[1 + \mathrm{a}\frac{\vec{p}_e \cdot \vec{p}_\nu}{E_e E_\nu} + \mathrm{A}\frac{\vec{\sigma} \cdot \vec{p}_e}{E_e} + \mathrm{B}\frac{\vec{\sigma} \cdot \vec{p}_\nu}{E_\nu} + \mathrm{D}\frac{\vec{\sigma} \cdot \vec{p}_e \times \vec{p}_\nu}{E_\nu E_e}]. \quad (1)$$

The parameters a, B, and D represent the neutrino-electron correlation, the spin-neutrino correlation, and the time-reversal violating correlation. The parameter A measures the parity violating correlation between the neutron spin and the electron momentum directions. The A correlation is related to the ratio of the fundamental nucleon vector and axial vector weak coupling constants, $\lambda = G_A/G_V$ through the relation,

$$A = \frac{-2\lambda(1+\lambda^2)}{1+2\lambda^2}. \qquad (2)$$

Neutron lifetime measurements are sensitive to a different mixture of G_V and G_A; the lifetime and A-correlation measurements may be combined to constrain these couplings separately. With the Fermi coupling constant, G_F, known from muon decay, the fundamental CKM matrix element V_{ud} may be extracted. Precision determinations of V_{ud} in this manner allow unitarity tests of the CKM matrix and possible access to new physics. A diagram of the current constraints on V_{ud} is shown in Figure 1.

A new experiment is proposed at the Los Alamos National Laboratory to study the parity-violating spin-electron correlation in neutron beta decay. The experiment relies on a superthermal deuterium source and a spallation proton target to generate a high flux of ultra-cold neutrons (UCNs). UCNs are defined as neutrons with velocity low enough that they undergo total external reflection from the surface of materials containing them. For the case of ^{58}Ni guides, $v_{neutron} < 8$ m/s.

UCN from pulsed spallation source offers significant advantages over other techniques of determining the neutron beta-asymmetry. The total external reflection property allows the UCNs to be transported far from the spallation source, through significant shielding. This separates the beta spectrometers from the high background environment at the spallation target. The timing structure of the pulsed beam allows one to reject prompt beam background during pulses; the produced UCNs may be stored to deliver nearly constant UCN flux in between pulses. Finally, a several Tesla magnetic field is sufficient to overwhelm the UCN kinetic energy, allowing a solenoid magnet to select one polarization state. Nearly 100% neutron polarization is achievable.

A SOLID DEUTERIUM SOURCE OF UCN

At the heart of the proposed experiment is new type of ultra-cold neutron source using a solid deuterium (SD$_2$) crystal above a spallation source of cold neutrons. Figure 2 shows a schematic diagram of the proposed source. Up to 100μC of 800 MeV protons from LANSCE will impinge on a tungsten spallation target. Produced neutrons will reach a thin film of SD$_2$ after being moderated in polyethylene at liquid nitrogen and liquid helium temperatures. These neutrons may phonon scatter in the SD$_2$ crystal to come nearly to rest. Above the SD$_2$, a structure of a ^{58}Ni coated guide surrounded by a cold beryllium reflector and a polyethylene moderator will serve both to contain the cold neutrons near the SD$_2$ and to transport produced UCN away. A storage volume above the SD$_2$ may be connected by an open valve

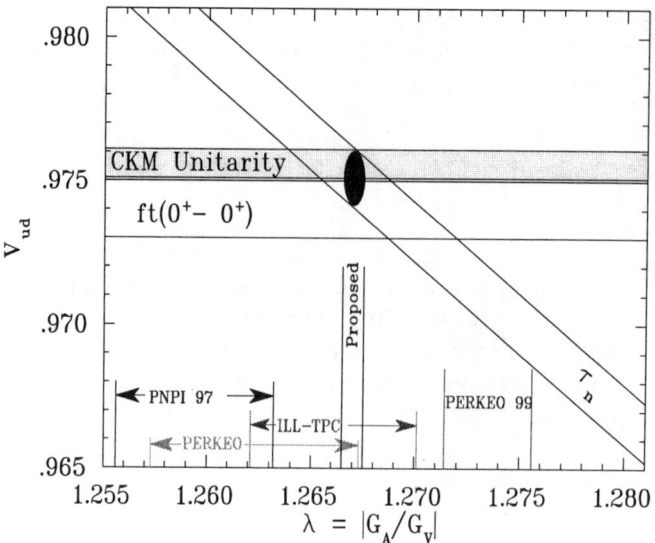

FIGURE 1. Constraints on the CKM matrix element V_{ud} for recent extractions of the parameter λ from the A-correlation. The λ values come from PNPI 97 [2], PERKEO [1], ILL-TPC [3], and PERKEO 99 [4]. The V_{ud} constraint from the ft measurement in superallowed beta decays comes from [5]. The dark oval represents the expected size of the 1σ allowed region when the results proposed experiment are combined with the world average neutron lifetime measurement.

to the UCN source during one second proton pulses; the valve will be closed for nine seconds in between pulses while the UCN are delivered to experiments.

In runs at Los Alamos over the past year, this design was tested with a prototype UCN SD_2 source connected to a ^{3}He spectrometer. The performance of the test source indicates that an equilibrium stored UCN density of 350 UCN/cm^3 is achievable for the A-correlation experiment.

THE SPECTROMETER

A schematic design of the A-correlation experiment is shown in Figure 3. UCNs will be transported from the storage to the experiment through diamond-coated guides. On their way, they will pass through a six Tesla solenoid field, which selects one neutron spin state with near 100% efficiency. The neutrons may then be rotated periodically to the opposite spin state by subsequent passage through an adiabatic fast passage spin flipper, with expected efficiency > 99.9%.

The polarized UCNs will then be fed into a 10-cm diameter bottle inside a one Tesla superconducting solenoid. The bottle is open ended and will store the neutrons for a average of five seconds. While less than 1% of the flowing UCNs decay

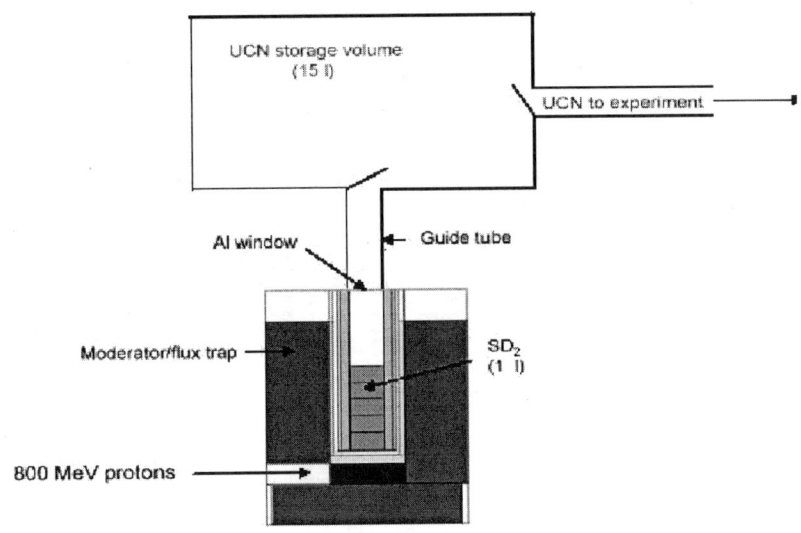

FIGURE 2. A schematic diagram of the UCN source.

inside the bottle, most are absorbed in ^{6}Li coated end chambers or flow out of the end region to neutron flux analyzing detectors.

When a UCN decays, its emitted beta will spiral in the solenoid field towards the end of the bottle; this field geometry will have nearly 4π beta collection efficiency. Before reaching electron detectors, the betas will encounter a magnetic field expansion region, which will both orient the trajectories more perpendicular to the detector face and act as a mirror to contain the electrons in the detector region. The impact of backscattering will then be reduced.

Two electron detector designs are being developed for the experiment. In one design a 15-cm diameter Time Projection Chamber(TPC) will cover a thin scintillator. The chamber allows a precision measurement of the electron trajectory for better fiducial volume definition and statistical sensitivity. The requirement of a TPC-scintillator coincidence in each event also eliminates backgrounds, such as from cosmic rays. Alternatively, a double-sided silicon strip detector may be placed in the experiment to provide excellent energy resolution. The combination of both detector technologies with differing sensitivities and systematics will allow cross-checks in the experiment.

A total of 108 Hz of neutrons will decay in the experiment, with 87 Hz analyzable after efficiencies and cuts are considered. Backgrounds in the experiment are essentially negligible. With the experiment comissioning by the end of 2001 the anisotropy in neutron beta decay is then expected to be measured to the 0.16% level statistically by the end of 2002.

The anticipated largest source of systematic uncertainty comes from knowledge

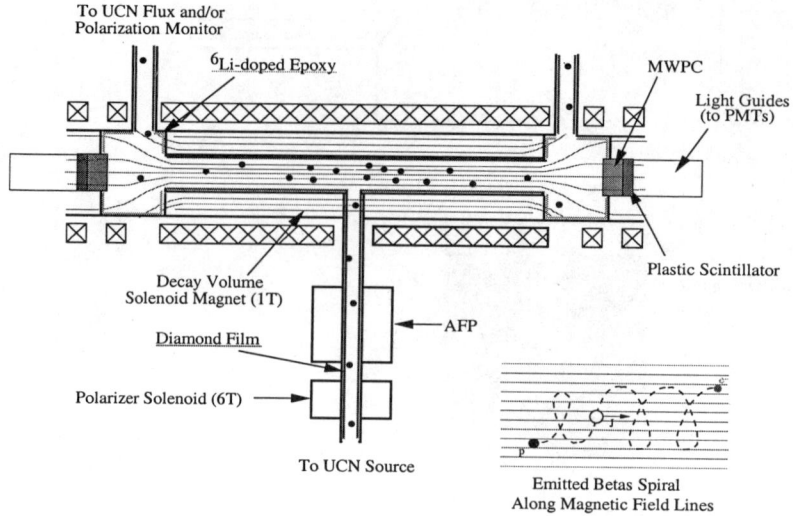

FIGURE 3. An overhead view of the spectrometer for the A-correlation experiment.

of the detector performances. The experimental goal is to keep the size of this systematic below 0.03% for a total systematic uncertainty of 0.042%.

An ambitious program has been begun to allow detailed study of the electron detectors after they are constructed. A combination of a custom built 150 keV electron gun and the 1 MeV Dynamotron at NASA's Jet Propulsion Laboratory provide continuous, stable, tunable electron beam from 5 keV to 1 MeV. A double focusing magnetic spectrometer has been constructed at Caltech to calibrate the electron beam energy to better than 0.3%. The combination of these devices will allow an extremely high precision characterization of the detector efficiency, calibration, linearity, resolution, and backscattering probability before and during the experiment.

CONCLUSION

A new experiment is proposed at Los Alamos to provide the world's best measurement of the A-correlation in neutron beta decay. The experimental collaboration plans to measure A to a precision of 0.2% by then end of 2002. This is an improvement of over a factor of four compared to the most recent A-correlation experiment [4]. The black dot in Figure 1 shows the impact of this measurement on the determination of V_{ud}, when combined with world data on the neutron lifetime.

ACKNOWLEDGMENTS

The author would like to acknowledge the support of the U.S. National Science Foundation.

REFERENCES

1. Bopp, P et. al., *Phys. Rev. Lett.* **56**, 919 (1986).
2. Erozoliminskii, B. G.et. al., *Phys. Lett.* **B236**, 33 (1991).
3. Schreckenbach, K. et. al., *Phys. Lett.* B349, 427 (1995).
4. Reich, J. et. al., *Phys. Lett.* **440**, 535 (2000).
5. Hardy, J.C., Towner, I.S., nucl-th/9807049.

Current Status of the CHORUS Experiment at CERN

Koichi Kodama

CHORUS collaboration

Aichi University of Education
Hirosawa 1, Igaya-cho, Kariya 448-8542, Japan

Abstract. Current status of the CHORUS experiment at CERN is presented with an emphasis on technical aspects. This experiment is aimed to search for $\nu_\mu \to \nu_\tau$ oscillation with a designed sensitivity of $sin^2(2\theta) \sim 2 \times 10^{-4}$ at large δm^2. Accumulation of neutrino interactions in the emulsion target with a data taking of electronic counters were performed from 1994 to 1997 at the CERN neutrino beam. Reading out track informations recorded in the emulsion target is then started using the Track Selector (a dedicated system which automatically reads out tracks recorded in emulsion). Successful development of the Track Selector enabled this experiment and opened a possibility for future experiments.

PHYSICS MOTIVATION

The CHORUS [1] experiment at CERN is designed to explore $\nu_\mu \to \nu_\tau$ oscillation in a region shown in Figure 1. The best limit in this region had been provided by the E531 experiment at FNAL [2]. The CHORUS experiment uses similar techneque as in E531 and designed to be about 10 times more sensitive to $sin^2(2\theta)$.

Since oscillation probability $P(\nu_\mu \to \nu_\tau)$ is related to mass difference between ν_μ and ν_τ as shown below, one can say that neutrino has non zero mass if the oscillaton could be observed.

$$P(\nu_\mu \to \nu_\tau) = sin^2(2\theta) sin^2(1.27 \delta m^2 \frac{L}{E})$$

θ mixing angle (rad)
δm^2 $m_{\nu_\tau}^2 - m_{\nu_\mu}^2$ (eV^2)
E neutrino energy (GeV)
L distance from a neutrino source (km)

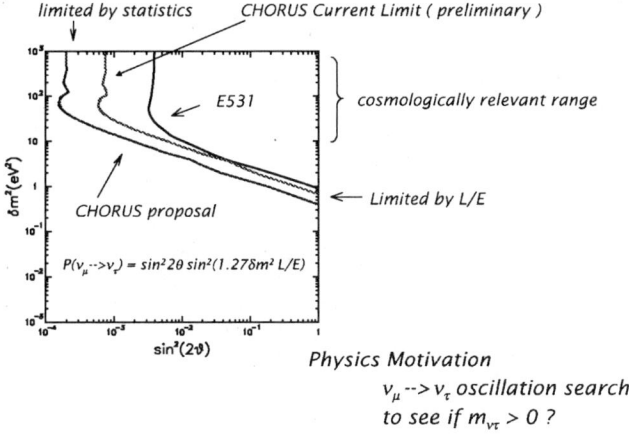

FIGURE 1. Exclusion plot showing a region where CHORUS is exploring $\nu_\mu \to \nu_\tau$ oscillation. Current preliminary limit is also shown.

NEUTRINO BEAM AND DETECTOR

The CERN WANF neutrino beam is suitable for the CHORUS experiment. Most of the beam flux has neutrino energy well above a charge current threshold of ν_τ and contamination of prompt ν_τ in the beam is very low, so that only 0.1 ν_τ charge current event is expected in CHORUS. Thus if one could observe a ν_τ charge current interaction, it is due to $\nu_\mu \to \nu_\tau$ oscillation. In other words, CHORUS is a ν_τ appearance experiment.

A schematic drawing of the CHORUS detector is shown in Figure 2. Detection of ν_τ is done by observing a decay topology of τ produced in a ν_τ charge current

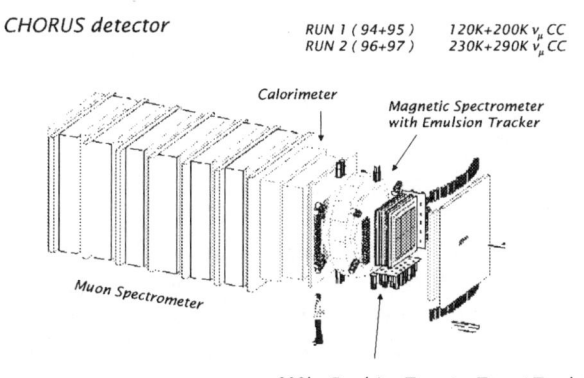

FIGURE 2. A schematic drawing of the CHORUS detector

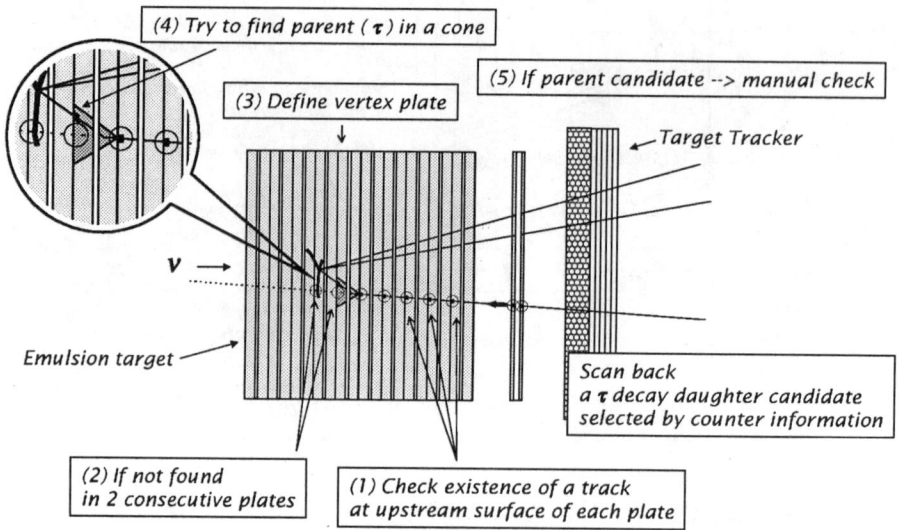

FIGURE 3. Phase 1 analysis procedure

interaction. Most of τ decays within $1mm$ from a primary vertex and excellent spacial resolution is required to observe it. Stacks of emulsion sheets used in CHORUS have spacial resolution of better than $1\mu m$ and is suitable for this purpose. But, since analysis of emulsion had taken much longer time compared to that of electronic counters, how to achieve 10 times more statistics than that of E531 was a big problem.

PHASE1 EMULSION ANALYSIS

To solve this problem, we developed segmented emulsion target followed by target trackers to expose 800kg emulsion, which is needed to collect enough neutrino interactions, and the Track Selector [3], which had been developed in Nagoya, to realize 10 times or more emulsion analysis power. Read-out speed of the Track Selector is about 0.3sec/view (1view $\sim 100\mu m \times 100\mu m$) for a specified track angle and efficiency to recognize tracks in emulsion is better than 98%.

Schematic analysis procedure in emulsion, assuming this version of the Track Selector, is shown in Figure 3, which we call Phase1 analysis. A track detected in target trackers and selected as a possible candidate of τ decay daughter (μ^- in charge current like events for example) is followed up to a τ decay vertex by checking an existence of a track at the suface in each emulsion sheet.

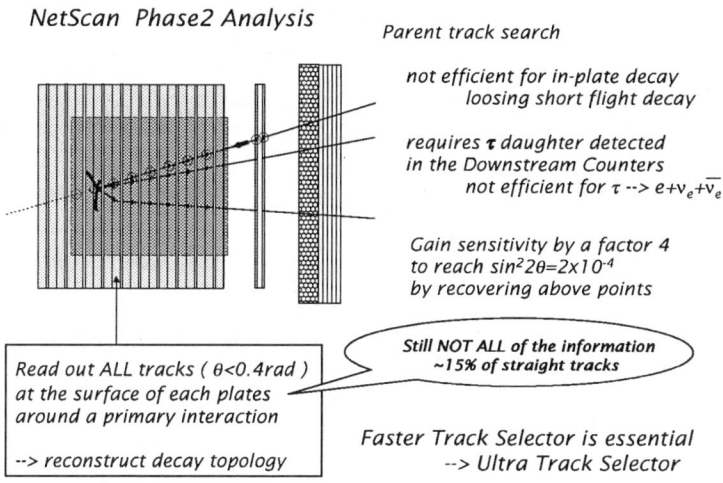

FIGURE 4. Phase 2 analysis procedure

A vertex plate is defined as the up-most sheet where the followed up track was observed when it is missed in 2 consecutive sheets. Then a search for parent track (τ) candidates is tried in the upstram of the vertex plate. Parent track candidates were found in about 3% of analyzed events. All of them were checked manually and no τ candidate was found in current statistics. A preliminary limit — $sin^2(2\theta) > 7.6 \times 10^{-4}$ is excluded at 90% C.L. for large δm^2 — is shown in Figure 1.

PHASE2 EMULSION ANALYSIS

Phase1 analysis procedure is not efficient for 1) short decay where no parent track is visible with current sampling step (i.e. at a surface of each sheet) and for 2) $\tau \to e$ decay mode because it is difficult to identify electrons only with electronic counters. To recover these inefficiencies and gain sensitiviy to oscillation limit, we developed a method called NETSCAN as described in Figure 4.

All track segments in a region around a neutrino interaction is read out and then a decay topology is reconstructed. Of course this method requires much faster Track Selector than used in Phase1 analysis and it has been developed successfully. A minimum distance resolution of any 2 tracks obtained in a test analysis is about $1.6\mu m$ for tracks with $\theta < 0.4 rad$ as shown in Figure 5, which means an impact parameter resolution of $0.8\mu m$. An example of charm decay observed in a test analysis is also shown in Figure 6. To indicate an ability to identify electron in emulsion, an example of cascade production of electron pairs observed in the CHORUS emulsion is shown in Figure 7.

A new emulsion data taking, which we call Phase2 analysis, is being started

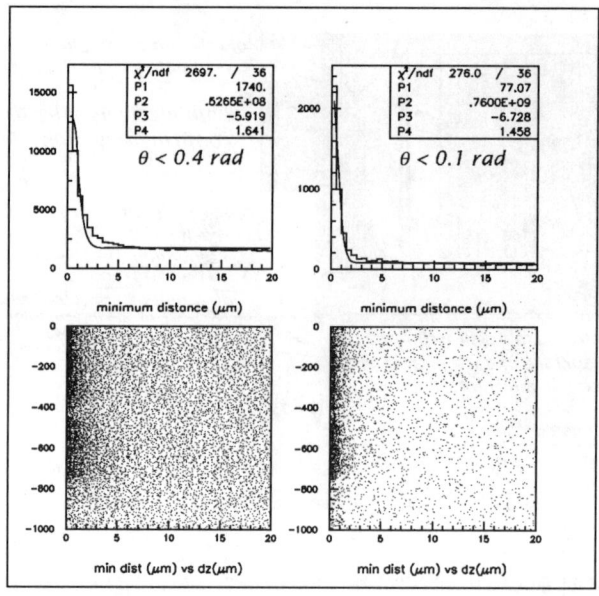

FIGURE 5. Impact paremeter resolution in NETSCAN

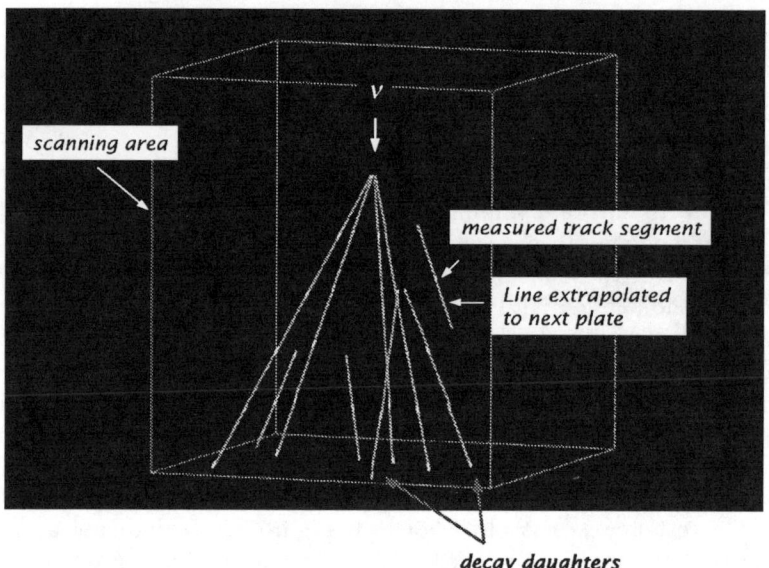

FIGURE 6. An example of charm decay found in NETSCAN

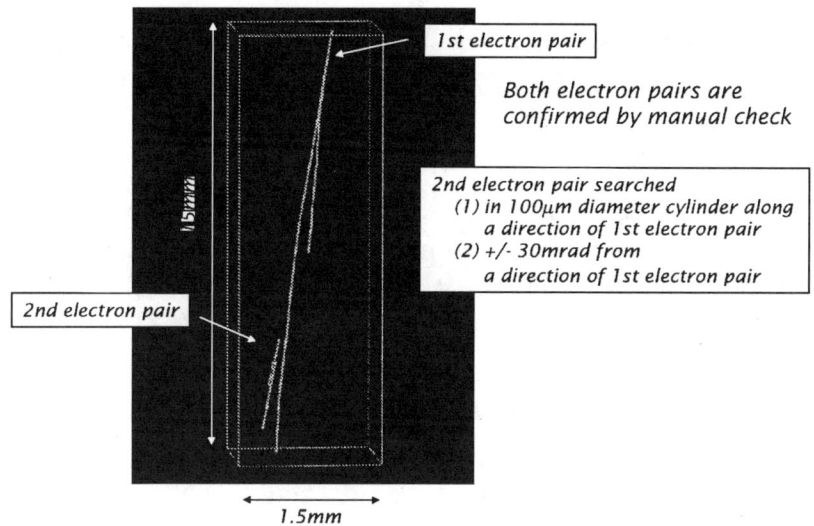

FIGURE 7. An example of electron pair cascade in NETSCAN

applying NETSCAN method to all interactions analyzed so far in Phase1 and to newly located neutrino interactions aiming to reach to a limit of $sin^2(2\theta) < 2\times 10^{-4}$.

CONCLUSION AND FUTURE PROSPECT

CHORUS experiment successfully applied a new emulsion techneque to a neutrino oscillation search in a region shown in Figure 1. Using statistics obtained so far, $sin^2(2\theta) > 7.6 \times 10^{-4}$ is excluded at 90% C.L. for large δm^2. A new emulsion data taking is being started applying NETSCAN method to reach to a limit of $sin^2(2\theta) < 2 \times 10^{-4}$.

An emulsion techneque based on the Track Selector is now entering a new era taking CHORUS experiment as a first application. Next application will be OPERA experiment which searches neutrino oscillation in a region suggested by the atmospheric neutrino results provided by Super-Kamiokande [4].

REFERENCES

1. E.Eskut et al., NIM A401(1997)7   http://choruswww.cern.ch
2. N.Ushida et al., Physical Review Letter 57(1986)2897
3. S.Aoki et al., Nuclear Tracks and Radiation Measurement 12(1986)249
4. Kajita T., *These proceedings.*

Parity Violation in p-p Scattering: the TRIUMF Experiment[1]

S.A. Page[b], A.R. Berdoz[a], J. Birchall[b], J.B. Bland[b], J.D. Bowman[c], J.R. Campbell[b], C.A. Davis[d], A.A. Green[e], P.W. Green[f], A.A. Hamian[g], D.C. Healey[d], R. Helmer[d], Y. Kuznetsov[h] *, L. Lee[b], C.D.P. Levy[d], R.E. Mischke[c], W.D. Ramsay[b], S.D. Reitzner[b], G. Roy[f], P.W. Schmor[d], A.M. Sekulovich[b], J. Soukup[f], G.M. Stinson[f], T. Stocki[f], V. Sum[b], N.A. Titov[h], W.T.H. van Oers[b], R.J. Woo[b], A.N. Zelenski[h]

Presented by S.A. Page

[a] *Carnegie Mellon University, Pittsburgh, PA 15213 USA*
[b] *University of Manitoba, Winnipeg, MB, Canada R3T 2N2*
[c] *Los Alamos National Laboratory, Los Alamos, NM 87545 USA*
[d] *TRIUMF, 4004 Wesbrook Mall, Vancouver, BC Canada V6T 2A3*
[e] *University of the Western Cape, Bellville 7535, South Africa*
[f] *University of Alberta, Edmonton, AB, Canada T6G 2N5*
[g] *University of Washington, Seattle, WA 98195-4290 USA*
[h] *Institute for Nuclear Research, 117312 Moscow, Russia*
* *deceased*

Abstract. Parity violation in proton-proton elastic scattering provides a unique window on the hadronic weak interaction. At low and intermediate energies, the coupling constants of an effective meson exchange model provide an appropriate parametrization. Measurements at different beam energies permit the lowest partial wave amplitudes to be determined in a model independent framework. The TRIUMF cyclotron provides a unique opportunity to measure the $^3P_2 - {}^1D_2$ contribution to proton-proton parity violation at 221 MeV beam energy. This partial wave in turn isolates the weak ρ meson exchange contribution, which is only known to within a factor of 2 from theoretical estimates. An experiment underway for many years at TRIUMF completed data taking in 1999, and the analysis is now nearing completion. Preliminary results, based on half of the acquired data sample, are consistent with the meson exchange model prediction at this beam energy.

---

[1] Research supported in part by NSERC Canada

INTRODUCTION

The parity violation experiment at TRIUMF [1] will determine the parity-violating longitudinal analyzing power $A_z = (\sigma^+ - \sigma^-)/(\sigma^+ + \sigma^-)$ in p-p elastic scattering where σ^+ and σ^- are the scattering cross sections for positive and negative helicity. The aim of the experiment is to measure A_z with a precision of $\pm 0.3 \times 10^{-7}$ at 221 MeV; theoretical predictions range from $(-0.2$ to $+2.5)\times 10^{-7}$ at this energy. The measurements are performed in transmission mode, with beam energy and detector geometries selected to ensure that only parity mixing in the $^3P_2$-$^1D_2$ partial wave amplitude contributes to the measured parity violating asymmetry [2]. This amplitude has never been studied experimentally, and the possibility is unique to the energy regime accessible with the TRIUMF cyclotron. In the context of the weak meson exchange model [3], this measurement will provide a direct determination of the weak ρ-nucleon coupling constant $h_\rho^{pp} = (h_\rho^0 + h_\rho^1 + h_\rho^2/\sqrt{6})$.

A major effort to minimize and understand systematic error contributions is required to successfully perform an experiment to this level of precision. The first significant data set for E497 was acquired in February 1997, with a statistical error of $\pm 0.4 \times 10^{-7}$ and most systematic errors at or below the 10^{-7} level. That result represented a major milestone for the experiment, the culmination of many years of effort to reduce both the helicity correlated beam modulations Δx_i and the sensitivities of the apparatus, $\frac{\partial A_z}{\partial x_i}$. Data taking was continued with another three month-long runs acquired during 1998 and 1999.

BEAMLINE AND INSTRUMENTATION

In addition to the measuring apparatus, the optically pumped polarized ion source (OPPIS), cyclotron, and transport beamlines are critical components of the experimental setup, as illustrated in Fig. 1. A 5 μA transversely polarized beam is transported to the cyclotron through an approximately 50 m long injection beamline. The ion source Wien filter is tuned to produce vertical polarization at the entrance to the cyclotron. A 200 nA beam at 75 - 80% vertical polarization is extracted at 221 MeV. Spin precession through a pair of solenoid and dipole magnets results in delivery of a longitudinally polarized beam to the 40 cm liquid hydrogen target, which scatters 4% of the beam. Transverse field parallel plate ion chambers TRIC1 and TRIC2 measure the beam current incident and transmitted through the target. The parity violation signal is derived from the helicity-correlated difference between the beam currents measured by the two TRICs. Upstream of the target are two polarization profile monitors (PPMs) to measure the distributions of transverse polarization $P_y(x)$ and $P_x(y)$ across the beam. Two intensity profile monitors (IPMs) measure the intensity distribution of beam current in x and y and are coupled to a pair of servo magnets which lock the beam path on the optimum

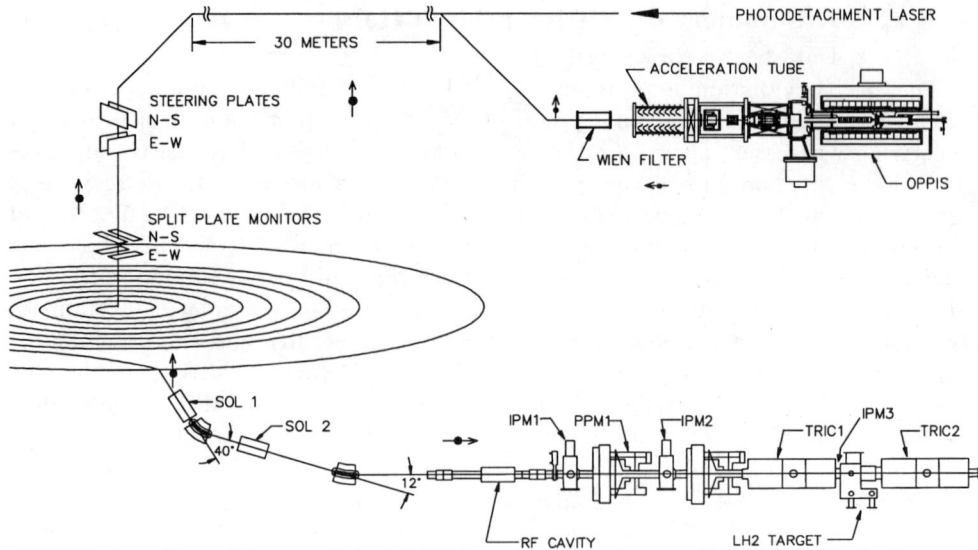

FIGURE 1. General layout of the TRIUMF parity experiment. (OPPIS: Optically Pumped Polarized Ion Source; SOL: Spin Precession Solenoid; IPM: Intensity Profile Monitor; PPM: Polarization Profile Monitor; TRIC: Transverse Field Ionization Chamber)

axis through the equipment. The beamline can be tuned to transport a spin 'up' beam from the cyclotron to the target in a state of either + or - helicity. Half the data are acquired in each of the beamline helicity tune states, which provides a check on systematic error contributions such as beam energy modulation which are not independently measureable at the parity apparatus.

DATA

The full parity data set now consists of four major data runs, taken in February 1997, in December 1997, July 1998, and June 1999. These runs have a raw statistical error of approximately $\pm 0.4 \times 10^{-7}$ each. Systematic error corrections limit the precision of the overall result in all cases. A major effort has been underway to analyze the recent data sets, and preliminary results are summarized in Table 1. The error quoted in the last column is a quadrature sum of statistical and systematic errors, which are roughly equal.

Table 1. Summary of existing parity data sets (preliminary).

Data Set	Raw σ (10^{-7})	Raw χ^2/d.f.	Corrected $A_z \pm \delta A_z (10^{-7})$
Feb. 1997	0.5	9	0.3 ± 0.65
Dec. 1997	0.4	30	0.8 ± 1.8
July 1998	0.3	5	0.9 ± 0.5
June 1999	0.4	6	(in progress)

The longitudinal analyzing power A_z is deduced from each 8-state data cycle by forming the helicity-correlated digitized TRIC difference signal, multiplied by a scale factor appropriate to the electronic and gas gains etc., and divided by the average incident beam current measured by TRIC1. Random beam and cyclotron instabilities are largely responsible for the width of this distribution (the normal 'counting statistics' contribution is negligible), which is minimized by aligning the beam along the symmetry or 'neutral axis' of the apparatus and ensuring that the two TRIC's are as identical as possible in their response to the beam. Helicity-correlated changes in beam properties other than longitudinal polarization give rise to systematic errors which shift the mean of the A_z distribution away from zero; these errors are studied in a series of calibration measurements in which small spin-state-correlated modulations of beam current, energy, position, angle, and transverse polarization are purposely introduced. Interspersed data acquired in the frequent spin-off cycles provide an important zero asymmetry check of the apparatus and electronics.

CORRECTIONS FOR SYSTEMATIC ERRORS

Table 2 summarizes helicity correlated beam properties and the net correction to A_z for the July 1998 data set, which is typical of the TRIUMF data. The only significant corrections to the data are for helicity correlated polarization moments $\langle xP_y \rangle$ and $\langle yP_x \rangle$, as discussed in detail below. The systematic error is due almost entirely to the uncertainty in the sensitivity to transverse polarization moments, which is difficult to determine given the finite resolution of the PPMs; it is in fact statistics dominated. This feature is common to all data sets.

Table 2: Helicity Correlated Beam Properties, July 1998 Run

Property	+ Helicity	- Helicity	ΔA_z (10^{-7})
P_x (%)	-0.12 ± 0.10	-.09 ± 0.01	0.06 ± 0.01
P_y (%)	0.08 ± 0.01	-.16 ± 0.01	-0.03 ± 0.01
yP_x (μm)	10.9 ± 0.4	6.7 ± 0.4	-0.34 ± 0.22
xP_y (μm)	2.0 ± 0.4	3.7 ± 0.4	1.05 ± 0.42
$\Delta I/I$ (10^{-5})	-4.1 ± 0.2	-6.8 ± 0.2	-0.01 ± 0.01
$\Delta\sigma_x$ (μm)	0.03 ± 0.04	0.04 ± 0.04	-0.04 ± 0.24
$\Delta\sigma_y$ (μm)	0.06 ± 0.04	0.04 ± 0.04	-0.03 ± 0.24
Δx (nm)	10 ± 13	12 ± 6	0.00 ± 0.01
Δy (nm)	8 ± 12	15 ± 9	-0.02 ± 0.04

The dominant correction to all data sets is due to the intrinsic moments of transverse polarization $\langle xP_y \rangle$ and $\langle yP_x \rangle$ resulting from a nonuniform distribution of transverse polarization within the beam envelope, as distinct from the corresponding extrinsic polarization moments $\langle x \rangle \langle P_y \rangle$ and $\langle y \rangle \langle P_x \rangle$ which arise when a beam with finite transverse polarization is displaced from the polarization neutral axis.

The polarization neutral axis, which defines $\langle x \rangle = 0$, $\langle y \rangle = 0$, is determined from routine calibration scans of the sensitivity to extrinsic polarization moments, several times per data run. These scans yield a set of coefficients which are used to correct the raw asymmetries for nonzero average polarization components $\langle P_y \rangle$ and $\langle P_x \rangle$. These extrinsic calibration coefficients also determine the optimum beam convergence at the LH$_2$ target to minimize the sensitivity to intrinsic polarization moments $\langle xP_y \rangle$ and $\langle yP_x \rangle$. The sensitivities of the apparatus to both intrinsic and extrinsic polarization moments are identical, so in principle only the extrinsic polarization scans are required to determine the sensitivity coefficients for both effects. However, since the resolution of the PPM's is significant compared to the intrinsic polarization moments that they measure, the corrections for intrinsic polarization moments made in this manner inflate the error bar on A_z significantly.

Corrections can also be made for intrinsic polarization moments based on a regression analysis applied to the parity data. The procedure used is to first correct the experimental asymmetries for all other systematic effects, and then use regression to determine the correlation between the semi-corrected asymmetries and the intrinsic polarization moments. When the data are bundled into samples of 5000 or more event pairs (approx. 1/2 hour acquisition time), the intrinsic moment sensitivities obtained from the regression analysis are in agreement with the sensitivities obtained from the extrinsic coefficients but are determined with better precision.

The corrected A_z values are consistent with both approaches to correct for the polarization moments; the spread in the data is dramatically reduced using the latter technique. Note that the regression analysis approach requires that the beam tune be constant over the entire set of runs, since it assumes a constant net intrinsic first moment sensitivity, while the approach based on extrinsic coefficients alone is beam tune independent. This approach could not be used for the December 1997 data set, which was acquired under a wide range of beam and cyclotron tune conditions as a result of several major failures in the cyclotron RF system that occurred during the run, and hence the uncertainty quoted in Table 1 is unusually large for that data set.

The agreement of data taken in different beamline helicity states and with different Wien filter settings, after being corrected for all measurable systematic errors, can be used to set upper limits on helicity correlated effects that are not measurable at the parity apparatus, such as energy modulation at the eV level in the extracted beam. The excellent consistency of the February 1997 data [6] after correcting for all known effects (χ^2/d.f. = 0.5) indicates that unmeasured systematic errors correlated with both the direction of spin at the ion source and the direction of spin in the cyclotron are consistent with zero at the level of precision in A_z, approximately $\pm 0.7 \times 10^{-7}$. Other data sets show a small shift between + and - helicity beamline tunes; this effect is under further investigation in the analysis.

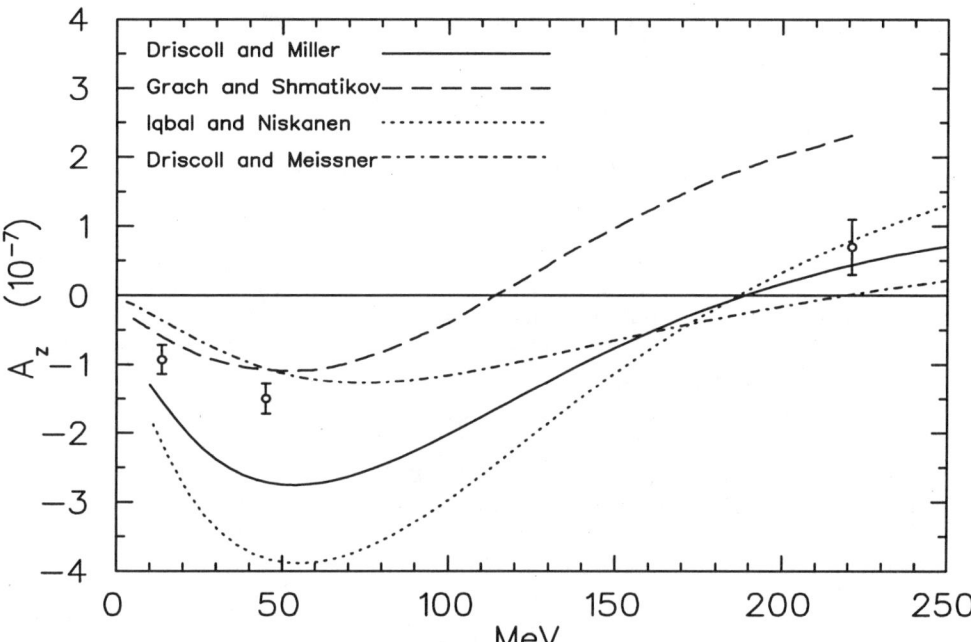

FIGURE 2. Theoretical predictions for A_z in pp scattering, as described in the text. The highest precision existing experimental data are also shown.

SUMMARY AND OUTLOOK

Preliminary analysis has already indicated that the existing full parity data set should be accurate at the $\pm 0.4 \times 10^{-7}$ level or better including corrections for all measureable systematic effects. Analysis is ongoing to determine the best upper limit on the systematic error due to beam energy modulation, which is inferred from control measurements interleaved with parity data taking, as well as from a comparison of corrected data for + and - helicity tunes of the parity beamline. The preliminary result from the February 1997 and July 1998 data is shown in Figure 2, together with recent theoretical predictions, calculated by Driscoll and Miller [7] using DDH predictions for the weak meson-nucleon coupling constants; also shown are quark model calculations by Grach and Shmatikov [8], meson exchange calculations by Iqbal and Niskanen [9] and a calculation by Driscoll and Meissner [10]. The highest precision existing experimental data are also shown [4,5].

The TRIUMF data point supports the meson exchange prediction at the level of this first result. The final result from E497 should be of significantly higher precision, and will serve to determine uniquely for the first time an experimental constraint on the weak meson nucleon coupling constant h_ρ^{pp}.

REFERENCES

1. S.A. Page, J. Birchall, W.T.H. van Oers et al., TRIUMF Experimental Proposal E497 (1987)
2. M. Simonius, Can. J. Phys. **66** 548 (1988)
3. B. Desplanques, J.F. Donoghue and B.R. Holstein, Ann. Phys. (N.Y.) **124** 449 (1980)
4. P.D. Eversheim et al., Phys. Lett. **B256**, 11 (1991); private communication (1994)
5. S. Kistryn et al., Phys. Rev. Lett. **58**, 1616 (1987)
6. A.A. Hamian, Ph.D. thesis: "The Measurement of Parity Violation in Proton-Proton Scattering at 221 MeV", Univ. of Manitoba, 1998.
7. D.E. Driscoll & G.A. Miller, Phys. Rev. **C39**, 1951 (1989), ibid **C40**, 2159 (1989)
8. I. Grach & M. Shmatikov, Phys. Lett. **B316**, 467 (1993)
9. M.J. Iqbal & J. Niskanen, Phys. Rev. **C42**, 1872, (1990)
10. D.E. Driscoll & U.G. Meissner, Phys. Rev. **C41**, 1303 (1990)

Asymmetric Quarks in the Proton

W. Melnitchouk

*Special Research Centre for the Subatomic Structure of Matter,
University of Adelaide, Adelaide 5005, Australia, and
Jefferson Lab, 12000 Jefferson Avenue, Newport News, VA 23606*

Abstract. Asymmetries in the quark momentum distributions in the proton reveal fundamental aspects of strong interaction physics. Differences between $\bar u$ and $\bar d$ quarks in the proton sea provide insight into the dynamics of the pion cloud around the nucleon and the nature of chiral symmetry breaking. Polarized flavor asymmetries allow the effects of pion clouds to be disentangled from those of antisymmetrization. Asymmetries between s and $\bar s$ quark distributions in the nucleon are also predicted from the chiral properties of QCD.

INTRODUCTION

Asymmetries in the proton's spin and flavor quark distributions provide direct information on QCD dynamics of bound systems. Differences between quark or antiquark distributions in the proton sea almost universally signal the presence of phenomena which require understanding of strongly coupled QCD. Their existence testifies to the relevance of long-distance dynamics (which are responsible for confinement) even at large energy and momentum transfers.

Over the past decade a number of high-energy experiments and refined data analyses have forced a re-evaluation of our view of the nucleon in terms of three valence quarks immersed in a sea of perturbatively generated $q\bar q$ pairs and gluons [1]. A classic example of this is the asymmetry of the light quark sea of the proton, dramatically confirmed in recent deep-inelastic and Drell-Yan experiments at CERN [2,3] and Fermilab [4]. Less firmly established, but equally intriguing, are asymmetries between quark and antiquark distributions for heavier flavors, such as s and $\bar s$, which can be measured in deep-inelastic neutrino scattering experiments [5].

LIGHT ANTIQUARK ASYMMETRY

Because gluons in QCD are flavor-blind, the sea generated through the perturbative process $g \to q\bar q$ is symmetric in the quark flavors. Differences can arise due to different quark masses, but because isospin symmetry is such a good symmetry in nature, one expects that the sea of light quarks generated perturbatively would be almost identical, $\bar{u}(x) = \bar{d}(x)$.

It was therefore a surprise to many when measurements by the New Muon Collaboration (NMC) at CERN [2] of the proton and deuteron structure functions suggested a significant excess of $\bar{d}$ over $\bar{u}$ in the proton. Indeed, it heralded a renewed interest in the application of ideas from non-perturbative QCD to deep-inelastic scattering analyses. While the NMC experiment measured the integral of the antiquark difference, more recently the E866 experiment at Fermilab has for the first time mapped out the shape of the $\bar{d}/\bar{u}$ ratio over a large range of x, $0.02 < x < 0.345$.

Specifically, the E866/NuSea Collaboration measured $\mu^+\mu^-$ Drell-Yan pairs produced in pp and pd collisions. If x_1 and x_2 are the light-cone momentum fractions carried by partons in the projectile and target, respectively, then in the limit $x_1 \gg x_2$ the ratio of pd and pp cross sections can be written [4]:

$$\frac{\sigma^{pd}}{2\sigma^{pp}} = \frac{1}{2}\left(1 + \frac{\bar{d}(x_2)}{\bar{u}(x_2)}\right) \frac{4 + d(x_1)/u(x_1)}{4 + d(x_1)/u(x_1) \cdot \bar{d}(x_2)/\bar{u}(x_2)} , \qquad (1)$$

where isospin symmetry has been used to relate quark distributions in the neutron to those in the proton. Corrections for nuclear shadowing in the deuteron [6], which are important at small x, are small in the region covered by this experiment.

The relatively large asymmetry found in these experiments, shown in Fig. 1, implies the presence of non-trivial dynamics in the proton sea which does not have a perturbative QCD origin. The simplest and most obvious source of a non-perturbative asymmetry in the light quark sea is the chiral structure of QCD. From numerous studies in low energy physics, including chiral perturbation theory, pions are known to play a crucial role in the structure and dynamics of the nucleon [7], and there is no reason why the long-range tail of the nucleon should not also play a role at higher energies.

As pointed out by Thomas [8], if the proton's wave function contains an explicit $\pi^+ n$ Fock state component, a deep-inelastic probe scattering from the virtual π^+, which contains a valence $\bar{d}$ quark, will automatically lead to a $\bar{d}$ excess in the proton. In the impulse approximation, deep-inelastic scattering from the πN component of the proton can then be understood in the infinite momentum frame (IMF) [9] as the probability for a pion to be emitted by the proton, folded with the probability of finding the a parton in the pion. For the antiquark asymmetry, this can be written as [10–13]:

$$\bar{d}(x) - \bar{u}(x) = \frac{2}{3}\int_x^1 \frac{dy}{y} f_{\pi N}(y) \, \bar{d}^{\pi^+}(x/y) , \qquad (2)$$

where $\bar{d}^{\pi^+}$ is the (valence) $\bar{d}$ quark distribution in the π^+, and the distribution of pions with a recoiling nucleon (or the $N \to \pi N$ splitting function) is given by [8,10–13]:

$$f_{\pi N}(y) = \frac{3g_{\pi NN}^2}{16\pi^3} \int_0^\infty \frac{d^2\mathbf{k}_T}{(1-y)y} \frac{\mathcal{F}_{\pi N}^2}{(M^2 - s_{\pi N})^2} \left(\frac{k_T^2 + y^2 M^2}{1-y} \right), \qquad (3)$$

where $s_{\pi N}$ is the invariant mass squared of the πN system, $s_{\pi N} = (k_T^2 + m_\pi^2)/y + (k_T^2 + M^2)/(1-y)$, and the πNN vertex form factor, $\mathcal{F}_{\pi N}$, plays the role of an ultraviolet cut-off.

Another contribution known to be important for nucleon structure is that from the $\pi\Delta$ component of the nucleon wave function [7]. For a proton initial state, the dominant Goldstone boson fluctuation is $p \to \pi^-\Delta^{++}$, which leads to an excess of $\bar u$ over $\bar d$. The relative contributions of the πN and $\pi\Delta$ components are determined partly by the πNN and $\pi N\Delta$ vertex form factors. The most direct way to fix these parameters is through a comparison of the axial form factors for the nucleon and for the N–Δ transition [14]. Within the framework of PCAC these form factors are directly related to the corresponding form factors for pion emission or absorption.

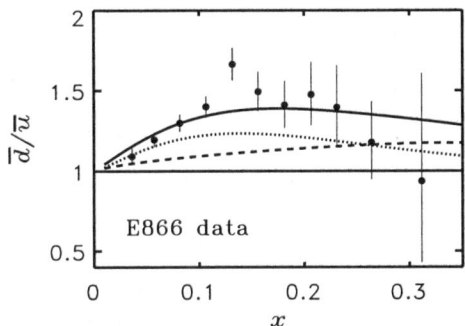

FIGURE 1. Flavor asymmetry of the light antiquark sea, including pion cloud (dashed) and Pauli blocking effects (dotted), and the total (solid) [12].

The resulting $\bar d/\bar u$ ratio, calculated from the pion cloud [12], is shown in Fig. 1 (dashed line). Data on the sum of the $\bar u$ and $\bar d$ (which is dominated by perturbative contributions) have been used to convert the calculated $\bar d - \bar u$ difference to the $\bar d/\bar u$ ratio. The results suggest that with pions alone one can account for about half of the observed asymmetry, leaving room for possible contributions from other mechanisms. One can fine tune the cut-off parameters, or include other, heavier mesons and baryons in the proton cloud [13] to obtain a better fit, however, the fact that an asymmetry exists follows directly from the chiral properties of QCD.

In particular, one can derive the non-analytic behavior of flavor asymmetries in the nucleon sea by considering the chiral ($m_\pi \to 0$) limit of Goldstone boson loops. The leading non-analytic (LNA) behavior of the excess number of $\bar d$ over $\bar u$ quarks in the proton has a chiral behavior typical of loop expansions in chiral effective theories, such as chiral perturbation theory [15]:

$$\int_0^1 dx \left(\bar{d}(x) - \bar{u}(x)\right)_{\text{LNA}} = \frac{2g_A^2}{(4\pi f_\pi)^2} m_\pi^2 \log(m_\pi^2/\mu^2), \qquad (4)$$

where μ is an ultraviolet cut-off mass. This result also generalizes to higher moments, each of which has a non-analytic component, so that the $\bar{d} - \bar{u}$ distribution itself, as a function of x, has a model-independent, LNA component. The presence of non-analytic terms indicates that Goldstone bosons play a role which cannot be canceled by any other physical process (except by chance at a particular value of m_π).

Another mechanism which could also contribute to the $\bar{d} - \bar{u}$ asymmetry is associated with the effects of antisymmetrization of $q\bar{q}$ pairs created inside the nucleon [16,22]. As pointed out originally by Field and Feynman [18], because the valence quark flavors are unequally represented in the proton, the Pauli exclusion principle will affect the likelihood with which $q\bar{q}$ pairs can be created in different flavor channels. Since the proton contains two valence u quarks compared with only one valence d quark, $u\bar{u}$ pair creation will be suppressed relative to $d\bar{d}$ creation. In the ground state of the proton the suppression will be in the ratio $\bar{d} : \bar{u} = 5 : 4$.

The shape of the Pauli contribution to the asymmetry is difficult to predict model-independently, but it is expected to have an x dependence similar to typical sea quark distributions. Phenomenologically, one finds a good fit for $x < 0.2$ if roughly half of the asymmetry is attributed to antisymmetrization [12]. At larger x it is difficult to reproduce the apparent trend in the data towards zero asymmetry, and possibly even an excess of $\bar{u}$ for $x > 0.3$. Unfortunately, the error bars are quite large beyond $x \sim 0.25$, and it is not clear whether new Drell-Yan data will be forthcoming in the near future to clarify this.

A solution may be available, however, through semi-inclusive scattering, in which one tags charged pions produced off protons and neutrons. The HERMES Collaboration has in fact recently measured this ratio [19], although there the rapidly falling cross sections at large x make measurements beyond $x \sim 0.3$ challenging. On the other hand, a high luminosity electron beam such as that available at Jefferson Lab, could allow a precise measurement of the asymmetry beyond $x \sim 0.3$.

The relative roles of pions and Pauli blocking can be further disentangled by measuring the polarized flavor distributions $\Delta \bar{d}$ and $\Delta \bar{u}$ in semi-inclusive scattering [20]. While the antiquark sea due to pions is necessarily unpolarized, the Pauli exclusion principle predicts quite a large asymmetry, $(\Delta \bar{u} - \Delta \bar{d})/(\bar{d} - \bar{u}) = 5/3$ [16]. By fixing the normalization of the Pauli effect from the polarized asymmetries, one could then determine the size of the pion cloud contribution to $\bar{d} - \bar{u}$.

STRANGENESS IN THE NUCLEON

A complication in studying the light quark sea is the fact that non-perturbative features associated with u and d quarks are intrinsically correlated with the valence core of the proton, so that effects of $q\bar{q}$ pairs can be difficult to distinguish from those of antisymmetrization or residual interactions of qurks in the core. The strange sector, on the other hand, where antisymmetrization between sea and valence quarks plays no role, is therefore more likely to provide direct information about the non-perturbative origin of the nucleon sea [21].

Generalizing the formal LNA analysis to the flavor SU(3) sector, one can show that the existence of an asymmetry between s and $\bar{s}$ in the nucleon is predicted on the basis of chiral SU(3) symmetry breaking, which gives rise to a kaon cloud through the fluctuation $N \to KY$, where the hyperon $Y = \Lambda, \Sigma, \cdots$. Since the $\bar{s}$ quark typically comes from the K and the s from the hyperon, the different K and Y masses and momentum distributions naturally lead to differences between s and $\bar{s}$ distributions in the nucleon [22,23]. For the case of the Λ, one has [22,24]:

$$s(x) - \bar{s}(x) = \int_x^1 \frac{dy}{y} \left(f_{YK}(y)\, s^Y(x/y) - f_{KY}(y)\, \bar{s}^K(x/y) \right), \tag{5}$$

where f_{KY} is the analog of the πN splitting function in Eq.(3), and $f_{YK}(y) = f_{KY}(1-y)$. Zero net strangeness in the nucleon implies the vanishing of the lowest moment of $s - \bar{s}$, although, higher moments in general do not vanish. In particular, the LNA behavior of the n-th moment of the $N \to K\Lambda$ splitting function is [15]:

$$\left. f_{K\Lambda}^{(n)} \right|_{\text{LNA}} = \frac{27}{25} \frac{M^2 g_A^2}{(4\pi f_\pi)^2} (M_\Lambda - M)^2 (-1)^n \frac{m_K^{2n+2}}{\Delta M^{2n+4}} \log(m_K^2/\mu^2), \tag{6}$$

where $\Delta M^2 = M_\Lambda^2 - M^2$. Since the LNA terms in the chiral expansion are model-independent, and in general not canceled by other contributions, this result establishes the fact that the process of dynamical symmetry breaking in QCD implies that the s and $\bar{s}$ distributions must have a different dependence on Bjorken x.

The s and $\bar{s}$ distributions can be individually measured in charm production in deep-inelastic ν and $\bar{\nu}$ scattering. Unfortunately, because of relatively large errors the available data from the CCFR collaboration [5] are unable to distinguish between zero asymmetry and a small amount of non-perturbative strangeness as would be expected from kaon loops [24]. More precise data would be valuable in determining the magnitude and even the sign of the asymmetry as a function of x, which depends rather strongly on the dynamics of the KNY interaction [24].

CONCLUSION

We have outlined a number of specific examples where measurement of asymmetries in the sea quark distributions of the nucleon can reveal hitherto hidden details of its non-perturbative structure. Asymmetries are predicted to exist on general grounds from the chiral properties of QCD, by examining the leading non-analytic chiral behavior of quark distributions associated with Goldstone boson loops.

For the $\bar{d}/\bar{u}$ ratio, it is important experimentally to confirm the downward trend of the ratio at large x, which may be feasible through semi-inclusive scattering.

Interestingly, Goldstone boson loops will not give rise to any flavor asymmetries for spin-dependent antiquark distributions, which can only arise from Pauli blocking effects in the proton.

For the strange content of the nucleon, the data from CCFR continue to be reanalyzed in view of possible nuclear shadowing corrections and charm quark effects [25] on s and $\bar{s}$. Together with complementary data on strange form factors currently being gathered in parity-violating elastic electron scattering experiments [26], this will provide valuable information about the role of non-perturbative strangeness in the nucleon.

REFERENCES

1. Watson, A., *Science* **283**, 472 (1999).
2. Amaudraz, P., et al., *Phys. Rev. Lett.* **66**, 2712 (1991).
3. Baldit, A., et al., *Phys. Lett.* **B 332**, 244 (1994).
4. Hawker, E.A., et al., *Phys. Rev. Lett.* **80**, 3715 (1998).
5. Bazarko, A.O., et al., *Zeit. Phys.* **C 65**, 189 (1995).
6. Melnitchouk, W. and Thomas, A.W., *Phys. Rev.* **D 47**, 3783 (1993).
7. Thomas, A.W., *Adv. Nucl. Phys.* **13**, 1 (1984).
8. Thomas, A.W., *Phys. Lett.* **B 126**, 97 (1983).
9. Drell, S.D., Levy, D.J. and Yan, T.M., *Phys. Rev.* **D 1**, 1035 (1970).
10. Sullivan, J.D., *Phys. Rev.* **D 5**, 1732 (1972).
11. Melnitchouk, W. and Thomas, A.W., *Phys. Rev.* **D 47**, 3794 (1993).
12. Melnitchouk, W., Speth, J. and Thomas, A.W., *Phys. Rev.* **D 59**, 014033 (1999).
13. Speth, J. and Thomas, A.W., *Adv. Nucl. Phys.* **24**, 83 (1998).
14. Jones, G.T., et al., *Z. Phys.* **C 43**, 527 (1989).
15. Thomas, A.W., Melnitchouk, W. and Steffens, F.M., hep-ph/0005043.
16. Schreiber, A.W., Signal, A.I. and Thomas, A.W., *Phys. Rev.* **D 44**, 2653 (1991).
17. Steffens, F.M. and Thomas, A.W., *Phys. Rev.* **C 55**, 900 (1997).
18. Field, R.D. and Feynman, R.P., *Phys. Rev.* **D 15**, 2590 (1977).
19. Ackerstaff, K., et al., *Phys. Rev. Lett.* **81**, 5519 (1998).
20. Dressler, B., Goeke, K., Polyakov, M.V. and Weiss, C., *Eur. Phys. J.* **C 14**, 147 (2000); Bhalerao, R.S., hep-ph/0003075.
21. Ji, X. and Tang, J., *Phys. Lett.* **B 362**, 182 (1995).
22. Signal, A.I. and Thomas, A.W., *Phys. Lett.* **B 191**, 206 (1987).
23. Geiger, P. and Isgur, N., *Phys. Rev.* **D 55**, 299 (1997).
24. Melnitchouk, W. and Malheiro, M., *Phys. Rev.* **C 55**, 431 (1997); *Phys. Lett.* **B 451**, 224 (1999).
25. Boros, C., Steffens, F.M., Londergan, J.T. and Thomas, A.W., *Phys. Lett.* **B 468**, 161 (1999).
26. Mueller, B. et al., *Phys. Rev. Lett.* **78**, 3824 (1997); Aniol, K.A. et al., *Phys. Rev. Lett.* **82**, 1096 (1999).

Status Report and preliminary results of the KLOE detector at the DAΦNE φ-factory

Graziano Venanzoni[1]

Institut für Experimentelle Kernphysik, Universität Karlsruhe
Postfach 3640, D-76021 Karlsuhe, Germany
e-mail: Graziano.Venanzoni@iekp.fzk.de

Abstract. The status of the KLOE detector at the Frascati φ-factory DAΦNE and preliminary results obtained with the data collected in 1999 ($\sim 2.4\,pb^{-1}$, corresponding to about 7 millions φ) are presented.

I THE DAΦNE MACHINE

DAΦNE [1] is an e^+e^- collider optimized to work at the φ mass with a target luminosity L= 5×10^{32} cm^{-2}s^{-1}. The strategy used to reach the target luminosity is to obtain the same single bunch luminosity as achieved at VEPP-2M and to increase the number of bunches (up to 120). To suppress multi-bunch instabilities

---

[1] for the KLOE Collaboration: M. Adinolfi, A. Aloisio, F. Ambrosino, A. Andryakov, A. Antonelli, M. Antonelli, F. Anulli, C. Bacci, A. Bankamp, G. Barbiellini, G. Bencivenni, S. Bertolucci, C. Bini, C. Bloise, V. Bocci, F. Bossi, P. Branchini, S.A. Bulychjov, G. Cabibbo, A. Calcaterra, R. Caloi, P. Campana, G. Capon, G. Carboni, A. Cardini, M. Casarsa, G. Cataldi, F. Ceradini, F. Cervelli, F. Cevenini, G. Chiefari, P. Ciambrone, S. Conetti, S. Conticelli, E. De Lucia, G. De Robertis, R. De Sangro, P. De Simone, G. De Zorzi, S. Dell'Agnello, A. Denig, A. Di Domenico, S. Di Falco, A. Doria, E. Drago, V. Elia, O. Erriquez, A. Farilla, G. Felici, A. Ferrari, M. L. Ferrer, G. Finocchiaro, C. Forti, A. Franceschi, P. Franzini, M. L. Gao, C. Gatti, P. Gauzzi, S. Giovannella, V. Golovatyuk, E. Gorini, F. Grancagnolo, W. Grandegger, E. Graziani, P. Guarnaccia, U.v. Hagel, H.G. Han, S.W. Han, X. Huang, M. Incagli, L. Ingrosso, Y. Y. Jiang, W. Kim, W. Kluge, V. Kulikov, F. Lacava, G. Lanfranchi, J. Lee-Franzini, T. Lomtadze, C. Luisi, C. S. Mao, M. Martemianov, M. Matsyuk, W. Mei, L. Merola, R. Messi, S. Miscetti, A. Moalem, S. Moccia, M. Moulson, S. Mueller, F. Murtas, M. Napolitano, A. Nedosekin, M. Panareo, L. Pacciani, P. Pagès, M. Palutan, L. Paoluzi, E. Pasqualucci, L. Passalacqua, M. Passaseo, A. Passeri, V. Patera, E. Petrolo, G. Petrucci, D. Picca, G. Pirozzi, C. Pistillo, M. Pollack, L. Pontecorvo, M. Primavera, F. Ruggieri, P. Santangelo, E. Santovetti, G. Saracino, R. D. Schamberger, C. Schwick, B. Sciascia, A. Sciubba, F. Scuri, I. Sfiligoi, J. Shan, T. Spadaro, S. Spagnolo, E. Spiriti, C. Stanescu, G.L. Tong, L. Tortora, E. Valente, P. Valente, B. Valeriani, G. Venanzoni, S. Veneziano, Y. Wu, Y.G. Xie, P.P. Zhao, Y. Zhou.

the electrons and the positrons circulate in two separate storage rings and collide at two interaction points with horizontal crossing angle of 12.5 mrad to minimize the effect of parasitic collisions. In one of the two interaction regions, equipped with two triplets of low-β quadrupoles, KLOE has been installed.

In March 1999, the first test run with the KLOE experiment was started. The machine was operated in the single and multi bunch mode and an integrated luminosity of ~ 100 nb^{-1} per day was reached, with a peak luminosity of $\sim 3 \times 10^{30}$ cm^{-2}s^{-1}.

At present an effort is underway by the accelerator division in order to increase the peak luminosity to the run I design value of ~ 1–3×10^{31} cm^{-2}s^{-1}.

II PHYSICS AT DAΦNE WITH KLOE

The main interest in a ϕ-factory is that the ϕ decays at rest into a K_S, K_L pair with BR $\sim 34\%$ of the time. The kaons are monochromatic, emitted back to back and they have $\beta \sim 0.2$, corresponding to mean free paths of 343 cm and 0.6 cm for the long and the short state, respectively. This allows for an efficient and clean tagging. In fact the presence of a *V-vertex* inside the beam pipe, with the mass and the momentum expected for the K_S two body decay, is the tag for a K_L in the other hemisphere, *independently from its decay process*. Similarly the identification of the K_L, either by its decay or by its interaction in the calorimeter, tags the presence of the K_S. In such a way a clean pure beam of K_S is obtained, unique with respect to fixed target machines.

The KLOE experiment at DAΦNE covers a wide range of possible physics items, both in *kaon physics* (CP violation studies, rare K_S decays, CPT studies through the semileptonic asymmetry, kaon form factors, regeneration measurements) and also in *non kaon physics*: as radiative ϕ decays ($\phi \to \eta\gamma, \eta'\gamma, f^0\gamma, a^0\gamma, ...$) and hadronic cross section measurement.

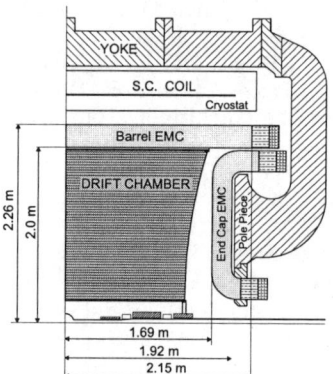

FIGURE 1. Schematic view of the KLOE detector.

III THE KLOE DETECTOR

The KLOE [2] detector (Fig.1) has been designed primarily with the goal of detecting CP violation in K^0 decays. It is a typical e^+e^- detector, of ~ 5 m diameter and ~ 4 m length, composed of a superconducting solenoid that provides a magnetic field of 6 kG, a central drift chamber, and an electromagnetic calorimeter installed inside the coil. The beam pipe at the interaction zone is a 10 cm (~ 16 λ_S) beryllium sphere 0.5 mm thick. This allows us to define a fiducial volume for K_S without complications from regeneration and to minimize multiple scattering and energy loss for charged particles. The two low-beta quadrupole triplets in the interaction region are instrumented to improve the photon acceptance. This helps in the rejection of $K_L \to \pi^0\pi^0\pi^0$.

A Electromagnetic Calorimeter Performance.

To discriminate between $K_L \to \pi^0\pi^0 / K_L \to \pi^0\pi^0\pi^0$ the calorimeter should have full efficiency for γ's in the energy range 20–280 MeV. In addition hermeticity and good energy resolution are required. The calorimeter should provide a fast trigger and have some capability for particle identification. Good timing performance is essential for reconstruction of the K_L neutral vertex with high resolution.

We have chosen to use a lead-scintillator sampling calorimeter [2] consisting of 0.5 mm lead layers in which 1 mm diameter scintillating fibers are embedded. The energy sampling fraction is 13 %.

The calorimeter has been fully calibrated in energy and in time using cosmic rays Bhabha and $\gamma\gamma$ events: an energy resolution $\sigma(E)/E \sim 5.7\%/\sqrt{E(GeV)}$ and a time resolution $\sigma_T \sim 80ps/\sqrt{E(GeV)}$ were measured, in good agreement with test beam results and the design performance.

B Tracking Chamber Performance.

The main requirements for the tracking system are:
- Large radius and uniform sampling to collect as many K_L decays as possible, taking into account that the K_L mean decay length is about 3.4 m.
- Optimized resolution for low momentum tracks by using low Z gas and thin walls to minimize multiple scattering, regeneration and γ conversion.

We have chosen a 2 m radius and 3.4 m long drift chamber [3] with a cell configuration that is almost square, with effective area of 2×2 cm^2 for the first 12 planes and 3×3 cm^2 for the outer 46 planes. The gas mixture is 90% helium and 10% isobutane with a total radiation length of ~ 900 m, including the $\sim 52,000$ wires. The mechanical structure is done entirely of carbon-fiber/epoxy and adds up to ~ 0.1 X_0.

The chamber is fully operational with very few dead channels (< 0.1 %). With the low thresholds of 4 mV the cell efficiency is ~ 99 %. The vertex resolution for Bhabha events is $\sigma_x \sim \sigma_y \sim$ 2mm, $\sigma_z \sim$ 5 mm, in good agreement with expectations. The same resolutions are found for lower momentum tracks as produced in the $\phi \to \rho\pi$ decay. The expected momentum resolution, entirely dominated by multiple scattering in our momentum range, is $\sigma_p/p \sim 0.4\%$, and agrees with the value we find for the high energy Bhabha events.

IV PRELIMINARY RESULTS ON NEUTRAL KAONS

In this section we present preliminary results on a sample of ~ 170,000 K_L K_S events tagged by a $K_S \to \pi^+\pi^-$ decay, which have been used to select all the relevant K_L decay modes and to isolate for the first time in KLOE the charged and neutral $\pi\pi$ CP-violating decays. This sample corresponds to an integrated luminosity of 400 nb^{-1}, acquired between August and October 1999.

A K_L charged decays

The most efficient and clean signature for the presence of a K_L is the observation of a K_S decay into $\pi^+ \pi^-$. This is detected by searching for two oppositely charged tracks belonging to a vertex contained in a 5 cm radius sphere around the interaction point with a cut on the invariant mass and momentum. With these selection criteria a clean mass peak is obtained with a sigma of 1 MeV/c^2, in agreement with the value expected from the drift chamber resolution (Fig. 2, left).

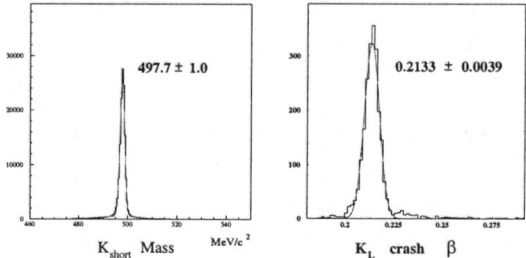

FIGURE 2. *Reconstructed invariant mass for charged pions from $K_S \to \pi^+\pi^-$ decay (left); the value of β of clusters belonging to a K_L interacting in the calorimeter (right).*

Given the $K_S \to \pi^+\pi^-$ decay vertex inside the beam pipe, the vertex of the K_L is searched in a 5 degree cone around the K_L direction, which is obtained by reversing the K_S momentum and correcting for the ϕ meson boost (see Fig. 3, left). However, not all the K_L mesons decay in the drift chamber. About 44% reach the calorimeter: these events can be easily identified by measuring the velocity

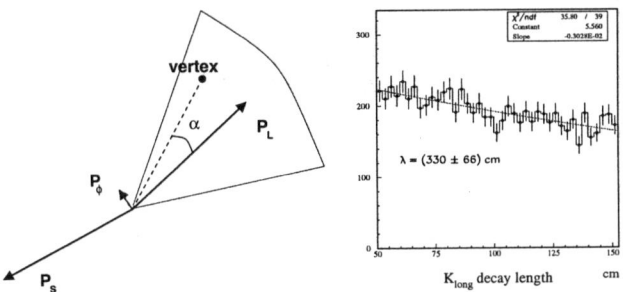

FIGURE 3. K_L charged decay selection (left); radial distribution of K_L charged vertices in a restricted region of the apparatus (right).

of the K_L using the arrival time of the cluster, corresponding to $\beta = 0.21$ (see Fig.2). For K_L decaying in the fiducial volume of the drift chamber, an efficiency of $(41.7 \pm 0.3)\%$ has been found (Monte Carlo predicts 40.6%). The K_L decay length has been fitted in a restricted region of the apparatus (50cm $< r <$ 150cm), where the reconstruction efficiency is reasonably uniform, giving $\lambda = (330 \pm 66)$ cm, in good agreement with expectation (Fig. 3, right).

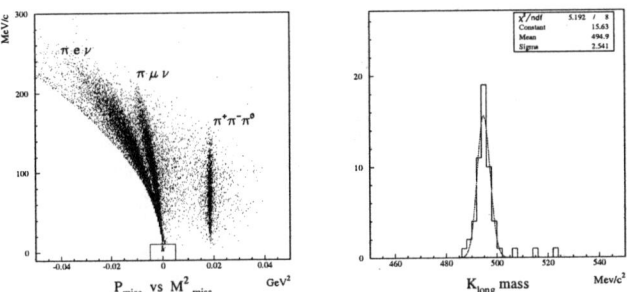

FIGURE 4. K_L decay modes. The box at small p_{miss} vs M^2_{miss} values defines the CP violation region of interest (left); mass peak for selected events in the box (right).

The different K_L charged decay modes ($\pi^+\pi^-\pi^0$, $\pi\mu\nu$, $\pi e\nu$, $\pi^+\pi^-$) can be separated by using the missing momentum (p_{miss}) and the missing invariant mass (M_{miss}) at the vertex, computed with the K_L momentum and the one measured for the charged tracks. The missing momentum is exactly zero for $\pi^+\pi^-$ decays and tends to be small for soft neutrino emission in semileptonic decays. The missing invariant mass is zero for $\pi^+\pi^-$ and semileptonic decays, and equal to m_{π^0} in the $\pi^+\pi^-\pi^0$ case. Fig. 4 (left) shows the correlation between p_{miss} and M^2_{miss} for the selected sample of charged vertices. The region of interest for the $\pi^+\pi^-$ signal is the one at small values of p_{miss} and M_{miss}: $p_{\text{miss}} < 10$ MeV/c and $M_{\text{miss}} < 70$ MeV/c^2. This selection shows a clear peak on the invariant mass of the two pions of 494.9 ± 2.5 MeV/c^2 (Fig. 4, right). The shift with respect to the nominal value of the K_L mass is due to a small systematic effect in the drift chamber calibration.

B K_L neutral decays

The $K_L \to 3\pi^0$ decay is unambiguously tagged by neutral secondary vertices with 6 associated clusters in the calorimeter. The distribution of the total energy of the 6 photons is shown in Fig. 5 (left): it is measured with the expected 10% resolution, although it peaks at (497 ± 44) MeV, pointing to a $\sim 3\%$ residual miscalibration, which still has to be corrected.

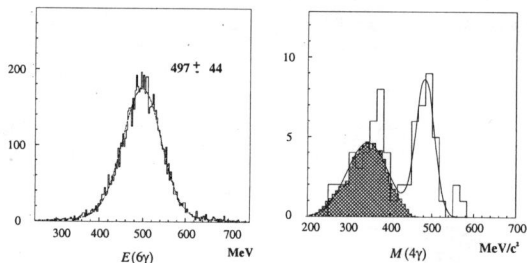

FIGURE 5. *Total energy of the neutral vertex for $K_L \to 3\pi^0$ decay (left); invariant mass for neutral K_L vertices with four γ's (right).*

Having such a clear signature for the $3\pi^0$ final state, a first attempt has been made to isolate the $2\pi^0$ CP-violating signal. For this purpose neutral vertices are selected with four clusters in the calorimeter. A loose collinearity cut is then applied between the K_L flight path and the momentum of the 4γ system. The final sample is selected requiring the presence of 2 properly reconstructed π^0's: this selection gives an invariant mass of the 4 photons as shown in Fig. 5 (right). Two peaks are clearly visible: the one centered close to K_L nominal value corresponds to the genuine $K_L \to \pi^0\pi^0$ decay, while the second one at lower values is due to $3\pi^0$ background events, in which two photons out of six have been lost. This is confirmed by the Monte Carlo distribution for such events, as shown in Fig.5 (hatched).

Fully neutral decay	Charged/neutral decay
$\phi \to \eta\gamma \to \gamma\gamma\gamma$	
$\phi \to \eta\gamma \to \pi^0\pi^0\pi^0\gamma$	$\phi \to \eta\gamma \to \pi^+\pi^-\pi^0\gamma$
$\phi \to \eta'\gamma \to \pi^0\pi^0\eta\gamma \to \pi^0\pi^0\gamma\gamma\gamma$	$\phi \to \eta'\gamma \to \pi^+\pi^-\eta\gamma \to \pi^+\pi^-\gamma\gamma\gamma$
$\phi \to f_0\gamma \to \pi^0\pi^0\gamma$	$\phi \to f_0\gamma \to \pi^+\pi^-\gamma$
Signature	
"n" photons no charged tracks in the DC	"n" photons 1 vertex with 2 charged tracks in the DC

TABLE 1. Final states and signature of ϕ radiative decays.

V PRELIMINARY RESULTS ON NON-KAON PHYSICS

In this section we report on the preliminary results for: (a.1) $\phi \to \eta\gamma$, (a.2) $\phi \to \eta'\gamma$, (b) $\phi \to \pi^0\pi^0\gamma$ and (c) $e^+e^- \to \pi^+\pi^-\gamma$ channels. (a.1) and (a.2) decays are important to constrain the gluonic content of the η' meson and to reveal possible $\eta - \eta'$ mixing, (b) decays allow clarification of the puzzle around the nature of the $f_0(980)$ meson, and (c) is important for two reasons: 1) to study the $f_0(980)$ meson; 2) to measure the hadronic cross section $\sigma(e^+e^- \to \pi^+\pi^-)$ with a tagged photon in the initial state. The following results were obtained with $1.8\,pb^{-1}$ of analysed data.

A ϕ radiative decay selection and signatures

The decays $\phi \to \eta\gamma, \eta'\gamma, \pi\pi\gamma$ [2] can be divided into two general classes according to the content of charged and neutral particles in their final state, as shown in Tab. 1. For these events a common selection procedure can be applied.

B $\phi \to \eta\gamma \to 3\gamma$ decay

Due to its clean signature, this decay can be used for normalization. An example is provided by the measurement of the ϕ resonance cross section:

$$\sigma(e^+e^- \to \phi) = \sigma(e^+e^- \to \phi \to \eta\gamma)/BR(\phi \to \eta\gamma) = N_{\eta\gamma}/(L \cdot \epsilon_{ana} \cdot BR(\phi \to \eta\gamma))$$

where L is the machine luminosity (measured with Bhabha events) and ϵ_{ana} is the overall $\phi \to \eta\gamma$ efficiency. By taking $BR(\phi \to \eta\gamma) = (0.49 \pm 0.02)\%$ from PDG [4],

---

[2] Different from the other channels, which are essentially produced by the ϕ decay, the final state $\pi^+\pi^-\gamma$ arises mostly from the annihilation reaction: $e^+e^- \to \gamma\gamma^* \to \gamma\pi^+\pi^-$

and an efficiency $\epsilon_{ana} = 0.643$ from Monte Carlo, we estimate:
$$\sigma(e^+e^- \to \phi)_{KLOE} = (3.19 \pm 0.02(stat.) \pm 0.26(syst.))\mu b$$
This results, though preliminary, agree well with the value obtained by the CMD-2 collaboration at the collider VEPP-2M [5]:
$$\sigma(e^+e^- \to \phi)_{CMD-2} = (3.114 \pm 0.034(stat.) \pm 0.048(syst.))\mu b$$

C $\phi \to \eta'\gamma/\eta\gamma : R_\phi$ measurement

R_ϕ is defined as the ratio of the branching ratios of the two decays $\phi \to \eta'\gamma$ and $\phi \to \eta\gamma$. It can be evaluted as:
$$R_\phi = \frac{BR(\phi \to \eta'\gamma)}{BR(\phi \to \eta\gamma)} = \frac{N^{\eta'}_{obs}\epsilon_{\eta\gamma}}{N^{\eta}_{obs}\epsilon_{\eta'\gamma}} R_{BR}$$

where $N^{\eta,\eta'}_{obs}$ are the number of observed decays, ϵ are the efficiencies and R_{BR} is related to the BR's of η and η' into the final state. By using for R_{BR} the values from PDG [4]:
$$R^{+-}_{BR} = 1.33 \pm 0.05 \quad (\pi^+\pi^-\gamma\gamma\gamma)$$
$$R^{00}_{BR} = 3.92 \pm 0.25 \quad (7\gamma)$$

we obtain the following values:
$$R^{+-}_\phi = (6.6 \pm 1.7(stat.) \pm 0.4(syst.)) \cdot 10^{-3} \quad \pi^+\pi^-\gamma\gamma\gamma \quad (15 \text{ events})$$
$$R^{00}_\phi = (6.3^{+2.8}_{-0.6}(stat.) \pm 1.0(syst.)) \cdot 10^{-3} \quad 7\gamma \quad (4 \text{ events})$$

The 4 fully neutral events give the first observation of the decay chain $\phi \to \eta'\gamma \to 7\gamma$.

D $\phi \to \eta'\gamma$ Branching Ratio and $\eta' - \eta$ mixing angle

By using R^{+-}_ϕ, BR$(\phi \to \eta'\gamma)$ was computed:
$$BR(\phi \to \eta'\gamma) = (8.4 \pm 2.1(stat.) \pm 0.5(syst.)) \cdot 10^{-5}$$
in agreement with the value obtained at the collider VEPP-2M [6,7]:
$$BR(\phi \to \eta'\gamma) = (8.2^{+2.1}_{-1.9}) \cdot 10^{-5} \quad (CMD-2 \; coll.)$$
$$BR(\phi \to \eta'\gamma) = (6.7^{+3.4}_{-2.9}) \cdot 10^{-5} \quad (SND \; coll.)$$

This value translates into a preliminary value for the pseudoscalar octet mixing angle of:
$$\theta_P = (-17.9^{+3.2°}_{-4.2°}(stat.) \pm 0.9°(syst.))$$

E $\phi \to f_0\gamma \to \pi^0\pi^0\gamma$

Tab. 2 shows the signal to background ratio expected for the main background channels to the $\pi^0\pi^0\gamma$ final state. These backgrounds can be considerably reduced

Background	S/B
$e^+e^- \to \omega\pi^0 \to \pi^0\pi^0\gamma$	0.52
$\phi \to \rho\pi^0 \to \pi^0\pi^0\gamma$	3.0
$\phi \to a_0\gamma \to \eta\pi^0\gamma \to 5\gamma$	3.1
$\phi \to \eta\gamma \to \pi^0\pi^0\pi^0\gamma$	0.02

TABLE 2. Signal to background for $\phi \to \pi^0\pi^0\gamma$ decay

by a cut on the photon direction in the dipion reference system; after a kinematical fit procedure 307 events are identified as $\phi \to \pi^0\pi^0\gamma$ with $M_{\pi^0\pi^0} > 700 MeV$, which allow to give a preliminary branching ratio:

$$BR(\phi \to \pi^0\pi^0\gamma) = (0.8 \pm 0.1(stat.) \pm 0.08(syst.)) \cdot 10^{-4}$$

This value should be compared with the one obtained at the collider VEPP-2M [8,7]:

$$BR(\phi \to \pi^0\pi^0\gamma) = (1.06 \pm 0.09(stat.) \pm 0.06(syst.)) \cdot 10^{-4} \ (CMD-2 \ coll.)$$
$$BR(\phi \to \pi^0\pi^0\gamma) = (1.14 \pm 0.1(stat.) \pm 0.12(syst.)) \cdot 10^{-4} \ (SND \ coll.)$$

F $\phi \to \pi^+\pi^-\gamma$ decay and $e^+e^- \to \pi^+\pi^-\gamma$ annihilation reaction

Different from the neutral decay, the decay into charged pions can be enhanced or suppressed due to the interference with the pure final state radiation process (FSR). An additional background is provided by the process of initial state radiation (ISR).

Fig. 6 shows a comparison between data (points) and Monte Carlo simulation (solid line) of pure QED processes [3] for $60° < \theta_\gamma < 120°$, $50 < \theta_\pi < 130°$. As can be seen the overall agreement is excellent. Preliminary study at $Q^2_{\pi^+\pi^-} > 0.8\,GeV^2$ points to destructive interference between $\phi \to \pi^+\pi^-\gamma$ decay and the final state radiation, as expected also from VEPP-2M data.

---

[3] Initial and final state radiation processes.

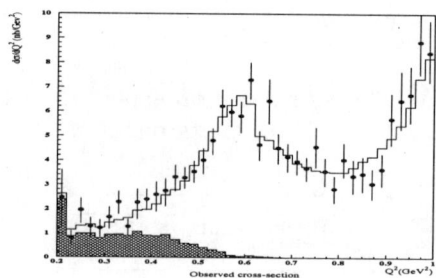

FIGURE 6. Invariant mass spectrum of the two pions in the reaction $e^+e^- \to \pi^+\pi^-\gamma$: points are data, the solid line is a Monte Carlo simulation of pure QED processes. Hatched region is for $\pi^+\pi^-\pi^0$ events. The enhancement around $0.6\,GeV^2$ is due to the return to the ρ resonance.

VI CONCLUSION

The status and the preliminary results of the KLOE detector have been presented. These results, though statistically limited, show a good agreement with Monte Carlo simulations and allow to extract first numbers for the ϕ radiative decays.

VII ACKNOWLEDGEMENTS

I would like to thank the organizers of $Symm$2000, particulary Prof. Anthony W. Thomas, for inviting me to such an interesting conference and a beautiful place. This work is supported by the EU Network EuroDAΦNE, contract FMRX-CT98-0169.

REFERENCES

1. G. Vignola, *Workshop on Physics and detector for DAΦNE 95*, ed. R. Baldini et al., SIS-Pubblicazioni, Frascati, 1995.
2. The KLOE Collaboration, The KLOE detector, Technical Proposal, LNF-93/002 (1993).
3. The KLOE Collaboration, The KLOE Central Drift Chamber, LNF-94/028 (1994).
4. Particle Data Group, G. Caso et al., Eur. Phys. J. **C 3** (1998) 1.
5. R. R. Akhmetshin et. al., Phys. Lett. **B 466** (1999) 385.
6. R. R. Akhmetshin et. al., Phys. Lett. **B 473** (2000) 337-342.
7. The SND Collaboration, *Experiments at VEPP-2M with SND detector*, hep-ex/9809013.
8. R. R. Akhmetshin et. al., Phys. Lett. **B 462** (1999) 380.

Neutrino Oscillation Results from CERN

Kevin Varvell*

*School of Physics, The University of Sydney, NSW 2006, Australia

Abstract. The study of neutrinos has formed an important part of the CERN experimental program for many years. During the 1990s, the CHORUS and NOMAD experiments were performed in order to search for neutrino oscillations governed by neutrino masses in the range of cosmological interest. In this presentation, the two experiments are described, and the current status of their oscillation searches reviewed.

INTRODUCTION

In the Standard Model of Particle Physics, the three flavours of neutrino form the neutral partners of the corresponding charged leptons, arranged into doublets of weak isospin.

$$\begin{pmatrix} \nu_e \\ e^- \end{pmatrix} \begin{pmatrix} \nu_\mu \\ \mu^- \end{pmatrix} \begin{pmatrix} \nu_\tau \\ \tau^- \end{pmatrix} + \text{associated antiparticles}$$

The absence of directly measured non-zero neutrino masses has led to Standard Model neutrinos being ascribed to be identically massless. If however, neutrinos do in fact possess a mass, and the different mass eigenstates are not degenerate, then neutrino oscillations will occur, pointing to physics beyond the Standard Model.

Neutrino oscillations are often described under the assumption that only two flavours are involved, the so-called "two flavour approximation". This simplification suffices for many situations. In this picture, oscillations occur because the weak eigenstates $\mid \nu_\alpha \rangle$, $\mid \nu_\beta \rangle$, which govern how neutrinos are produced and interact, differ from the mass eigenstates $\mid \nu_1 \rangle$, $\mid \nu_2 \rangle$, governing how neutrinos propagate in free space. The two bases are related by a mixing matrix.

$$\begin{pmatrix} \mid \nu_\alpha \rangle \\ \mid \nu_\beta \rangle \end{pmatrix} = \begin{pmatrix} \cos\theta & \sin\theta \\ -\sin\theta & \cos\theta \end{pmatrix} \begin{pmatrix} \mid \nu_1 \rangle \\ \mid \nu_2 \rangle \end{pmatrix}$$

The transition probability for a neutrino born as flavour $\mid \nu_\alpha \rangle$ to be detected as flavour $\mid \nu_\beta \rangle$ after having travelled a distance L is given by

$$P(\nu_\alpha \to \nu_\beta) = \sin^2 2\theta \cdot \sin^2\left(1.27 \frac{\Delta m^2}{eV^2} \frac{L}{E} \frac{GeV}{km}\right)$$

In this equation, θ is the mixing angle, L the distance the neutrino has travelled (in km), E the neutrino energy (in GeV), and $\Delta m^2 = |m_2^2 - m_1^2|$ the difference in the squares of the masses (in eV^2) of the two mass eigenstates involved.

If $\Delta m^2 \sim$ few eV^2, neutrinos become cosmologically interesting [1]. For example, if there exists a mass hierarchy $m_3 \gg m_2 \sim m_1$, with $| \nu_\tau \rangle \approx | \nu_3 \rangle$, then this would make ν_τ a significant component of dark matter.

The possibility of observing neutrino oscillations under this scenario and contributing to the understanding of the dark matter problem prompted two experiments to be proposed for the West Area at CERN. CHORUS [2] was approved in 1991 and took data from 1994 to 1997, while NOMAD [3] was approved in 1992 and took data from 1995 to 1998. Both experiments had the goal of exploring the oscillation mode $\nu_\mu \to \nu_\tau$. The source detector distances involved ($\sim$ 0.6 km) and neutrino energies available from the West Area beam (mean $E_\nu \sim 20 - 25$ GeV) place the sensitivity of these experiments in the relevant range of Δm^2.

THE EXPERIMENTS

The CHORUS and NOMAD experiments share the same experimental goal, namely to search for $\nu_\mu \to \nu_\tau$ oscillations by detecting the interaction

$$\nu_\tau + N \to \tau^- + X$$

through decay of the τ^- (here N is a nucleon and X an inclusive hadronic final state). They are, however, complementary in the detection methods used, as will be outlined below following a description of the neutrino beam.

The CERN West Area Neutrino Facility

At the CERN West Area Neutrino Facility (WANF), neutrinos are obtained mainly from the decays of secondary pions and kaons which themselves are produced when a proton beam of momentum 450GeV/c extracted from the Super Proton Synchrotron (SPS) strikes a beryllium target. Some focussing and sign-selection of the short-lived charged mesons prior to decay is achieved using a system of magnetic lenses known as the horn and reflector. A 290m evacuated decay tunnel is followed by earth and iron shielding in order to absorb the majority of muons and hadrons accompanying the neutrinos. CHORUS and NOMAD are located approximately 830m from the beryllium target, with the mean flight distance of the neutrinos from the decay tunnel to the detectors approximately 600m.

The majority of neutrinos comprising the WANF neutrino beam are of flavour ν_μ. Figure 1 gives the predicted spectra for the species ν_μ, $\bar{\nu}_\mu$, ν_e and $\bar{\nu}_e$ as a function of neutrino energy, averaged in this case over an area of 2.6m × 2.6m at the position of the NOMAD detector and normalised to 10^9 protons incident on the beryllium target. The spectra for the CHORUS experiment are similar, differing in detail however due to the smaller fiducial area of the emulsion target and more upstream position of the detector.

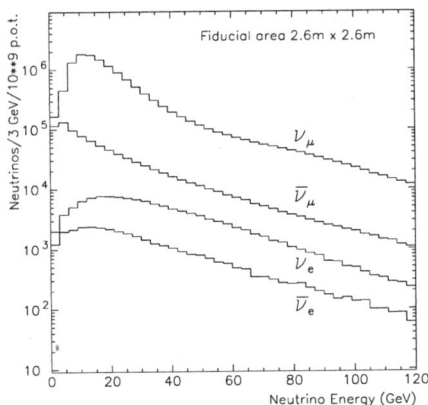

FIGURE 1. The predicted spectra as a function of neutrino energy of the neutrino species comprising the WANF neutrino beam, at the position of the NOMAD detector. Reprinted from *Nucl. Instr. and Meth.* **A404**, J. Altegoer et al., "The NOMAD experiment at the CERN SPS", page 101, Copyright (1998), with permission from Elsevier Science.

The mean energies of each flavour of neutrino in the beam, along with the relative abundance with respect to the dominant ν_μ species, are given in Table 1, again for the NOMAD detector. Also given are the same quantities for charged current (CC) interactions in the detector, taking into account the different neutrino and antineutrino interaction cross sections, and the differing radial distributions of each component.

TABLE 1. Composition of WANF neutrino beam at NOMAD.

Type	Flux		CC interactions	
	$\langle E_\nu \rangle$ [GeV]	rel. abundance	$\langle E_\nu \rangle$ [GeV]	rel. abundance
ν_μ	23.5	1.0	43.8	1.0
$\bar\nu_\mu$	19.2	0.061	42.8	0.025
ν_e	37.1	0.0094	58.3	0.015
$\bar\nu_e$	31.3	0.0025	54.5	0.0016

Crucial to the ability of CHORUS and NOMAD to reach the necessary level of sensitivity is the absence of a significant ν_τ component in the neutrino beam, which would if present mask the oscillation signal. ν_τ in the beam arise primarily from decay of charmed $D_s^\pm$ mesons, with the relative abundance of ν_τ with respect to ν_μ having been calculated to be at the level of a few parts in 10^6 [4]. This is not significant for the present experiments.

The CHORUS Experiment

CHORUS employs a hybrid detector based on a nuclear emulsion target (770 kg in total) to give a high resolution reconstruction of the emerging tracks from the interaction vertex, followed by electronic detectors to provide tracking, calorimetry and muon identification. The detector is fully described in [5]. A side view of the detector is reproduced in Figure 2.

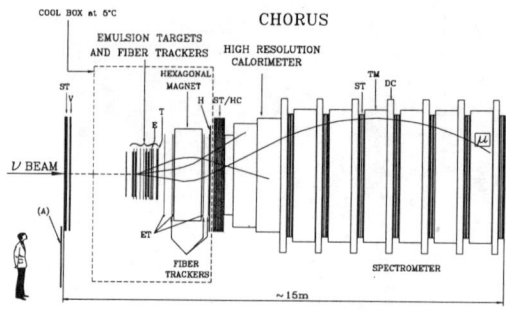

FIGURE 2. Side view of the CHORUS detector. Reprinted from *Nucl. Instr. and Meth.* **A401**, E. Eskut et al., "The CHORUS experiment to search for $\nu_\mu \to \nu_\tau$ oscillation", page 10, Copyright (1997), with permission from Elsevier Science.

Despite the short decay path of the τ^- (typically $\sim$ 1mm), the excellent spatial resolution of the nuclear emulsion ($\sim 1\mu$m) enables direct observation of the decay vertex of the τ^-. The method employed is illustrated in Figure 3.

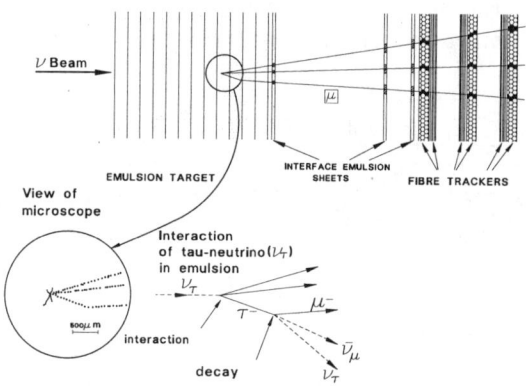

FIGURE 3. The CHORUS detection method. Reprinted from *Nucl. Instr. and Meth.* **A401**, E. Eskut et al., "The CHORUS experiment to search for $\nu_\mu \to \nu_\tau$ oscillation", page 10, Copyright (1997), with permission from Elsevier Science.

Fully automatic scanning microscopes are used for vertex location, based on predictions provided by extrapolation of tracks detected in the tracking detectors back into the emulsion. In scanning performed to date, two event samples with vertices in the emulsion are searched:

- events with an identified μ^- the so-called "1 μ sample". This sample will contain candidates for the decay $\tau^- \to \mu^- \bar{\nu}_\mu \nu_\tau$

- events with no identified μ, the so-called "0 μ sample". In this sample the following decays would be expected: $\tau^- \to h^- \nu_\tau +$ anything, $\tau^- \to e^- \bar{\nu}_e \nu_\tau$, and $\tau^- \to \mu^- \bar{\nu}_\mu \nu_\tau$ where the μ^- is not identified.

Detection efficiencies are evaluated using neutrino induced charm (dimuon) interactions. The ability of CHORUS to detect τ^- decay kinks is very well illustrated by the event reproduced in Figure 4, which is interpreted as the diffractive production of a D_s^{+*} meson which decays to $D_s^+ \gamma$, followed by the semileptonic decay of the D_s^+ to a τ^+ and ν_τ and of the τ^+ to a μ^+ and two neutrinos. Further details on this event can be found in [6].

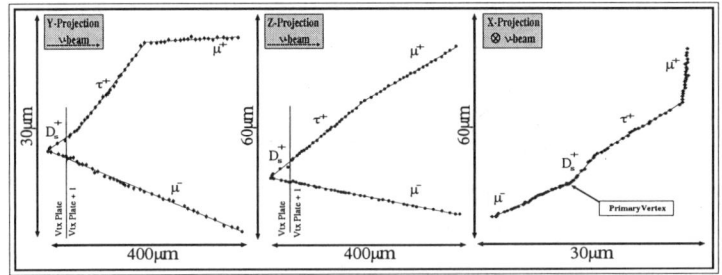

FIGURE 4. Detail of diffractive D_s^{+*} event in the CHORUS detector. Reprinted from *Phys. Lett.* **B435**, P. Annis et al., "Observation of neutrino induced diffractive D_s^{*+} production and subsequent decay $D_s^{*+} \to D_s^+ \to \tau^+ \to \mu^+$", page 461, Copyright (1998), with permission from Elsevier Science.

The ability to detect the decay kink of the τ^- makes CHORUS a very low background experiment. The principal background to the 1 μ sample comes from antineutrino induced charm,

$$\bar{\nu}_\mu(\bar{\nu}_e) + \text{nucleon} \to \mu^+(e^+) + D^- + \text{anything}$$

where the D^- decays to a μ^- and neutral system, and the μ^+ or e^+ is not identified. For the 0 μ sample, the main background comes from so-called "white kinks", which are interactions of one of the outgoing hadrons from a neutrino interaction producing a kink but no visible recoil of the struck nucleon. The total background from all sources has been estimated to be at a level of less than one event over the entire data sample.

A summary of the current status of the CHORUS scanning effort is given in Table 2. No τ^- candidate has been observed to date. If N_τ is defined as the

TABLE 2. Current status of scanning in CHORUS.

		1994 - 1997	
Protons on target		5.06×10^{19}	
Emulsion triggers		2.305×10^6	
ν_μ CC events in target		840,000	
$1\mu^-$ events	to be scanned	458,601	
	scanned so far	350,834	($\sim 75\%$)
	located in emulsion	133,616	
0μ events	to be scanned	116,049	
	scanned so far	67,107	($\sim 60\%$)
	located in emulsion	7,514*	

* 1994 and 1995 data only included to date

number of τ^- events expected to be observed if all ν_μ were to oscillate to ν_τ i.e. $P(\nu_\mu \to \nu_\tau) \equiv 1$, then, based on the above dataset, Table 3 gives the values of N_τ presented at this conference [7] for the decay channels of the τ^- to which CHORUS is sensitive.

TABLE 3. Predicted event numbers in CHORUS for $P(\nu_\mu \to \nu_\tau) \equiv 1$ [7].

Channel	N_τ
$\tau^- \to \mu^-$	5130
$\tau^- \to h^-$	998
$\tau^- \to e^-$	100
$\tau^- \to \mu^-$ (μ^- unidentified)	51

The absence of observed events and the values of N_τ allow a limit to be set on the probability of $\nu_\mu \to \nu_\tau$ oscillations of [7]

$$P(\nu_\mu \to \nu_\tau) < 3.8 \times 10^{-4}, \quad 90\% \text{C.L.}$$

This limit calculation takes into account an estimated 17% systematic uncertainty in the experimental sensitivity, following the prescription described in [8].

It is possible to reinterpret the absence of a signal in terms of $\nu_e \to \nu_\tau$ oscillations, assuming that the less abundant ν_e beam component rather than the ν_μ component is oscillating. CHORUS have previously quoted such a limit based on a slightly smaller data set than that used for the above ν_μ limit [9]

$$P(\nu_e \to \nu_\tau) < 3.0 \times 10^{-2}, \quad 90\% \text{C.L.}$$

Exclusion plots based on these results will be shown below. CHORUS is presently undertaking a major reanalysis of their emulsion data, employing improved scanning techniques and a volume scan around located primary vertices, which will lead to a significant improvement in the sensitivity of the experiment [7].

The NOMAD Experiment

A side view of the NOMAD detector is given in Figure 5. The core of the detector is a set of large-area drift chambers located within a 0.4T transverse magnetic field, providing both target and tracking. These are followed by a transition radiation detector, preshower, electromagnetic calorimeter, hadron calorimeter and muon chambers. Details of the detector operation and performance can be found in [10].

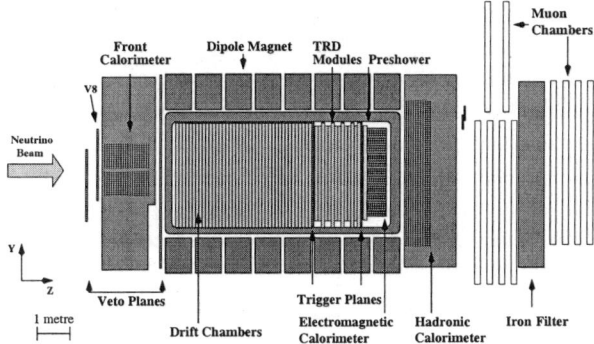

FIGURE 5. Side view of the NOMAD detector. Reprinted from *Nucl. Instr. and Meth.* **A404**, J. Altegoer et al., "The NOMAD experiment at the CERN SPS", page 98, Copyright (1998), with permission from Elsevier Science.

The vertex resolution achievable in the large volume instrumented drift chamber target of NOMAD precludes direct observation of the kink associated with the decay of a τ^-. By contrast with CHORUS, the approach here is to detect the presence of ν_τ induced events through kinematical methods. The following decay channels of the τ^- are explored, constituting in total 82.8% of the total branching ratio.

$$\tau^- \to e^- \bar{\nu}_e \nu_\tau$$
$$\tau^- \to \pi^- \pi^- \pi^+ \nu_\tau + n\pi^0$$
$$\tau^- \to h^- \nu_\tau + n\pi^0$$
$$\tau^- \to \rho^- \nu_\tau \to \pi^- \pi^0 \nu_\tau$$

The predicted data sample for NOMAD over the lifetime of the experiment, assuming 100% detection efficiency, is summarised in Table 4.

Full details of the kinematical methods employed by NOMAD may be found in [11]. A much simplified discussion will be presented here for the $\tau^- \to e^-$ channel, in order to illustrate the main components of the analysis.

Figure 6 gives an example of the key point of the kinematical method, the different event characteristics expected for signal and background events. In the case

TABLE 4. Predicted event numbers in NOMAD.

NOMAD data set	($\times 10^6$ events)
$\nu_\mu + N \to \mu^- + X$	1.354
$\bar{\nu}_\mu + N \to \mu^+ + X$	0.034
$\nu_e + N \to e^- + X$	0.020
$\bar{\nu}_e + N \to e^+ + X$	0.002
$\nu(\bar{\nu}) + N \to \nu(\bar{\nu}) + X$	0.485

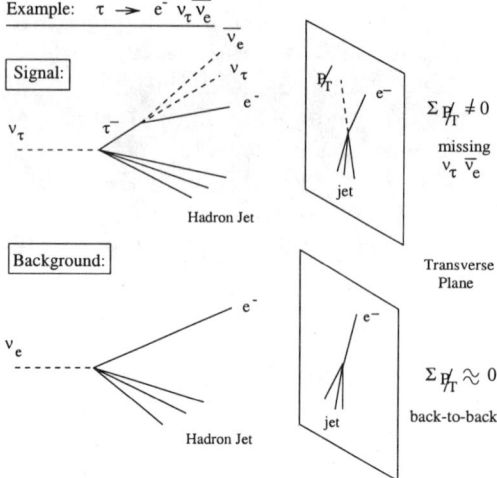

FIGURE 6. Signal/background discrimination in NOMAD.

of a signal event where the τ^- decays to an electron and two neutrinos, missing transverse momentum is expected in the plane perpendicular to the neutrino line of flight, due to the undetected neutrinos. A background charged current event should be well balanced in transverse momentum up to the resolution of the detector, fermi motion and nuclear effects, since the electron will recoil against the produced jet of hadrons. Thus missing transverse momentum should be a good discriminating variable.

In practice, in order to achieve the necessary degree of signal/background discrimination, a more sophisticated analysis than this simple example is needed. Approximate likelihood functions are employed, constructed from a number of different kinematical variables describing the events.

NOMAD observes that discrepancies exist between data and Monte Carlo in the properties of the hadronic jet (notably missing transverse momentum) which would lead to errors in estimation of efficiencies for signal and background. For this reason these efficiencies are obtained from the data themselves through a *data simulator* technique, in which the identified muon in a measured ν_μ charged current event is replaced by a simulated lepton of appropriate type to obtain a fake sample of events. More details may be found in [11].

In order to avoid biases in event selection which may result from studying data events in the signal region, and hence incorrect background estimates, a "blind analysis" strategy is employed. This involves defining a "box" around the signal region, in which data may not be examined. Selection criteria and cuts are defined using Monte Carlo events, with a view to obtaining the best *sensitivity* for the measurement. Sensitivity is defined as the average upper limit on oscillations that would be obtained from an ensemble of experiments with the same expected background and no true signal. Once criteria are in place, background predictions are compared with data outside the box region. Additionally, the procedure can be checked by searching for τ^+ events in data, since due to the heavily suppressed antineutrino components in the beam, no oscillation signal would be expected here given the sensitivity of the experiment. Once the integrity of the search procedure has been demonstrated, the "box" is opened and examined for a signal.

The main background to the $\tau^- \to e^-$ channel comes from ν_e charged current (CC) interactions. Figure 7 shows the predicted log likelihood ratios for signal (unshaded histogram, for 100% oscillation probability, reduced by a factor of 20) and ν_e CC background (shaded histogram). The "box" in this variable was chosen to be the region to the right of log likelihood ratio of 3.0. In practice, a second likelihood ratio to discriminate against neutral current background is used in addition, making the actual "box" two dimensional. Furthermore, the box is itself subdivided into several independent bins, in order to enhance the sensitivity of the search by exploiting the predicted shape of the event distribution in the box. The points with error bars show the distribution in log likelihood ratio for the complete NOMAD data set, to be compared with the predictions for signal and background.

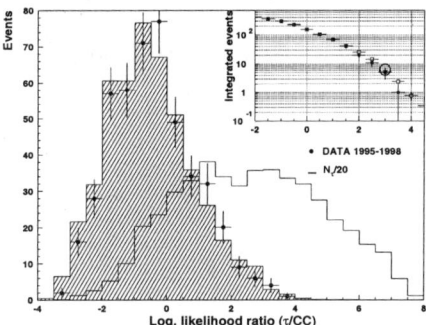

FIGURE 7. Event distributions as a function of log likelihood ratio for signal to charged current background for the τ^- search in NOMAD [12].

The inset in Figure 7 shows a comparison between Monte Carlo and data for the integrated event distribution, i.e. the distribution obtained when, for each value of log likelihood ratio, the total number of events with log likelihood ratio greater than that value is calculated. The circle indicates the edge of the box in this variable. A similar comparison for the τ^+ search, not reproduced here, shows good agreement across the whole range with, as expected, no τ^+ signal evident.

Clearly there is no evidence for an excess to be interpreted as a signal for τ^- events. Table 5 summarises the results for all of the channels searched for in the NOMAD analysis [12]. N_τ is as defined for the CHORUS analysis, and ϵ the selection efficiency for the given channel. Note that for each channel the sample has been divided into a "Deep Inelastic" (DIS) and "Low Multiplicity" (LM) sample, divided according to whether the total momentum of the hadron system in an event is greater than or less than 1.5 GeV/c. This division improves the sensitivity of the search by separately exploiting the simpler topologies and higher selection efficiencies obtainable in the low multiplicity sample.

TABLE 5. Summary of NOMAD results for the τ^- search [12].

Channel	est. backg.	N_{obs}	N_τ	$\epsilon(\%)$
$\tau^- \to e^- \nu_\tau \bar{\nu}_e$ DIS	$5.3^{+0.8}_{-0.5}$	5	4110	3.6
$\tau^- \to e^- \nu_\tau \bar{\nu}_e$ LM	5.4 ± 0.9	6	859	6.3
$\tau^- \to \rho^- (n\pi^0) \nu_\tau$ DIS	9.5 ± 2.5	7	3307	1.04
$\tau^- \to h^- (n\pi^0) \nu_\tau$ DIS	6.8 ± 2.1	5	2022	0.70
$\tau^- \to h^-/\rho^- (n\pi^0) \nu_\tau$ DIS	$0.0^{+0.74}_{-0.0}$	1	210	0.07
$\tau^- \to \rho^- (n\pi^0) \nu_\tau$ LM	5.2 ± 1.8	7	458	2.0
$\tau^- \to h^- (n\pi^0) \nu_\tau$ LM	6.7 ± 2.3	5	357	0.84
$\tau^- \to \pi^- \pi^- \pi^+ (n\pi^0) \nu_\tau$ DIS	9.6 ± 2.4	9	1820	1.9
$\tau^- \to \pi^- \pi^- \pi^+ (n\pi^0) \nu_\tau$ LM	3.5 ± 1.2	5	288	1.8

Agreement between predicted and observed event numbers is observed across all channels considered. From this, NOMAD calculates a limit on the oscillation probability for $\nu_\mu \to \nu_\tau$ of [12]

$$P(\nu_\mu \to \nu_\tau) < 2.2 \times 10^{-4}, \quad 90\% \text{C.L.}$$

with a sensitivity, as defined earlier, of 4.3×10^{-4}. For $\nu_e \to \nu_\tau$ oscillations, this becomes

$$P(\nu_e \to \nu_\tau) < 1.1 \times 10^{-2}, \quad 90\% \text{C.L.}$$

with sensitivity 2.0×10^{-2}. Note that the prescription used by NOMAD in calculating these limits, that of Feldman and Cousins [13], differs slightly from that used by CHORUS.

THE CHORUS AND NOMAD RESULTS IN CONTEXT

In Figure 8, the exclusion regions set by CHORUS and NOMAD based on their current results are given. Also shown are the limits set by other experiments [14]. It can be seen that the stated goal of the experiments, to improve the (then) present

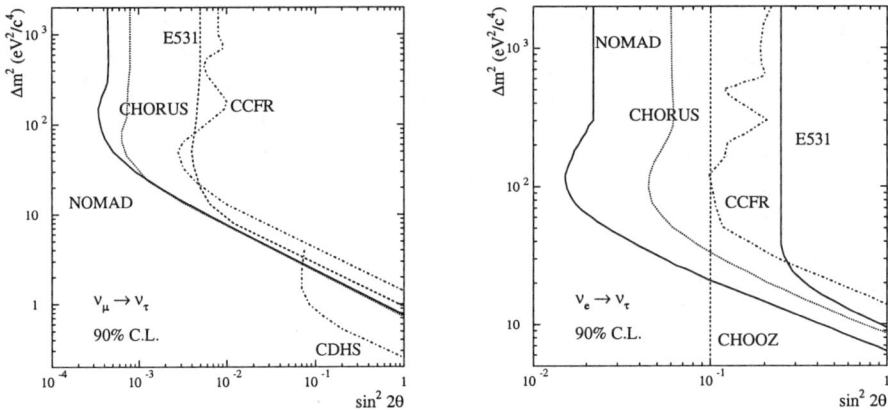

FIGURE 8. Current limits on $\nu_\mu \to \nu_\tau$ (left) and $\nu_e \to \nu_\tau$ (right) oscillations for the CHORUS and NOMAD experiments. Results from previous experiments are also shown [14].

oscillation limits by an order of magnitude if oscillations were not observed, is close to being realised.

Figure 9 shows schematically where the CHORUS and NOMAD results fit in to the overall picture as far as neutrino oscillations are concerned.

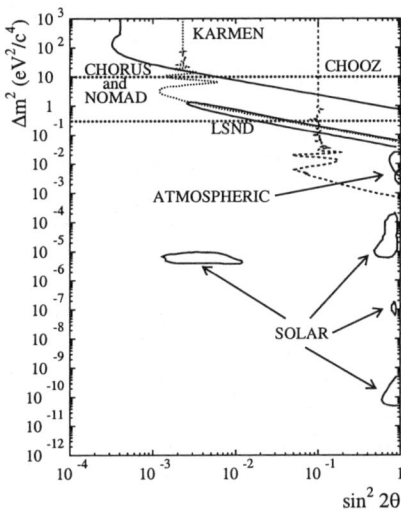

FIGURE 9. Schematic of neutrino oscillation landscape. The horizontal dotted lines indicate the approximate bounds in Δm^2 of the cosmologically interesting region.

THE FUTURE OF NEUTRINO PHYSICS AT CERN

The CHORUS and NOMAD experiments, decommisioned in 1999, constitute the last of several generations of neutrino experiments performed in the West Area at CERN, stretching back two decades. The focus of neutrino physics at CERN now turns to the program to send a neutrino beam to the Gran Sasso Laboratory in Italy, over a distance of 730 km. Approved by CERN Council in December 1999, this program, which requires construction of a new neutrino beam line, plans to commence operation in May 2005. The aim, as with CHORUS and NOMAD, is to search for ν_τ appearance, now with a much longer baseline in order to probe to significantly lower mass-squared differences; the region of interest suggested by the atmospheric neutrino results [15]. At the time of writing, two experiments are proposed for Gran Sasso, OPERA and ICANOE.

REFERENCES

1. Harari H., *Phys. Lett.* **B216**, 413 (1989).
2. CHORUS Collaboration. *CERN-SPSC/90-42*, (1990).
3. NOMAD Collaboration. *CERN-SPSLC/91-21*, (1991). *CERN-SPSLC/91-48*, (1991). *CERN-SPSLC/91-53*, (1991).
4. Van de Vyver B., *Nucl. Instr. and Meth.* **A385**, 91 (1997). Gonzalez-Garcia, M.C. and Gomez-Cadenas J.J., *Phys. Rev.* **D55**, 1297 (1997).
5. Eskut E. et al., *Nucl. Instr. and Meth.* **A401**, 7 (1997).
6. Annis P. et al., *Phys. Lett.* **B435**, 458 (1998).
7. Kodama K., *These proceedings*.
8. Cousins R.D., and Highland, V.L., *Nucl. Instr. and Meth.* **A320**, 331 (1992).
9. CHORUS Collaboration. *hep-ex/9907015*. Contributed paper to the XIX International Symposium on Lepton and Photon Interactions at High Energies, Stanford, August 9-14 1999.
10. Altegoer J. et al., *Nucl. Instr. and Meth.* **A404**, 96 (1998).
11. Altegoer J., et al., *Phys. Lett.* **B431**, 219 (1998). Astier P. et al., *Phys. Lett.* **B453**, 169 (1999).
12. NOMAD Collaboration, De Santo A., Talk given at the XXXV Recontres des Moriond, Les Arcs, March 11-18 2000. Astier P. et al., Accepted for publication in *Phys. Lett. B*.
13. Feldman G.J., and Cousins R.D., *Phys. Rev.* **D57**, 3873 (1998).
14. E531 Collaboration, Ushida N. et al., *Phys. Rev. Lett.* **57**, 2897 (1986).
    CCFR Collaboration, McFarland K.S. et al., *Phys. Rev.* **D75**, 3993 (1995). Naples D. et al., *Phys. Rev.* **D59**, 031101 (1998).
    CHOOZ Collaboration, Apollonio M. et al., *Phys. Lett.* **B466**, 415 (1999).
    CDHS Collaboration, Dydak F. et al., *Phys. Lett.* **B134**, 281 (1984).
15. Kajita, T., *These proceedings*.

Rare Kaon Decays

Toshio Numao

TRIUMF, 4004 Wesbrook Mall, Vancouver, B.C., Canada V6T 2A3
email: toshio@triumf.ca

Abstract. Rare kaon decays via Flavor Changing Neutral Currents are discussed in the context of the CKM unitarity triangle with a particular interest in the rare kaon decays $K^+ \to \pi^+ \nu \bar{\nu}$ and $K_L^0 \to \pi^0 \nu \bar{\nu}$. New results and the status of these experiments are reported.

INTRODUCTION

The study of rare kaon decays has a glorious history. At the level of 10^{-1} in branching ratio, parity violation was first "discovered" as the θ-τ puzzle [1], CP violation was discovered at the level of 10^{-3} [2], and the GIM mechanism [3] was suggested by the strong suppression of the decay $K_L^0 \to \mu^+ \mu^-$, or the absence of the Flavor Changing Neutral Currents (FCNC)— the decay mode was subsequently found at the level of 10^{-8}. Further searches for rare kaon decays at lower branching ratio undoubtably involve rich physics and may yet offer another surprise— e.g. they are likely to elucidate the origin of CP violation, and provide some clues for physics beyond the Standard Model (SM). In this talk, the decay modes via FCNC with the expected branching ratios of less than 10^{-9} are discussed. Since exotic decays violating the lepton flavor conservation law have been discussed by the previous speaker [4], the focus of this talk is on the rare kaon decays which are allowed in the SM.

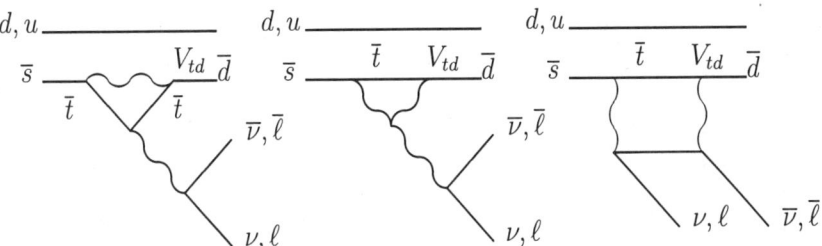

FIGURE 1. Typical Feynman diagrams.

Typical leading Feynman diagrams of these decays are shown in Fig. 1, where d and u at the top lines are for K^0 and K^+ decays, respectively, and ℓ indicates the electron or the muon. Because of the GIM mechanism, the lowest order diagrams come from second order weak interactions with a u-type quark in the loop diagrams, in which the top-quark contributions dominate because of the mass. This makes these decay modes very sensitive to V_{td}, the least constrained coupling-constant of the top and down quarks in the Cabibbo-Kobayashi-Maskawa (CKM) matrix. Due to one of the unitarity conditions of the CKM matrix $V_{ud}V_{ub}^* + V_{cd}V_{cb}^* + V_{td}V_{tb}^* = 0$ and with the approximation $V_{ud} \sim V_{tb} \sim 1$, V_{td} and V_{ub}^* form two sides of a unitarity triangle as shown in Fig. 2. The height of the triangle $Im(V_{td})$ is an indication of "direct" CP violation, or CP violation through the decay amplitude. For the decay $K^+ \to \pi^+ \nu \bar{\nu}$, the branching ratio is roughly proportional to $|V_{td}|^2$ and for $K_L^0 \to \pi^0 \nu \bar{\nu}$ to $Im^2(V_{td})$. These measurements alone can determine the unitarity triangle, and when they are combined with measurements from B-decays it will provide over-constrained information that is sensitive to a presence of supersymmetry and other physics beyond the SM [5,6].

DECAYS $K_L^0 \to \pi^0 \ell^+ \ell^-$ AND $K_L^0 \to \ell^+ \ell^-$

If the final states include a charged lepton pair, there are additional contributions from diagrams with virtual photons that usually prevent clear interpretations of measurements [7].

The decays $K_L^0 \to \pi^0 \ell^+ \ell^-$ have a "direct" CP violating component, which is expected to occur, if it were the only component, at a branching ratio of $\sim 5 \times 10^{-12}$. There are two other contributions, however, at the same level to this process; one arises from the mixing of the CP even state in K_L^0, and the other from two-virtual-photon intermediate states that conserve CP. At present, there are theoretical am-

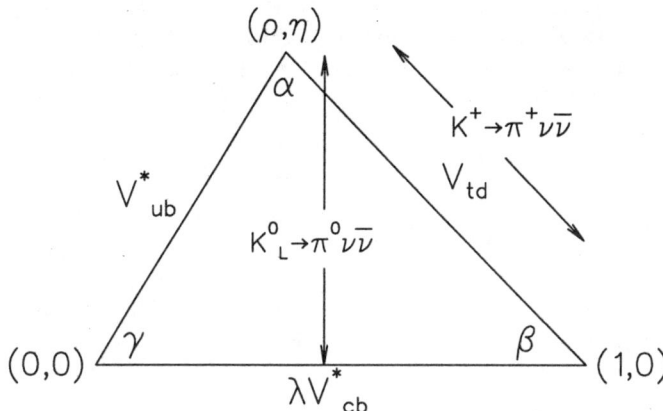

FIGURE 2. Unitarity triangle. The coordinates of the vertices are for the rescaled triangle[8].

biguities in the estimations of these contributions, which may eventually be sorted out by measurements of $K_S^0 \to \pi^0 \ell^+ \ell^-$ and other radiative decays. To make the matter a bit more complicated, there is a physical background coming from the radiative decay $K_L^0 \to \ell^+ \ell^- \gamma\gamma$, which has the identical event topology. With the tightest cuts, the background level is estimated to be still at a 10^{-11} level for $\ell = e$ [9]. The situation is similar in the case of $\ell = \mu$. The present upper limits (90 % c.l.) are $B(K_L^0 \to \pi^0 e^+ e^-) \leq 5.6 \times 10^{-10}$ [10] and $B(K_L^0 \to \pi^0 \mu^+ \mu^-) \leq 3.8 \times 10^{-10}$ [11].

The decay $K_L^0 \to e^+ e^-$ is very similar to $K_L^0 \to \mu^+ \mu^-$, which is sensitive to the ρ parameter [8], but it is further suppressed by the helicity mechanism. This decay is sensitive to pseudo-scalar interactions coming from physics beyond the SM. Four events from the decay $K_L^0 \to e^+ e^-$ have been reported recently which correspond to a branching ratio, $B(K_L^0 \to e^- e^+) = 8.7^{+5.7}_{-4.1} \times 10^{-12}$ [12], being consistent with the unitarity bound expected from the long-distance contributions.

DECAY $K^+ \to \pi^+ \nu \bar{\nu}$

In the SM calculation of the decay $K^+ \to \pi^+ \nu \bar{\nu}$, the dominant contribution comes from second order loop diagrams with a virtual top quark (Fig. 1). The hadronic matrix element can be extracted from the decay $K \to \pi^0 e^+ \nu$ and theoretical uncertainties in the calculation due to long distance contributions and other effects are small [5]. The signature of the decay $K^+ \to \pi^+ \nu \bar{\nu}$ is a single pion with no other observable decay-products. Definitive observation of this signal requires that all possible backgrounds are suppressed well below the signal level. Major background sources are: a muon from the copious decay $K^+ \to \mu^+ \nu$ ($K_{\mu 2}$) which is misidentified as a pion, a pion from the decay $K^+ \to \pi^+ \pi^0$ ($K_{\pi 2}$) when two photons from the π^0 decay are unobserved, a beam pion scattered by the target into the detector, and charge exchange reactions of K^+'s which result in decays $K_L^0 \to \pi^+ \ell^- \bar{\nu}$, where ℓ (e or μ) is undetected. In order to suppress the backgrounds, good particle identification, efficient photon veto, and detection of incoming particles are essential.

The E787 experiment at Brookhaven National Laboratory (BNL) as shown in Fig. 3 is designed to effectively distinguish these backgrounds from the signal. Kaons of about 700 MeV/c at a rate of $(4-7) \times 10^6$ per 1.6-s spill are detected and identified by a Čerenkov counter and hodoscopes, degraded by BeO and stopped in an active target, primarily consisting of 413 5-mm square scintillating fibers. The momentum (P), kinetic energy (E) and range (R) of decay products are measured using the target, a central drift chamber, 21 layers of 1.9-cm thick plastic scintillator (Range Stack) and two layers of straw chambers, all contained in a 1-T magnetic field. The $\pi^+ \to \mu^+ \to e^+$ decay sequence of the decay products in the Range Stack scintillator is observed by 500-MHz transient digitizers for particle identification. Photons are detected by a 4π-sr calorimeter consisting of a 14-radiation-length-thick lead/scintillator barrel detector, 13.4-radiation-length-thick end caps of CsI crystals, and a 3.5-radiation-length-thick lead-glass Čerenkov counter which also works as an active beam degrader.

The E787 experiment reported an observation of one clean event at a branching ratio $B(K^+ \to \pi^+ \nu \bar{\nu}) = 4.2^{+9.7}_{-3.5} \times 10^{-10}$ using the data sample taken in 1995 [13]. Since then, additional data samples taken in 1996 and 1997, together with those taken in 1995, have been analyzed with improved algorithms, which have resulted in less non-Gaussian tails in P, R and E measurements, and a $\sim$30 % higher acceptance than that in Ref. [13]. Also, in the 1996–7 runs, lowering the incident K^+ beam momentum resulted in a larger fraction of kaons stopping in the target, which reduced accidental hits originated from nuclear reactions in the beam degrader. The higher proton intensity at the production target compensated the kaon yield loss at lower momentum.

In order to avoid a possible bias in the analysis, the signal region is kept untouched until the background estimations as well as optimization of acceptance have been done. In the background study, the data sample after applying all the final cuts except two orthogonal (uncorrelated) cut groups to be studied, e.g. the kinematical cuts and those related to the observation of the decay sequence $\pi^+ \to \mu^+ \to e^+$ in the stopping counter for the estimation of the $K^+ \to \mu^+ \nu$ background, is used to obtain the suppression factor for each cut group. The correlation between the two cut groups, which may invalidate the method, is studied by varying the cuts being studied or by enhancing certain types of background events.

The total background for the entire 1995–1997 exposure with the final analysis cuts is estimated to be 0.08 ± 0.02 events. The acceptance for $K^+ \to \pi^+ \nu \bar{\nu}$, $A = 0.0021 \pm 0.0001(stat) \pm 0.0002(syst)$ is calculated based on data and Monte Carlo calculations. The largest uncertainty comes from the uncertainty in pion-nucleus interaction. The measurement of the branching ratio for $K^+ \to \pi^+ \pi^0$

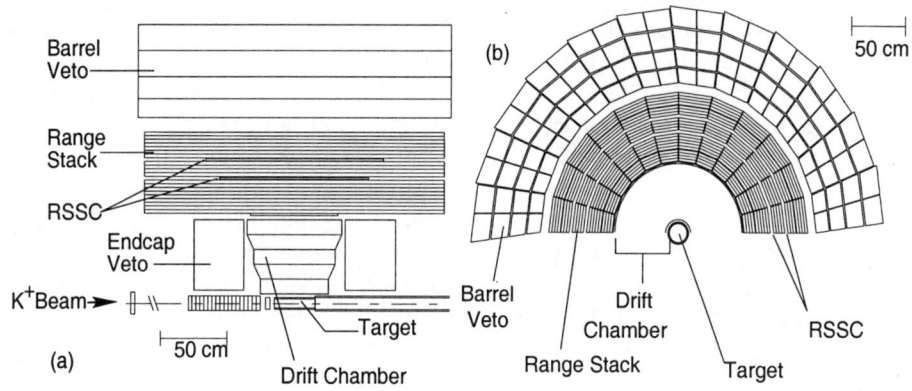

FIGURE 3. Upper half of the BNL E787/949 detector: (a) side view and (b) end view.

within a few % of Ref. [14] confirms the acceptance calculation. Analysis of the full data sample as shown in Fig. 4 has yielded only the single event previously reported. Based on the acceptance A and the total exposure of $N_{K^+} = 3.2 \times 10^{12}$ kaons, the new branching ratio is $B(K^+ \to \pi^+ \nu \bar{\nu}) = 1.5^{+3.4}_{-1.2} \times 10^{-10}$ [15]. This provides a constraint, $0.002 < |V_{td}| < 0.04$.

The goal of a new experiment E949 at BNL is to improve the sensitivity by an order of magnitude. Since the AGS in the RHIC era will be used for two hours a day to feed heavy ions into the RHIC ring, the rest of 22 hours can be used for the high energy program. The operation is expected to provide a more stable and longer running period. Exploiting a higher beam intensity, the incident K^+ momentum can be further lowered to reduce accidental coincidence. The photon veto capability will be improved by additional lead/scintillator layers in the barrel region and by additional active degrader in the beam region. In the region below the $K^+ \to \pi^+ \pi^0$ peak at 205 MeV/c, where nuclear interactions result in a large momentum-tail, the additional photon veto capability may suppress the background by more than an order of magnitude as low as to the signal level, doubling the phase space in the search.

Decay $K_L^0 \to \pi^0 \nu \bar{\nu}$

The decay $K_L^0 \to \pi^0 \nu \bar{\nu}$ violates the CP conservation law through decay amplitude. This process is expected to occur at $B(K_L^0 \to \pi^0 \nu \bar{\nu}) \sim 3 \times 10^{-11}$ [5]. The contribution from CP mixing in K_L^0 is expected to be around 10^{-15} [16]. Since no

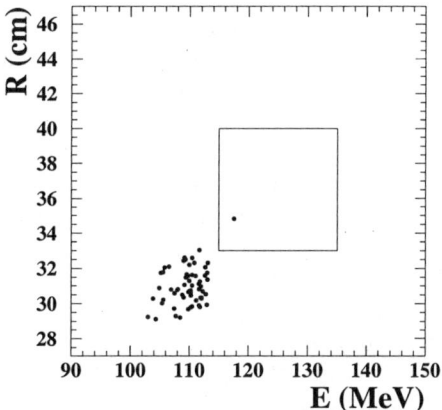

FIGURE 4. Range (cm in plastic scintillator) vs Kinetic energy plot of the 95–97 data. The concentration at E=107 MeV is due to $K_{\pi 2}$ events.

charged leptons are involved in this decay, the process is free from virtual photon contributions and clean for the study of the origin of CP violation. The theoretical ambiguity is only ~ 1 % except those in the CKM matrix elements. Conversely, the observation of this decay mode uniquely determines $Im(V_{td})$ or η in the Wolfenstein parametrization. The goal of the experiment is to determine $Im(V_{td})$ with a 10–15 % accuracy, which corresponds to a single event sensitivity of 10^{-12}. The present upper limit of this decay is 5.7×10^{-7} [17].

The decay $K_L^0 \to \pi^0 \nu \bar{\nu}$ is a three-body decay that involves only neutral particles. The signature of this decay is two photons from the π^0 and no other activity in the detector. In this decay mode, available kinematical parameters are very limited; relatively easy ones to measure are the positions and energies of γ-rays. Also, the direction and position of the K_L^0 beam can be limited by tightly collimating the beam at a cost of beam intensity. The decay vertex can be reconstructed from these two constraints with the assumption of the pion mass. The KEK experiment [18] and the FNAL approach [19] are classified in this category.

The BNL experiment E926 [20] attempts to measure more kinematic values; the directions of γ-rays and the momentum of the K_L^0. Measurements of γ-ray directions allow full reconstruction of the π^0 kinematics without the assumption of the π^0 mass, providing more constraints with redundancy, which is necessary to suppress the background to the level well below the signal. The time-of-flight (TOF) of a K_L^0 between the production target and the decay vertex can be measured for low momentum K_L^0's if the incident proton beam is bunched. This measurement allows calculation of the missing mass and kinematical reconstruction in the center of mass system, which is effective to eliminate backgrounds from two-body decays. The major decay modes of K_L^0 are $K_L^0 \to \pi^0 \pi \pi$ and $K_L^0 \to \pi^\pm \ell^\mp \nu$. Suppression of most backgrounds is achieved by high-efficiency hermetic photon and charged-particle detector system surrounding the decay volume, and kinematical constraints.

In the BNL E926 experiment, a low energy K_L^0 beam with an average momentum around 650 MeV/c is produced by irradiating a target with a 24-GeV proton beam from the AGS and extracted at $40°$ with respect to the incident beam. The proton beam is bunched to form a ≤ 200 ps wide bucket at a rate of 25 MHz. About 16 % of K_L^0's decay in the 4-m long decay volume, which is evacuated to a level of 10^{-7} Torr to suppress the background from neutron-induced π^0 production. The decay region is surrounded by a charged particle veto system and a photon veto system of 18-radiation-length-thick lead/scintillator sandwiches in the barrel region. The two photons from the π^0 decay are converted into pairs of a positron and an electron in a 2-radiation-length preradiator next to the vacuum region for the measurement of the directions of the γ-rays. The preradiator consists of sandwiches of 2-mm thick scintillator, a copper plate as a mechanical support and radiator, and a tracking chamber. This is followed by an 18-radiation-length calorimeter for the measurement of γ-ray energies and for vetoing additional photons.

The estimates of sensitivity for $K_L^0 \to \pi^0 \nu \bar{\nu}$ are tightly coupled to the cuts required for background suppression, particularly for the $K_L^0 \to \pi^0 \pi^0$ and $K_L^0 \to \pi^0 \pi^+ \pi^-$ backgrounds. An acceptance of ~ 0.015 % for the case S/N=0.5 comes from the combination of factors; 0.58 for the fiducial region and usable kaon momentum region, 0.33 for the solid angle, 0.5 for the efficiency of the preradiator, and the remaining factor for the π^0 mass cut and other cuts. Assuming three years of running with 6×10^6 kaons per 2-s beam spill (50 % duty factor), the expected number of events is 65.

The toughest background is the CP violating $K_L^0 \to \pi^0\pi^0$ decay when two photons are undetected. Backgrounds from $K_L^0 \to \pi^0\pi^0$ arise when two photons are detected in the forward detector and the other two are undetected anywhere. These backgrounds can be classified into two categories; even pairing when two photons come from the same π^0, and odd pairing when two photons come from different π^0's. Events in the even pairing category form a two-body-decay peak in the momentum spectrum of π^0 in the CM system, while the two γ-rays in the odd pairing category do not reproduce the π^0 mass. The above requirements and photon veto essentially suppress $K_L^0 \to \pi^0\pi^0$ backgrounds to 0.2 of the signal level. Similarly, $K_L^0 \to \pi^0\pi^+\pi^-$ backgrounds can be suppressed to 0.1 of the expected signal. Backgrounds from other K_L^0 decay modes are estimated to be less than 0.1 of the signal. Because of the large angle extraction, the cross sections for producing Λ's are small and they completely decay before reaching the decay volume. Backgrounds could arise from Λ's produced by halo neutrons and K_L^0's, but they are estimated to be negligible. Neutrons with $P_n \geq 800$ MeV/c can react on the remaining atoms in the vacuum region to produce π^0's. This background level is again estimated to be 0.01. Accidental backgrounds are caused by beam halo neutrons and γ-rays which create a π^0 signal in the preradiator. The background level is estimated to be 0.02 of the expected signal.

The BNL E926 was proposed in 1997 but the group is still waiting for full funding. In the present scenario, the detector construction is expected to start in 2002. The

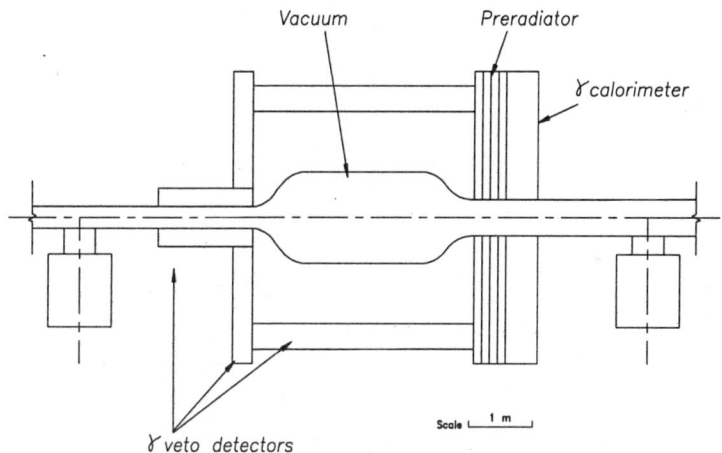

FIGURE 5. A typical detector system for $K_L^0 \to \pi^0 \nu \bar{\nu}$ experiments.

KEK experiment has partially been approved, but the sensitivity goal is only around the SM level—the new Japan Hadron Facility is expected to improve the sensitivity.

CONCLUSION

Studies of rare kaon decays have contributed to the discoveries of several symmetry violations. The decays $K^+ \to \pi^+ \nu \bar{\nu}$ and $K_L^0 \to \pi^0 \nu \bar{\nu}$ are expected to "complete" the measurement of the CKM matrix in the next decade independently from the B-decay system, and to elucidate the origin of CP violation.

REFERENCES

1. T.D. Lee and C.N. Yang, Phys. Rev. **104**, 254 (1956).
2. J.H. Christenson et al., Phys. Rev. Lett. **13**, 138 (1964).
3. S.L. Glashow, J. Iliopoulos and L. Maiani, Phys. Rev. **D2**, 1285 (1970).
4. K. Jungmann, these proceedings.
5. A.J. Buras, these proceedings; G. Buchalla and A.J. Buras, Phys. Rev. **D54**, 6782 (1996).
6. Y. Nir and M.P. Worah, Phys. Lett. **B423**, 319 (1998).
7. L.S. Littenberg and G. Valencia, Ann. Rev. Nucl. Sci. **43**, 729 (1993).
8. L. Wolfenstein, Phys. Rev. Lett. **51**, 1945 (1983).
9. H.B. Greenlee, Phys. Rev. **D42**, 3724 (1992).
10. J. Whitemore for KTEV, to be published in the proceedings of KAON'99 Conference at FERMILAB, 1999.
11. KTEV Collaboration, hep-ex/0001006 (2000).
12. BNL E871 Collaboration, Phys. Rev. Lett. **81**, 4309 (1998).
13. BNL E787 Collaboration, Phys. Rev. Lett. **79**, 2204 (1997).
14. Particle Data Group, Review of Particle Physics, Z. Phys. **C3**, 1 (1998).
15. BNL E787 Collaboration, Phys. Rev. Lett. **84**, 3768 (2000).
16. L.S. Littenberg, Phys. Rev. **D39**, 3322 (1989).
17. E799-II/KTEV Collaboration, Phys. Rev. **D61**, 072006 (2000).
18. KEK E391A Collaboration, KEK Internal 96-13 (1996).
19. KAMI Collaboration, Expression of Interest, hep-ex/9709026 (1997).
20. BNL E926 Collaboration, AGS proposal E926 (1997).

Experiments Searching for Lepton Number Violation

Klaus P. Jungmann

Physikalisches Institut, Universität Heidelberg
Philosophenweg 12, D-69129 Heidelberg, Germany

Abstract. Most sensitive searches for lepton number or lepton flavour violation can be carried out at present in kaon decays and in muon decays. Important information can be extracted from searches for neutrinoless double β-decay and muonium to antimuonium conversion. Although there is no confirmed signal reported yet from any of these systems, stringent limits for parameters in speculative extensions to the standard model can be set. Some of these theoretical approaches could recently be strongly disfavoured.

I INTRODUCTION

According to all confirmed experimental data acquired to date lepton numbers appear to be conserved quantities. This observations can be summarized by several different empirical laws [1–5], some of which are additive and some obey multiplicative, parity-like, schemes. There is no experimental indication yet which would favour any of them. In the standard model states every lepton flavour is separately conserved with an additive quantum number. Based on a suggestion by Lee and Young in 1956 [6] we have in modern physics a strong believe that a strict conservation of baryon and lepton number is without a foundation unless it can be associated with a local gauge invariance and with new long-distance interactions which are yet excluded by experiments [7]. In particular, since no symmetry related to lepton numbers could be revealed yet, the observed conservation laws have no status in physics [8].

Mixings between different generations are well known and understood in the quark sector of the standard theory where the Cabbibo-Kobayashi-Maskawa (CKM) matrix [9] relates the weak quark eigenstates with their mass eigenstates. A familiar and well known example are the K^0-$\overline{K^0}$ oscillations. At present we are left puzzled why leptons do not show any similar mixing and the analogon to the CKM matrix appears just as the unity matrix. Recent exciting experimental hints for neutrino oscillations in the Super-Kamiokande experiment have a high potential for changing this situation; they are not covered here (see e.g. [10]).

TABLE 1. Recently obtained upper limits on lepton number violating processes (90% C.L.).

decay		limit	reference	experiment	future possib.
Z^0	$\to \mu e$	$1.7 \cdot 10^{-6}$	[12]	LEP	$\approx 10^{-7}$
K_L	$\to \mu e$	$4.7 \cdot 10^{-12}$	[13]	BNL E871	$\approx 10^{-13}$
K_L	$\to \pi^0 \mu e$	$3.1 \cdot 10^{-9}$	[14]	KTeV	$\approx 10^{-13}$
K^+	$\to \pi^+ \mu e$	$4.8 \cdot 10^{-11}$	[15]	BNL E865	$\approx 10^{-13}$
μ^+	$\to e^+ \nu_\mu \bar{\nu}_e$	$2.5 \cdot 10^{-3}$	[16]	KARMEN	
μ	$\to eee$	$1 \cdot 10^{-12}$	[17]	SINDRUM I	$\approx 10^{-13}$
μ	$\to e\gamma$	$1.2 \cdot 10^{-11}$	[18]	MEGA	10^{-14}
$\mu^- \text{Ti}$	$\to e^- \text{Ti}$	$6.1 \cdot 10^{-13}$	[19]	SINDRUM II	$5 \cdot 10^{-17}$ for Al
$\mu^- \text{Ti}$	$\to e^+ \text{Ca}$	$1.7 \cdot 10^{-12}$	[20]	SINDRUM II	
B^0	$\to \mu e$	$5.9 \cdot 10^{-6}$	[22]	CLEO	
$\mu^+ e^-$	$\to \mu^- e^+$	$G_{M\bar{M}} < 3 \cdot 10^{-3} G_F$	[21]	MACS	
τ	$\to e\gamma$	$2.7 \cdot 10^{-6}$	[23]	CLEO	$\approx 10^{-7}$
τ	$\to \mu\gamma$	$3.0 \cdot 10^{-6}$	[23]	CLEO	$\approx 10^{-7}$
^{76}Ge	$\to ^{76}\text{Se}\ e^- e^-$	$T_{1/2} > 1.2 \cdot 10^{25} y$	[24]	HD-MOSCOW	$> 6 \cdot 10^{28} y$
		$m_{\nu_e}(Maj.) < 0.2 eV$	[24]		$< 6 meV$

A variety of extensions to the standard model have been suggested which try to explain further some of its not well understood features, which include e.g. parity violation in weak interaction or the values of the fundamental fermion masses. Such facts are introduced by hand into this remarkable theoretical framework which appears to serve as an extremely robust description of all confirmed particle physics. However, the lack of understanding of some basic features may be related to the large number of independent parameters needed in the standard model.

Lepton flavor violation (LFV) appears naturally in most of the speculative extensions to standard theory. They include Left-Right-Symmetry, Supersymmetry, Technicolor, Grand Unification, String Theories, Compositeness, and many others. The predictions of these models are a very strong stimulus for experimental searches in a large range of energies and with a variety of experimental approaches.

In particular, with some low energy experiments New Physics can be probed at mass scales far beyond the reach of present accelerators or such planned for the future and at which predicted new particles could be produced directly. To present date highest sensitivity has generally been reached in dedicated search experiments particularly such on Kaons (K) and on muons (μ) (Table 1), where we have also a high discovery potential for new physics [11]. Further there is a significant window for New Physics discovery for non accelerator experiments, e.g. in searches for neutrinoless double β-decays. The decays of heavier elementary particles, however, which can be created in high energy collisions can be observed with lower accuracy in general and their potential to limit speculative models (or verify their predictions) is mostly restricted to theories in which particle masses enter with high powers. The progress in the K and μ (see sec. IV and V) field is indicated in Fig.1 which shows more than 10 decades of improvement since the first experiments in the late 1940's. The highest gain in sensitivity has been achieved recently is for muonium

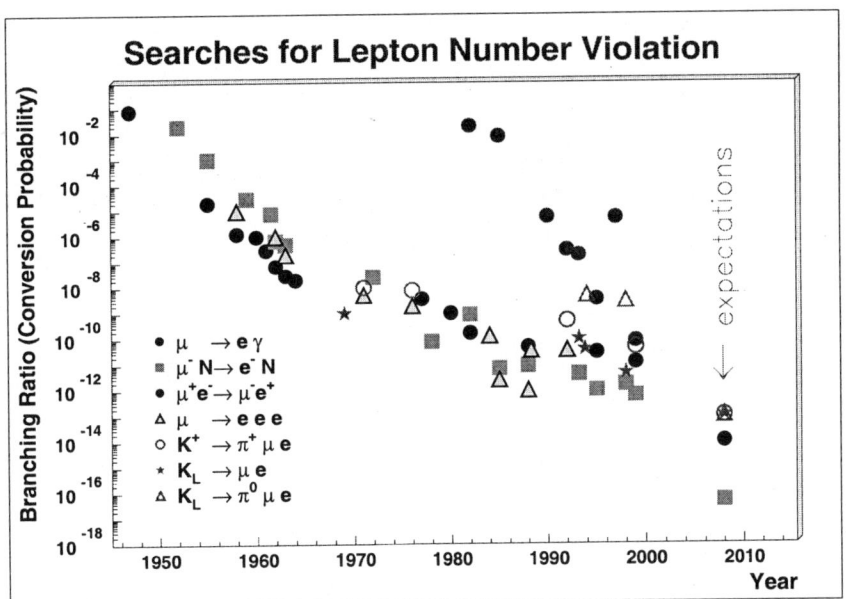

FIGURE 1. Dedicated searches for lepton number violating processes involving muons (μ) and kaons (K). Recent K experiments and $\mu^+e^- - \mu^-e^+$ conversion happen to show the most significant gain in sensitivity. Entries beyond the year 2000 are projections of possibilities by the respective experimenters.

(M=μ^+e^-) to antimuonium ($\overline{M}=\mu^-e^+$) conversion due to the exploitation of a new signature (see sec. V C) which was yet unused in experiments.

II NEUTRINOLESS DOUBLE β-DECAY

Lepton number would be violated by two units in a β-decay of a nucleus which involves two electrons only and no neutrinos. This neutrinoless double β-decay has been suggested in many models which involve neutrinos of the Majorana type. A variety of experiments has been set up to search for this process (see Table 2) employing ^{48}Ca, ^{76}Ge, ^{82}Se, ^{100}Mo, ^{116}Cd, ^{130}Te and ^{136}Xe nuclei. Among those the Heidelberg-Moscow Germanium experiment yields the most stringent half life limit of $T_{1/2} \geq 5.7 \cdot 10^{25}$ y (90% C.L.) [24]. It uses isotopically enriched material with 86% ^{76}Ge as a semiconductor detector to watch its own nuclei decay. It is situated in a clean and carefully against background radiation shielded environment in the Gran Sasso underground laboratory. The experimantal care includes purging with purified nitrogen as well as the use of copper material in the cooling system in the vicinity of the actual detector which had been selected for low intrinsic radiation. Remaining background counts were further suppressed by pulse shape analysis. The result achieved in 24 kg years with the 11.5 kg detector can be used to impose

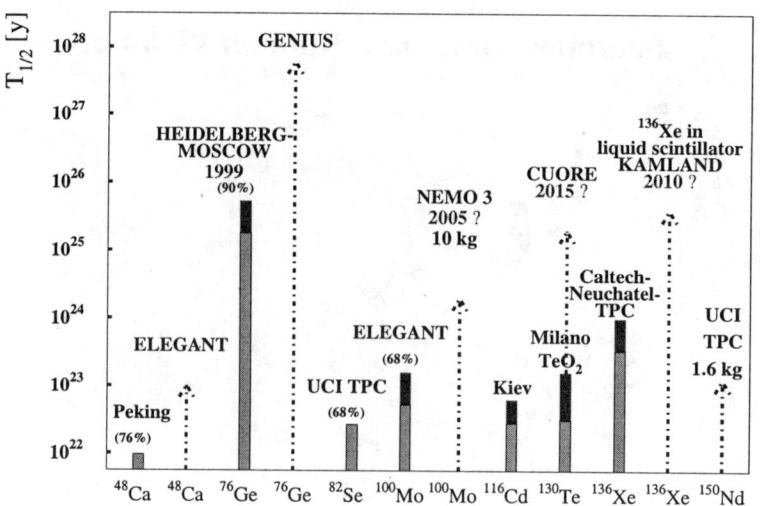

FIGURE 2. Lifetime limits from experiments searching for neutrinoless double β-decay. Most sensitive appear enriched ^{68}Ge detectors. The dark areas represent the current status and the lighter colour indicates near future possibilities. The dashed arrows are long term future plans. Among the most intriguing projects ranks a 1 ton Ge detector with which two orders of magnitude could be gained in sensitivity (from ref. [24]).

an upper limit on the electron neutrino Majorana mass of 0.2 eV, which is well below the electron neutrino mass limit of 2.5 eV established in model independent general direct searches using tritium decay [25].

With 1 ton enriched ^{76}Ge distributed in 288 individual detectors, as suggested by the GENIUS proposal, and after 10 years running time a limit of $T_{1/2} > 6 \cdot 10^{28}$ y could be expected. This would correspond to a Majorana neutrino mass limit of below 6 meV/c^2 [24]. It is a appealing feature of most search experiments for neutrinoless double β-decay that they are also sensitive to cold dark matter, particularly to weakly interacting massive particles (WIMPS) in mass regions above ≈ 20 GeV/c^2.

III EXPERIMENTS AT ELECTRON-POSITRON COLLIDERS - Z^0 AND $W^\pm$ BOSONS AND τ LEPTONS

The general purpose detectors installed at the large high energy electron-positron colliders provide the possibility heavy elementary particles and gauge bosons like the τ lepton or the $W^\pm$ and Z^0 bosons to be observed for rare decays and particularly for lepton number violating effects. Their high mass offers for each particle a large number of different possible purely leptonic and semileptonic decay channels. Z^0 bosons were produced in large quantities at the LEP storage ring of CERN and the Stanford Linear Collider. With the LEP200 upgrade a significant number of $W^\pm$ bosons became available. For τ's the CLEO detector at the Cornell CESR facility provided a significant amount of the available data particularly on neutrinoless τ decays [23,26] as well as on B^0 and D^0 decays [27,22].

The sensitivity of all analyses for lepton number violating (LNV) decays have a principle limit set by statistics. For a particular decay channel further restrictions arise from finite acceptances for the final state particles which explain the course differences in the upper bounds reported for the different channels although starting from the same initial state (Table 1). The limits on branching ratios are in general much higher than the ones obtained in dedicated experiments on K's and μ's. In the near future no sensitivity better than 10^{-7} for any τ decay mode can be expected.

However, such bounds are of great value for discriminating theoretical models where mass scaling runs with a high power of the mass ratios. In the framework of superstring models, for example, the decay $l \to e\gamma$, where l stands for μ or τ, scales with the fifth power of the lepton mass m_l. In this particular case the upper limit of $2.7 \cdot 10^{-6}$ for $\tau \to e\gamma$ can compete with the present $3.8 \cdot 10^{-11}$ limit on $\mu \to e\gamma$ due to the $1.3 \cdot 10^6$ enhancement factor from the mass ratio $m_\tau/m_\mu \approx 16.8$. However in general the mass scaling is expected to be less dramatic.

IV   EXPERIMENTS ON KAONS

Sensitive searches for lepton number violating Kaon decays became possible because of two major developments, i.e. with the availability of intense Kaon sources at the Fermi National Accelerator Laboratory (FNAL) and the Brookhaven National Laboratory (BNL) and with novel experimental techniques which allow to operate at high data rates.

The experiments BNL-871 searching for $K^+ \to \pi^+ e\mu$ is finished now and yielded an upper limit of $4.7 \cdot 10^{-12}$ for the branching ratio of this forbidden process. The experiment BNL-865 searching for $K_L \to \mu e$ and the Fermilab KTeV effort FNAL-799II investigating $K_L \to \pi^0 \mu e$ promise significant improvements (see Table 2). Lepton flavour violating decays are searched here along with measurements on very rare K decay channels. Among the new physics that could be revealed are new heavy gauge bosons with masses up to order 50-200 TeV/c^2, far beyond the reach of even any planned accelerator [28]. At the projected Japanese Hadron Facility (JHF) one could expect significant improvements beyond the present status; new experiments could be expected to reach a sensitivity level at 10^{-13}.

V   EXPERIMENTS INVOLVING MUONS

Shortly after the muon's nature as a heavy electron-like particle became apparent, the decay $\mu \to e\gamma$ was the first which has been searched for. However, the muon

TABLE 2. Three presently ongoing searches for lepton flavour violating K decays.

	past limit	present limit (2000)	expected near future result
$K^+ \to \pi^+ e\mu$	$2 \cdot 10^{-10}$ BNL-777	$3.2 \cdot 10^{-11}$ BNL-865	factor of 3 expected upon completion
$K_L \to \mu e$	$3 \cdot 10^{-11}$ BNL-791	$4.7 \cdot 10^{-12}$ BNL-871	finished; 10^{-13} possible in principle, no plans
$K_L \to \pi^0 \mu e$	$3.1 \cdot 10^{-9}$ FNAL-799	$3.1 \cdot 10^{-9}$ FNAL799II	$\approx 10^{-11}$ expected since no background

turned not out to be just an excited electron. Since then searches for rare and forbidden muon decays have been among the most precise experiments in physics [29]. They were always of special interest in the context of unified gauge theories, as they can provide accurate tests of speculative models. Owing to the achievable high experimental precision they may be able to discriminate between such theoretical approaches [30]. Recently forbidden muon decays have attracted attention, when their possible sensitivity to effects arising in minimal supersymmetry (SUSY) were discussed in theoretical studies [31]. It was pointed out that for values of $\tan\beta$ (the ratio of the vacuum expectation values of the two Higgs fields involved) which exceed about 3, the branching ratio should be above $\approx 10^{-14}$ for a decay $\mu \to e\gamma$ and above $\approx 10^{-16}$ for $\mu \to e$ conversion on a Ti nucleus, almost independent of all other parameters in the model. This has stimulated promising proposals to the Paul Scherrer Institute (PSI), Switzerland, and to BNL to search for the respective processes. Both experiments are on their way.

In the field of searching for SUSY effects, particularly such with moderately high values of $\tan\beta$, in low energy experiments rare decay experiments are in some competition with the just started new precision measurement [32] of the muon magnetic anomaly a_μ where the contribution from SUSY is of order $a_\mu(SUSY) = 140 \cdot 10^{-11} \tan\beta * (100 GeV/\tilde{m})^2$ with $\tilde{m}$ the mass of the lightest SUSY particle (see [33]). The measurement goal is $\Delta a_\mu(exp) = 40 \cdot 10^{-11}$ and should be reached around the year 2001/2002.

It depends strongly on the model to be tested, which process is best suited. There can be no general rule given, as long as we have no hint for lepton number violating processes. Fortunately many authors allow to compare different processes in terms of their usefulness [34,35]. Provided the sub eV neutrino square mass differences suggested by neutrino oscillation experiments can be verified, then loop processes could induce lepton number violation. For the process $\mu \to e\gamma$ a branching ratio at the level $10^{-39} \cdot (m_{\nu_1}^2 - m_{\nu_2}^2)/400 eV^2$ has been estimated [41]. This must be considered far below any observability in the near future. However, many speculative theories offer much larger effects.

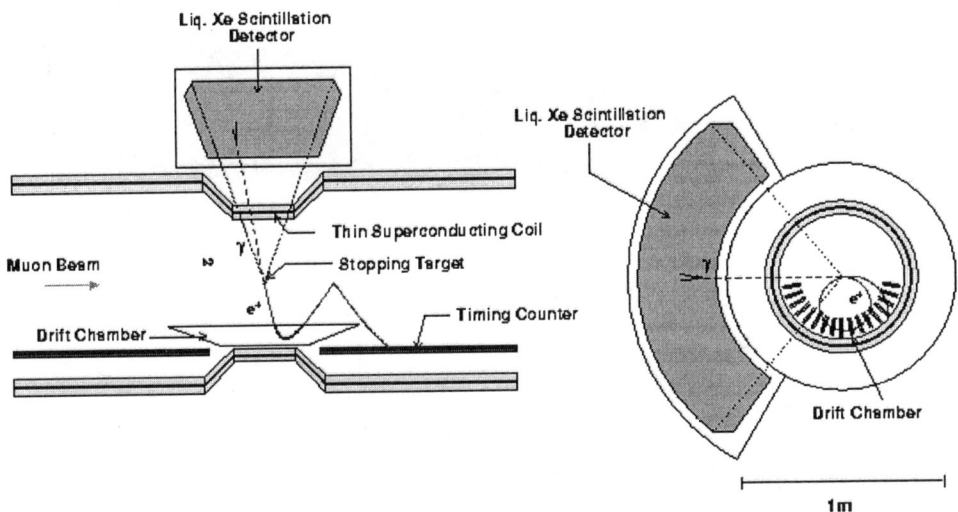

FIGURE 3. The approved $\mu e \gamma$ experiment at PSI. (see [37]).

A $\mu \to e\gamma$ Decay

The signature of a $\mu \to e\gamma$ event is a 52.8 MeV positron emitted back to back with a 52.8 MeV photon. The MEGA experiment at the late Los Alamos Meson Physics Facility (LAMPF) consisted of a magnetic spectrometer to observe the charged final state particle and three pair spectrometers for detecting the photon through its e^+e^- pair creation in lead converters. Random coincidences at high rates are reported as major background in addition to electronic cross talk in part of the recorded data. Data taking is completed and some 16% of the data could be analyzed leading to an upper limit on the branching ratio of $1.2 \cdot 10^{-11}$ [18] which improves the previously best value of $4.9 \cdot 10^{-11}$ established in a crystal box detector also at LAMPF [36].

At PSI new efforts are being discussed to reach a sensitivity of about 10^{-14} for this decay mode within the next few of years. The suggested instrument uses a liquid Xe calorimeter (see Fig. 3) for the detection of the γ potentially released in the process [37]. The associated electron will be detected in drift chambers.

It should be noted that the tightest bounds on bileptonic gauge bosons, which are common to many speculative standard model extensions, come from $\mu \to e\gamma$, if flavour democracy is assumed [35].

B $\mu \to e$ Conversion

Many constraints for speculative models arise from the present experimental bound on the conversion process $\mu + Z \to e + Z$ (Table 1), which is the tightest for all studied LNV decays. The variety of theoretically possible processes that can

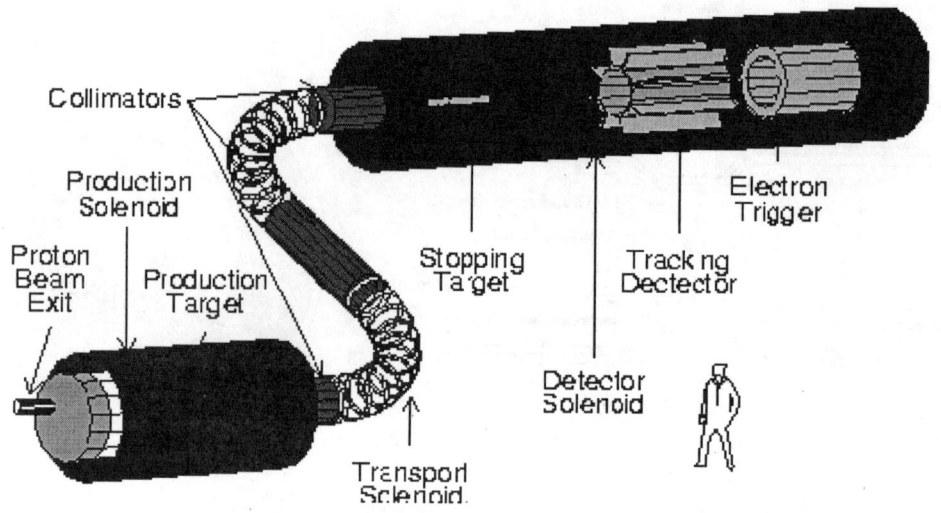

FIGURE 4. The MECO experiment at BNL (see [39]).

be tested includes, e.g. supersymmetric loop graphs, heavy neutrinos, leptoquarks, compositeness, Higgs bosons and heavy Z' bosons with anomalous couplings. Generally it is more sensitive to new Physics than $\mu \to e\gamma$ in a wide class of models where the process is generated at the one loop level [38].

The $\mu \to e$ conversion process needs to involve a nucleus to assure that elementary conservation laws can be obeyed. If the nucleus is left behind in its ground state, a conversion event is signaled through the release of a 105 MeV electron, which is uniquely distinguishable from normal muon decay electrons ranging up to 53 MeV. Among the physically relevant intrinsic background processes is μ decay in the atomic orbit after a muonic atom has been formed, which can release much higher energetic electrons, and radiative muon capture, where the photokinematic end point can be close to the signal electron energy.

The ongoing SINDRUM II experiment uses the worldwide brightest continuous muon channel π E5 at PSI. Their new results limit the branching ratios $\mu^-\text{Ti} \to e^+\text{Ca}^{gs}$ to below $1.7 \cdot 10^{-12}$ for the Ca nucleus in the ground state [20], $\mu^-\text{Ti} \to e^+\text{Ca}^{GDR}$ to below $3.6 \cdot 10^{-11}$ leaving Ca with giant dipole resonance excitation [20], $\mu^-\text{Ti} \to e^-\text{Ti}$ to below $6.1 \cdot 10^{-13}$ for Ti in the ground state [19], $4.6 \cdot 10^{-12}$ for ground state Pb [40], and $1.2 \cdot 10^{-11}$ for ground state Au [40]. For the ground state processes the nucleons interact coherently which enhances the possible effect. In order to boost accuracy in the near future the SINDRUM II collaboration wants to take advantage in the gain of muon flux through a $\pi - \mu$ converter, a novel superconducting device in the beam line which collects π's and releases only μ's with very low π contamination. The latter point is essential as π's are a source of potential background due to nuclear reactions. The projections of the collaboration for the achievable limit in the coherent $\mu^-\text{Ti} \to e^-\text{Ti}$ case are in the 10^{-14} region.

The new Muon Electron Conversion (MECO) experiment at BNL [39] (see Fig.4) is very close in its design to the MELC proposal by Lobashev and collaborators for the Moscow Meson Factory. The setup consists of a target station for π/μ production which uses a proton beam from the AGS accelerator, an S-shaped transport and purification section and a detector the basic idea of which is to let electrons from normal muon decay pass without being seen and to observe only the 105 MeV signal electrons. The goal is a $5 \cdot 10^{-17}$ level in sensitivity, which will stringently test supersymmetric models; there is an anticipated ultimate sensitivity of 10^{-18}.

C Muonium-Antimuonium Conversion

Two leptons from different generations form the hydrogen-like muonium atom. in which the close confinement of the bound state offers excellent opportunities to study precisely fundamental electron-muon interactions [42,43]. Since the effect of all known fundamental forces in this atom can be calculated mainly in the framework of quantum electrodynamics (QED) to the required accuracies, this system renders the possibility to search sensitively for yet unknown interactions between both particles. An M-$\overline{\text{M}}$-conversion would violate additive lepton family number conservation and is forseen in many speculative theories (see Fig. 5). It would be an analogy in the lepton sector to the well established K^0-$\overline{K^0}$ oscillations.

The setup at PSI (Fig. 6) [21] has been designed to employ the signature developed in a predecessor experiment at LAMPF, which requires the coincident identification of both constituents of the antiatom in its decay [49,21]. Muonium atoms in vacuum with thermal velocities, which are produced from a SiO_2 powder target, are observed for antimuonium decays. Energetic electrons from the decay of the μ^- in the antiatom can be observed in a magnetic spectrometer. The positron in the atomic shell of the antiatom is left behind after the decay with 13.5 eV average kinetic energy [51]. It can be accelerated and guided in a magnetic transport system onto a position sensitive microchannel plate detector (MCP). Annihilation radiation can be observed in a segmented pure CsI calorimeter around it.

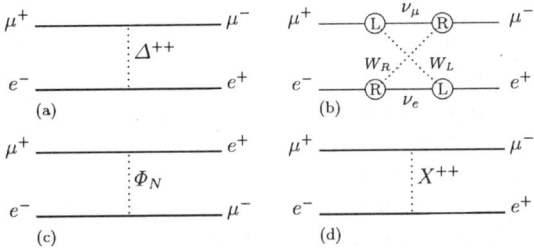

FIGURE 5. Muonium-antimuonium conversion in theories beyond the standard model. The interaction could be mediated by (a) a doubly charged Higgs boson Δ^{++} [44,45], (b) heavy Majorana neutrinos [44], (c) a neutral scalar Φ_N [46], e.g. a supersymmetric τ-sneutrino $\tilde{\nu}_\tau$ [8,47], or (d) a bileptonic gauge boson X^{++} [48].

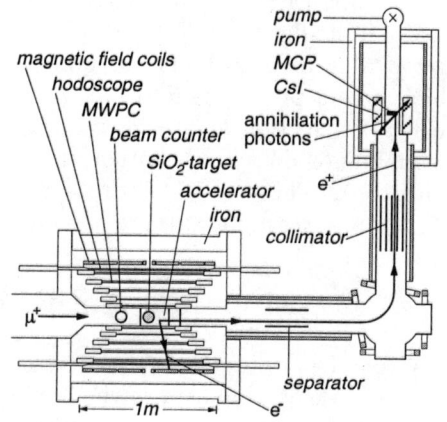

FIGURE 6. Top view of the MACS (Muonium - Antimuonium - Conversion - Spectrometer) apparatus at PSI to search for $M-\overline{M}$ - conversion [21].

The relevant measurements were performed during in total 6 month distributed over 4 years during which $5.7 \cdot 10^{10}$ muonium atoms were in the interaction region. One event fell within a 99% confidence interval of all relevant distributions (Fig. 7). The expected background due to accidental coincidences is 1.7(2) events. Thus an upper limit on the conversion probability of $P_{M\overline{M}} \leq 8.2 \cdot 10^{-11}/S_B$ (90% C.L.) was found, where S_B accounts for the interaction type dependent suppression of the conversion in the magnetic field of the detector due to the removal of degeneracy between corresponding levels in M and $\overline{M}$. The reduction is strongest for $(V\pm A)\times(V\pm A)$, where $S_B=0.35$ [52,53]. This yields for the traditionally quoted upper limit on the coupling constant in effective four fermion interaction $G_{M\overline{M}} \leq 3.0 \cdot 10^{-3} G_F (90\% C.L.)$ with G_F the weak interaction Fermi constant.

This new result, which exceeds bounds from previous experiments [49,54] by a factor of 2500 and the one from an early stage of the experiment [21] by 35, has some impact on speculative models. A certain Z_8 model is ruled out which has more than 4 generations of particles and where masses could be generated radiatively with heavy lepton seeding [55]. A new lower limit of $m_{X^{\pm\pm}} \geq 2.6$ TeV/c^2 $*g_{3l}$ (95% C.L.) on the masses of flavour diagonal bileptonic gauge bosons in GUT models can be extracted which lies far beyond the value derived from direct searches, measurements of the muon magnetic anomaly or high energy Bhabha scattering [48,35]. Here g_{3l} is of order 1 and depends on the details of the underlying symmetry. For 331 models this translates into $m_{X^{\pm\pm}} \geq 850$ GeV/c^2 which disfavours their minimal Higgs version in which an upper bound of 600 GeV/c^2 has been extracted from an analysis of electroweak parameters [56,57]. The 331 models may still be viable in some extended form involving a Higgs octet [58]. In the framework of R-parity violating supersymmetry [47,8] the bound on the coupling parameters could be lowered by a factor of 15 to $| \lambda_{132}\lambda^*_{231} | \leq 3*10^{-4}$ for assumed superpartner masses of 100 GeV/c^2. Further the achieved level of sensitivity allows to narrow slightly the interval of allowed heavy muon neutrino masses in minimal left-right

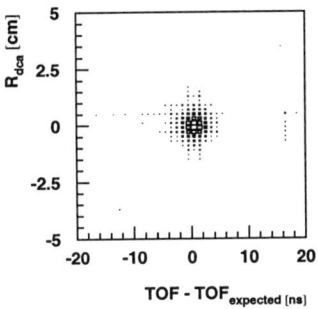 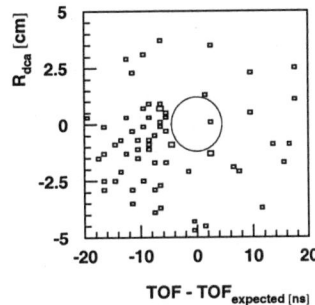

FIGURE 7. Time of flight (TOF) and vertex quality for a muonium measurement (left) and the same for all data of the final 4 month search for antimuonium (right). One event falls into the indicated 3 standard deviations area.

symmetry [45] (where a lower bound on $G_{M\overline{M}}$ exists, if muon neutrinos are heavier than 35 keV) to ≈ 40 keV/c^2 up to the present experimental bound at 170 keV/c^2. In minimal left right symmetric models, in which $M\overline{M}$ conversion is allowed, the process is intimately connected to the lepton family number violating muon decay $\mu^+ \to e^+ + \nu_\mu + \bar{\nu}_e$. With the limit achieved in this experiment this decay is not an option for explaining the excess neutrino counts in the LSND neutrino experiment at Los Alamos [59,60]. The consequences for atomic physics of muonium are such that the expected level splitting in the ground state due to $M - \overline{M}$ interaction is below 1.5 Hz/$\sqrt{S_B}$ reassuring the validity of fundamental constants determined in muonium spectroscopy [61].

A future $M - \overline{M}$ experiment could take advantage of high intense pulsed muon sources. In contrast to other LNV muon decays, the conversion through its nature as particle - antiparticle oscillation, has a time evolution in which the probability for finding $\overline{M}$ in the ensemble increases quadratically in time, giving the signal an advantage growing in time over major exponentially decaying background [50].

VI LONG TERM FUTURE POSSIBILITIES

It appears that the availability of particles limits the ability to find very rare processes or to impose significantly improved limits in continuation of the search program of dedicated experiments. Therefore any measure to boost the respective particle fluxes is a very important step forward. The $\pi - \mu$ converter at PSI or the dedicated tailored muon production of the planned MECO experiment at BNL are examples of novel attempts to overcome this problem. In principle, we need significantly more intense accelerators, such as they are discussed at various places. In the intermediate future the Japanese Hadron Facility (JHF) as well as a planned intense proton machine at GSI are important options for intense beams. Also the discussed Oak Ridge neutron spallation source or a European Spallation Source (ESS) could in principle accommodate intense muon beams. The most promising

TABLE 3. Muon fluxes of some existing and future facilities, Rutherford Appleton Laboratory (RAL), Japanese Hadron Facility (JHF), European Spallation Source (ESS), Muon collider (MC).

	RAL(μ^+)	PSI(μ^+)	PSI(μ^-)	JHF(μ^+)	ESS(μ^+)	MC (μ^+, μ^-)
Intensity (μ/s)	3×10^6	3×10^8	1×10^8	4.5×10^{11}	4.5×10^7	7.5×10^{13}
Momentum bite $\Delta p_\mu/p[\%]$	10	10	10	10	10	5-10
Spot size (cm × cm)	1.2×2.0	3.3×2.0	3.3×2.0	1.5×2.0	1.5×2.0	few×few
Pulse structure	82 ns 50 Hz	50 MHz continuous	50 MHz continuous	300 ns 50 Hz	300 ns 50 Hz	50 ps 15 Hz

facility would be, however, a muon collider [62], the front end of which could provide muon rates 5-6 orders of magnitude higher than present beams (see Table 3). An intense neutrino factory, which may be raelistic in the near future, would be a good starting point already. Particularly attractive are new muon beam designs, where relatively more muons are collected at the production target compared to present beam lines and where new techniques like phase rotation will be employed. Among those the design of the PRISM beam for JHF is very appealing [63].

It was noted already in the early 60ies that, e.g. the process $e^-e^- \to \mu^-\mu^-$ is closely related to muonium-antimuonium conversion [64]. Indeed such scattering experiments were carried out at the Princeton-Stanford storage rings at Stanford yielding the at the time best limit on the coupling constant $G_{M\overline{M}}$ [65]. Today, similar proposals have been made for scattering of high energy e^- on e^-, e^- on e^+, μ^- on μ^- and μ^- on μ^+ [66–68]. They were mainly discussed in connection with bileponic gauge bosons. Even a lower limit for the cross section of the process $e^-e^- \to \mu^-\mu^-$ was found, provided the sum of the light neutrino masses exceeds $\approx$ 90 eV [68]. Pronounced resonances have been predicted particularly for such experiments at the Next Linear Collider or the high energy end of a muon collider.

VII CONCLUSIONS

Although lepton flavour conservation remains a mystery and searches for its violation have not yet led to a successful observation, both the theoretical and experimental work in this connection have led to a deeper understanding of particle interactions. The experiments contribute continuously towards guiding theoretical developments by excluding various speculative models or by limiting their parameters. In some cases by the creation of new models could be stimulated.

ACKNOWLEDGMENTS

The organizers of the Symmetries 2000 conference deserve our gratitude for creating a wonderful atmosphere for a stimulating conference. The author is grateful to M. Cooper, H. Klapdor-Kleingrothaus, Y. Kuno, W. Molzon and A.v.d. Schaaf for discussions and updates, respectively latest results from their experiments and unpublished insights.

REFERENCES

1. Y.B. Zeldovitch, Dan. SSR **86**, 505 (1952)
2. B. Pontecorvo, Sov.Phys.-JETP **37**, 1751 (1959) and Sov. Phys. JETP **6**, 381 (1958)
3. N. Cabbibo and R. Gatto, Phys.Rev.Lett. **5**, 114 (1960); N. Cabbibo, Nuovo Cim. **19**,612 (1961)
4. E.J. Konopinski and H.M. Mahmoud, Phys.Rev.**92**, 1045 (1953)
5. G. Feinberg and S. Weinberg, Phys.Rev.Lett. **6**, 381 (1961)
6. T.D. Lee and C.N. Yang, Phys. Rev. **98**,1501 (1956)
7. J.L. Chkareuli, et al.,Phys.Rev. **D62**, 015014 (2000) and J.L. Chkareuli and C.D. Froggatt, hep-ph/0004090 (2000)
8. A. Halprin and A. Masiero, Phys.Rev.D**48**, 2987 (1993)
9. M. Kobayashi and T. Maskawa, Prog. Theor. Phys. **49**, 652 (1973)
10. T. Kajita, this conference
11. R.N. Mohapatra, Prog.Part.Nucl.Phys. **31**, 39 (1993)
12. O. Adriani et al. Phys. Lett. B **316**, 427 (1993); L. Bugge et al., in: Proc. Europhysics Conference on High-energy Physics, Brussels, J. Lemonne et al. (eds.), World Scientific, Singapore (1996); P. Abreu et al., Z. Phys. C **73**, 243 (1997)
13. D. Ambrose et al., Phys. Rev. Lett. **81**, 5734 (1998)
14. J. Belz, Proc. Intersections between Particle and Nuclear Physics, 6th conf, T.W. Donnelly (ed.), AIP Press, New York, p.763 (1997); R. Ray, JHF98 workshop, Tsukuba (1998)
15. M. Zeller, priv. com.;see also ref. [?] and S. Eilerts, loc. cit. [14], p. 779 (1997)
16. K. Eitel, doctoral thesis, University of Karlsruhe (1995)
17. W. Bertl et al., Nucl.Phys. B**260**, 1 (1985)
18. M.L. Brooks et al., Phys. Rev. Lett. **83**, 1521 (1999); M.D. Cooper, priv. com. (1999)
19. S. Eggli et al., publication in preparation (1998)
20. J. Kaulard et al., Phys.Lett. **B 422**, 334 (1998)
21. L. Willmann et al., Phys.Rev.Lett. **82** 49 (1999)
22. R. Ammar et al., Phys.Rev.D **49**, 5701 (1994)
23. K. Edwards et al., Phys.Rev.D **55**, 3919 (1997)
24. L. Baudis, Phys. Rev. Lett. **83**, 41 (1999) H.V. Klapdor-Kleingrothaus, Proc. Beyond the Desert Conference, Institute of Physics Publishing, Bristol, p.485 (1998)
25. V.M. Lobashev et al., Phys. Lett. **B460**, 227 (1999); see also C. Weinheimer et al. Phys. Lett. **B460**, 219 (1999) and Phys. Lett. **B464**, 352 (1999)
26. D. Bliss et al., Phys.Rev.D **57**,5903 (1998)
27. A. Freyberger et al., Phys.Rev.Lett.**76**, 3065 (1996)
28. S.H. Kettell, hep-ex/0002011 (2000) and references therein
29. Y. Kuno and Y. Okada, hep-ph/9909265 (1999)
30. T.S. Kosmas, G.K. Leontaris, J.D. Vergados, Prog.Part.Nucl.Phys. **33**, 397 (1994)

31. R. Barbieri, L. Hall and A. Strumia, Nucl.Phys. **B445**, 219 (1995)
32. R.M. Carey et al., Phys. Rev. Lett. **82**, 1632 (1999); K. Jungmann et al., nucl-ex/0002005 (2000)
33. U. Chattopadhyay and P. Nath, Phys.Rev. **D53**, 1648 (1996)
34. J. Bordes et al., hep-ph/9909321 (1999)
35. F. Cuypers and S. Davidson, Eur.Phys.J. **C2**, 503 (1998)
36. T. Bolton et al., Phys. Rev **D38**, 2077 (1988)
37. T. Mori et al., PSI experiment R-99-05 (1999)
38. M. Raidal and A. Santamaria, hep-ph/9710389 (1997)
39. W. Molzon et al., Proposal to BNL E-940 (1997)
40. A. van der Schaaf, priv.com. (2000)
41. A.A. Godzev et al., Phys. Lett **B338**, 212 (1994)
42. V.W. Hughes and G. zu Putlitz, in: *Quantum Electrodynamics*, World Scientific, Singapore, T. Kinoshita (ed.), p. 822 (1990)
43. K. Jungmann, in: *Atomic Physics 14* (New York: AIP Press), D. Wineland et al. (ed.), p. 102 (1994)
44. A. Halprin, Phys.Rev.Lett. **48**, 1313 (1982)
45. P. Herczeg and R.N. Mohapatra, Phys.Rev.Lett. **69**, 2475 (1992)
46. W.S. Hou and G.G. Wong, Phys.Rev. **D53** 1537 (1996)
47. R.N. Mohapatra, Z.Phys. **C56**, S117 (1992)
48. H. Fujii et al., Phys.Rev. D **49** 559 (1994)
49. B.E. Matthias et al., Phys.Rev.Lett. **66**, 2716 (1991)
50. L. Willmann and K. Jungmann, Lecture Notes in Physics, Vol. 499, (1997)
51. L. Chatterjee et al., Phys. Rev. **D46**, 46 (1992)
52. K. Horrikawa and K. Sasaki, Phys. Rev. **D53**, 560 (1996)
53. G.G. Wong and W.S. Hou, Phys.Lett.**B357**, 145 (1995)
54. V.A. Gordeev et al, JETP Lett. **59**, 589 (1994)
55. G.G. Wong and W.S. Hou, Phys.Rev.**D50**, R2962 (1994)
56. P. Frampton , Phys.Rev.Lett**69**, 1889 (1994); see also: hep-ph/97112821 (1997)
57. P. Frampton and S. Harada, hep-ph/9711448 (1997) and hep-ph/0002017 (2000)
58. P. Frampton, priv. comm. (1998); V. Pleitez, Phys.Rev. **D61**, 057903 (2000)
59. P. Herczeg, Conference "Beyond the Desert 97", Castle Ringberg (1997); S. Bergmann and Y. Grossman, Phys.Rev.**D59**, 093005 (1999)
60. C. Athanassopoulos et al., Phys.Rev. C54, 2685 (1996); see also: nucl-ex/9709006
61. W. Liu et al., Phys. Rev. Lett **82**, 711 (1999); see also: K. Jungmann, physics/9910023 (1999)
62. R.B. Palmer and J.C. Gallardo, physics/9802002 (1998); R.B. Palmer, physics/9802005 (1998)
63. Y. Kuno, priv. com. (1998)
64. S. Glashow, Phys.Rev.Lett. **6**, 196 (1961)
65. W.C. Barber et al, Phys.Rev.Lett. **22**, 902 (1969)
66. P. Frampton , Phys.Rev. **D45**, 4240 (1992)
67. W.S. Hou, Nucl. Phys. **B51A**, 40 (1996)
68. M. Raidal, Phys.Rev.**D57**, 2013 (1998)

LIST OF PARTICIPANTS

Name	Institution	E-mail contact
Dr. Brad B Abbott	BaBar	bkabbott@lbl.gov
Mr Jim Bashford	CSSM	jbashfor@physics.adelaide.edu.au
Mr Sundance Bilson-Thompson	CSSM	sbilson@physics.adelaide.edu.au
Mr Francois Bissey	CSSM	bissey@in2p3.fr
Dr. Robert Bluhm	Colby College	rtbluhm@colby.edu
Mr Frederic Bonnet	CSSM	fbonnet@physics.adelaide.edu.au
Dr. Csaba Boros	CSSM	cboros@physics.adelaide.edu.au
Mr Patrick Bourke	CSSM	pbourke@physics.adelaide.edu.au
Mr Patrick Bowman	CSSM	pbowman@physics.adelaide.edu.au
Prof. Andrzej J. Buras	Technical University Muenchen	buras@feynman.t30.physik.tu-muenchen.de
Dr. Paolo Christillin	Universita di Pisa	christ@difi.unipi.it
Prof. Robert Delbourgo	University of Tasmania	bob.delbourgo@utas.edu.au
Prof. Pierre Depommier	University of Montreal	pom@LPS.UMontreal.ca
Dr. Bertrand Desplanques	Inst. des Sciences Nucleaires	desplanq@isn.in2p3.fr
Mr Will Detmold	CSSM	wdetmold@physics.adelaide.edu.au
Dr. Vladimir Dzuba	University of NSW	dzuba@phys.unsw.edu.au
Prof. Magda Ericson	University. of Lyon/CERN	magda.ericson@cern.ch
Mr Torleif Ericson	TSL Uppsala	torleif.ericson@cern.ch
Mr. Craig Everton	University of Melbourne	everton@liszt.ph.unimelb.edu.au
Prof. Wulf Fetscher	ETH Zurich	Wulf.Fetscher@psi.ch
Prof. Victor Flambaum	University of NSW	flambaum@phys.unsw.edu.au
Assist. Prof. Susan S Gardner	University of Kentucky	gardner@pa.uky.edu
Mr. Ascelin Gordon	University of Melbourne	gordon@physics.unimelb.edu.au
Dr. Xin-Heng Guo	CSSM	xhguo@physics.adelaide.edu.au
Dr. Vadim Guzey	CSSM	vguzey@physics.adelaide.edu.au
Ms Emily Hackett-Jones	CSSM	ehackett@physics.adelaide.edu.au
Mr. Nick Hastings	University of Melbourne	hastings@physics.unimelb.edu.au
Dr. Fred Hawes	CSSM	fhawes@adelaide.edu.au
Dr. Eric Heenan	University of Melbourne	heenan@physics.unimelb.edu.au

Prof. Ernest Henley	University of Washington	henley@nucthy.phys.washington.edu
Dr. Yee Bob Hsiung	Fermilab	hsiung@fnal.gov
Prof. W-Y Pauchy Hwang	National Taiwan University	wyhwang@phys.ntu.edu.tw
Dr. Klaus P. Jungmann	University of Heidelberg	jungmann@physi.uni-heidelberg.de
Dr. Takaaki Kajita	University of Tokyo	kajita@icrr.u-tokyo.ac.jp
Mr Waseem Kamleh	CSSM	wkamleh@physics.adelaide.edu.au
Dr. Ayse Kizilersu	CSSM	akiziler@physics.adelaide.edu.au
Dr. Koichi Kodama	Aichi University f Education	kkodama@auecc.aichi-edu.ac.jp
Prof. Yoshitaka Kuno	IPNS, KEK	yoshitaka.kuno@kek.jp
Ms Sarah Lawley	CSSM	slawley@physics.adelaide.edu.au
Dr. Derek Leinweber	CSSM	dleinweb@physics.adelaide.edu.au
Mr Olivier Leitner	CSSM	oleitner@physics.adelaide.edu.au
A/Prof. Max Lohe	CSSM	mlohe@physics.adelaide.edu.au
Prof. Timothy J. Londergan	Indiana University	tlonderg@niobe.iucf.indiana.edu
Prof. Kim Maltman	York University	maltman@fewbody.phys.yorku.ca
Prof. Bruce McKellar	University of Melbourne	mckellar@physics.unimelb.edu.au
Dr. Wally Melnitchouk	CSSM	wmelnitc@jlab.org
Dr. Glenn Moloney	University of Melbourne	glenn@physics.unimelb.edu.au
Prof. Alfonso Mondragon	UNAM	mondra@ft.ifisicacu.unam.mx
Dr. David Muller	Stanford Linear Accelerator Center	muller@slac.stanford.edu
Dr. Toshio Numao	TRIUMF	toshio@triumf.ca
Assis. Prof. Allena K. Opper	Ohio University	opper@ohiou.edu
Dr. Hitoshi Ozaki	KEK	hitoshi@bmail.kek.jp
Prof. Shelley A. Page	University of Manitoba	spage@cc.umanitoba.ca
Dr. Mark L. Pitt	Virginia Tech	pitt@vt.edu
Mr Shaun Ryan	CSSM	sryan@physics.adelaide.edu.au
Dr. Koichi Saito	Tohoku College of Pharmacy	ksaito@nucl.phys.tohoku.ac.jp
Prof. Lukas Schaller	University of Fribourg	lukas.schaller@unifr.ch
Dr. Andreas Schreiber	CSSM	aschreib@physics.adelaide.edu.au
Dr. Kamal Seth	Northwestern University	kseth@nwu.edu
Dr. Martin Sevior	University of Melbourne	msevior@physics.unimelb.edu.au

Dr. Bill Y. N. Shin	TRIUMF	shin@triumf.ca
Dr. William M. Snow	Indiana University	snow@iucf.indiana.edu
Dr. Jerzy Sromicki	ETH	jerzy.sromicki@psi.ch
Dr. Sergey Sukhoruchkin	Petersberg Nuclear Physics Institute	sergeis@hep486.pnpi.spb.ru
Mr David Tellis	CSSM	dtellis@physics.adelaide.edu.au
Prof. Anthony W Thomas	CSSM	athomas@physics.adelaide.edu.au
Dr. Feodor Tikhonin	Inst. For High Energy Physics	tikhonin@mx.ihep.su
Dr. Bryan Tipton	Kellogg Radiation Laboratory	tipton@krl.caltech.edu
Dr. Kazuo Tsushima	CSSM	ktsushima@physics.adelaide.edu.au
Prof. German Valencia	Iowa State University	valencia@iastate.edu
Dr. Pierre Van Hove	Universite Catholique de Louvain la Neuve	vanhove@fynu.ucl.ac.be
Dr. Willem van Oers	TRIUMF	vanoers@triumf.ca
Dr. Kevin Varvell	University of Sydney	kev@physics.usyd.edu.au
Dr. Graziano Venanzoni	University of Karlsruhe	graziano.venanzoni@iekp.fzk.de
Dr. Heinrich Wahl	CERN	wahl@mail.cern.ch
Mr. Michael Walker	Australian National University	mlw105@rsphysse.anu.edu.au
Dr. Ron Walsworth	Harvard Smithsonian Center for Astrophysics	rwalsworth@cfa.harvard.edu
Dr. John Wilkerson	University of Washington	jfw@u.washington.edu
A/Prof. Anthony Williams	CSSM	awilliams@physics.adelaide.edu.au
Mr. Stewart Wright	CSSM	swright@physics.adelaide.edu.au
Dr. Zhi-Zhong Xing	University Munich	Xing@theorie.physik.uni-muenchen.de
Dr. Shuqian Ying	Fudan University	sqying@fudan.edu.cn
Mr Ross Young	CSSM	ryoung@physics.adelaide.edu.au
Mr James Zanotti	CSSM	jzanotti@physics.adelaide.edu.au

AUTHOR INDEX

A

Abbott, B., 197
Alduschenkov, A., 286
Alexandrou, C., 73
Asahi, K., 286

B

Bear, D., 119
Berdoz, A. R., 298
Birchall, J., 298
Bland, J. B., 298
Bluhm, R., 109
Bodek, K., 167, 280
Böni, P., 280
Boros, C., 255
Bowles, T., 286
Bowman, J. D., 298
Buchmann, A. J., 148
Budzanowski, A., 167
Buras, A. J., 13

C

Campbell, J. R., 298

D

Danneberg, N., 167, 261, 280
Davis, C. A., 298
Delbourgo, R., 265
Desplanques, B., 130
Deutsch, J., 261
Dzuba, V. A., 136

E

Egger, J., 261

F

Fetscher, W., 167, 261, 280
Filippone, B., 286
Flambaum, V. V., 92, 136
Foroughi, F., 261
Fowler, M., 286

G

Gardner, S., 41
Geltenbort, P., 286
Ginges, J. S. M., 136
Goldman, T., 60
Govaerts, J., 261
Green, A. A., 298
Green, P. W., 298
Guo, X.-H., 153

H

Hadri, M., 261
Hamian, A. A., 298
Hartmann, F., 286
Healey, D. C., 298
Helmer, R., 298
Henley, E. M., 148
Hilbes, C., 167, 261, 280
Hill, R., 286
Hime, A., 286
Hino, M., 286
Hoedl, S., 286
Hogan, G., 286
Hsiung, Y. B., 1
Humphrey, M., 119
Hwang, P. W.-Y., 160

I

Ito, T., 286

J

Jarczyk, L., 167
Jones, C., 286
Jungmann, K. P., 341

K

Kajita, T., 31
Kawai, T., 286
Kharitonov, A., 286
Kirch, K., 167, 261, 286
Kistryn, S., 167, 280
Kitagaki, T., 286
Klement, J., 167
Knowles, P., 261
Kodama, K., 292
Köhler, K., 167
Kozela, A., 167
Kuno, Y., 177
Kuznetsov, Y., 298

L

Lamoreaux, S., 286
Lang, J., 167, 261, 280
Lassakov, M., 286
Lee, L., 298
Levy, C. D. P., 298
Liu, C.-Y., 286
Liu, D., 265
Llosá Llácer, G., 167
Londergan, J. T., 207, 255
Lüthy, M., 280

M

Makela, M., 286
Maltman, K., 48
Markiewicz, M., 167, 261, 280
Martin, J., 286
Mattison, E. M., 119
McKellar, B. H., 60
McKeown, R., 286
Medve, R., 261
Melnitchouk, W., 305
Mischke, R. E., 298

Mondragón, A., 86
Morelle, X., 167, 261
Morris, C., 286
Muller, D., 99

N

Naviliat, O., 261
Ninane, A., 261
Numao, T., 333

O

Opper, A. K., 238

P

Page, S. A., 298
Phillips, D. F., 119
Pichlmaier, A., 286
Pitt, M. L., 228, 286
Prieels, R., 261
Pusenkov, A., 280

R

Ramsay, W. D., 298
Reitzner, S. D., 298
Rodríguez-Jáuregui, E., 86
Rosenfelder, R., 73
Roy, G., 298
Rudnev, Y., 286

S

Saito, K., 245
Saunders, A., 286
Schaller, L. A., 187
Schebetov, A., 280
Schmor, P. W., 298
Schreiber, A. W., 73
Schweizer, T., 167, 261
Seestrom, S., 286
Sekulovich, A. M., 298
Serebrov, A., 280, 286

Smith, D., 286
Smyrski, J., 167
Soukup, J., 298
Soyama, K., 286
Sromicki, J., 167, 261, 280
Steffens, F. M., 255
Stephan, E., 167
Stephenson, Jr., G. J., 60
Stinson, G. M., 298
Stocki, T., 298
Stoner, R. E., 119
Strzałkowski, A., 167
Sukhoruchkin, S. I., 142
Sum, V., 298

T

Thomas, A. W., 153, 255
Tikhonin, F. F., 273
Tipton, B., 217, 286
Titov, N. A., 298

U

Utsuro, M., 286

V

Valencia, G., 80
van Hove, P., 261
van Oers, W. T. H., 66, 298

Varvell, K., 321
Vasilev, A., 286
Venanzoni, G., 311
Vessot, R. F. C., 119
Vogelaar, B., 286

W

Walstrom, P., 286
Walsworth, R. L., 119
Wilhelmy, J., 286
Wolfe, C. E., 48
Woo, R. J., 298

X

Xing, Z., 54

Y

Young, A. R., 286
Yuan, J., 286

Z

Zejma, J., 167
Zelenski, A. N., 298